DISTORTION IN MUSIC PRODUCTION

Distortion in Music Production offers a range of valuable perspectives on how engineers and producers use distortion and colouration as production tools. Readers are provided with detailed and informed considerations on the use of non-linear signal processing, by authors working in a wide array of academic, creative, and professional contexts.

Including comprehensive coverage of the process, as well as historical perspectives and future innovations, this book features interviews and contributions from academics and industry practitioners. *Distortion in Music Production* also explores ways in which music producers can implement the process in their work and how the effect can be used and abused through examination from technical, practical, and musicological perspectives.

This text is one of the first to offer an extensive investigation of distortion in music production and constitutes essential reading for students and practitioners working in music production.

Gary Bromham is a composer, artist, music producer and mix engineer. Also a researcher and published author who recently completed a PhD in Computer Science, specifically in music and timbre perception, at Queen Mary, University London. He began his career as a Fairlight programmer and a tape op at Trident Studios in London. He subsequently began his songwriting and music production career in Iceland. During this time, he co-produced the 1988 Icelandic Eurovision entry, made an album with jazz-funk band Mezzoforte, and worked with progressive artist, Bjork. Whilst collaborating with Andrew Ridgeley, of Wham, on his solo album, he was also fortunate enough to work with George Michael on some of his album, Listen Without Prejudice. As a result of his collaboration with George Michael and Andrew Ridgeley he was offered his first publishing contract with Warner Chappell Music in 1990. This proved to be a great success. He has written with many varied and successful artists and songwriters including Sheryl Crow, Bow Wow Wow, The Tubes, Robbie Neville, Eddie Money and The Fixx to name but a few, spending a large period of his time working in New York, Nashville and Los Angeles. Later collaborations include Editors, Graham Coxon from Blur, The Maccabees, Lulu, Delta Goodrem and again Sheryl Crow on her single, 'Soak up the Sun'. He was also signed to EMI Records in 1996 as a recording artist in the band The Big Blue. They released two singles and achieved moderate success. He has just completed his PhD at Queen Mary University London where he is researching the role of retro aesthetics in music production and how some of these features are used in intelligent music production systems. He has several publications in this ever-expanding field. He is currently recording as an artist using the project name Convergence.

Austin Moore is a senior lecturer and course leader of sound engineering and music production courses at the University of Huddersfield, UK, where he is also the Music and Audio Production (MAP) research group leader. He completed a PhD in music production and technology, which investigated the use of non-linearity in music production, with a focus on dynamic range compression and the 1176 compressor. His research interests include music production sonic signatures, Atmos in music production, and semantic audio. He has a background in the music industry and spent many years producing and remixing various forms of electronic dance music under numerous artist names and working as an engineer in studios.

Perspectives on Music Production
Series Editors: Russ Hepworth-Sawyer
York St John University, UK

Jay Hodgson
Western University, Ontario, Canada

Mark Marrington
York St John University, UK

This series collects detailed and experientially informed considerations of record production from a multitude of perspectives, by authors working in a wide array of academic, creative and professional contexts. We solicit the perspectives of scholars of every disciplinary stripe, alongside recordists and recording musicians themselves, to provide a fully comprehensive analytic point-of-view on each component stage of music production. Each volume in the series thus focuses directly on a distinct stage of music production, from pre-production through recording (audio engineering), mixing, mastering, to marketing and promotions.

Mastering in Music
Edited by Russ Hepworth-Sawyer and Jay Hodgson

Innovation in Music
Future Opportunities
Edited by Russ Hepworth-Sawyer, Justin Paterson and Rob Toulson

Recording the Classical Guitar
Mark Marrington

The Creative Electronic Music Producer
Thomas Brett

3-D Audio
Edited by Justin Paterson and Hyunkook Lee

Understanding Game Scoring
The Evolution of Compositional Practice for and through Gaming
Mackenzie Enns

Coproduction
Collaboration in Music Production
Robert Wilsmore and Christopher Johnson

Distortion in Music Production
The Soul of Sonics
Edited by Gary Bromham and Austin Moore

For more information about this series, please visit: www.routledge.com/Perspectives-on-Music-Production/book-series/POMP

Distortion in Music Production
The Soul of Sonics

Edited by Gary Bromham and Austin Moore

LONDON AND NEW YORK

Designed cover image: Tony Barnard

First published 2023
by Routledge
4 Park Square, Milton Park, Abingdon, Oxon OX14 4RN

and by Routledge
605 Third Avenue, New York, NY 10158

Routledge is an imprint of the Taylor & Francis Group, an informa business

© 2023 selection and editorial matter, Gary Bromham and Austin Moore; individual chapters, the contributors

The right of Gary Bromham and Austin Moore to be identified as the authors of the editorial material, and of the authors for their individual chapters, has been asserted in accordance with sections 77 and 78 of the Copyright, Designs and Patents Act 1988.

All rights reserved. No part of this book may be reprinted or reproduced or utilised in any form or by any electronic, mechanical, or other means, now known or hereafter invented, including photocopying and recording, or in any information storage or retrieval system, without permission in writing from the publishers.

Trademark notice: Product or corporate names may be trademarks or registered trademarks, and are used only for identification and explanation without intent to infringe.

British Library Cataloguing-in-Publication Data
A catalogue record for this book is available from the British Library

Library of Congress Cataloging-in-Publication Data
Names: Bromham, Gary, editor. | Moore, Austin (Music producer) editor.
Title: Distortion in music production : the soul of sonics / edited by Gary Bromham and Austin Moore.
Description: Abingdon, Oxon ; New York : Routledge, 2023. | Series: Perspectives on music production | Includes bibliographical references and index.
Subjects: LCSH: Sound recordings—Production and direction. | Electric distortion.
Classification: LCC ML3790 .D498 2023 (print) | LCC ML3790 (ebook) | DDC 781.49—dc23/eng/20221129
LC record available at https://lccn.loc.gov/2022056486
LC ebook record available at https://lccn.loc.gov/2022056487

ISBN: 978-0-367-40587-8 (hbk)
ISBN: 978-0-367-40585-4 (pbk)
ISBN: 978-0-429-35684-1 (ebk)

DOI: 10.4324/9780429356841

Typeset in Times New Roman
by Apex CoVantage, LLC

Contents

List of Contributors viii
Foreword xii
Preface xiii

PART I
Technology of Distortion
1

1 A History of Distortion in Music Production 3
CLIVE MEAD, GARY BROMHAM, AND DAVID MOFFAT

2 The Development of Audio Software with Distortion 13
ERIC TARR

3 A Browser-based WebAudio Ecosystem to Dynamically Play with Real-time Simulations of Historic Guitar Tube Amps and Their Typical Distortions 28
MICHEL BUFFA AND JEROME LEBRUN

4 Non-linearity and Dynamic Range Compressors 47
AUSTIN MOORE

5 Low Order Distortion in Creative Recording and Mixing 65
ANDREW BOURBON

PART II
Perception and Semantics of Distortion
91

6 Understanding the Semantics of Distortion 93
GARY BROMHAM

7 An Ecological Approach to Distortion in Mixing Audio: Is
   Distortion an Expected, Rather than an Unwanted Artefact?   109
   LACHLAN GOOLD

8 Towards a Lexicon for Distortion Pedals   128
   TOM RICE AND AUSTIN MOORE

PART III
Retrospective Perspectives of Distortion   145

9 Hit Hardware: Vintage Processing Technologies and the Modern
   Recordist   147
   NIALL COGHLAN

10 Even Better than the Real Thing: A Comparison of Traditional
   and Software-Emulated Distortion in the Contemporary Audio
   Production Workflow   160
   DOUG BIELMEIER

11 'It Just Is My Inner Refusal': Innovation and Conservatism
   in Guitar Amplification Technology   174
   JAN-PETER HERBST

12 A Saturated Market   185
   ASK KÆREBY

PART IV
Musicology of Distortion   199

13 The Studio's Function in Creating Distortion Related
   Compositional Structures in Hard Rock and Heavy Metal   201
   CIRO SCOTTO

14 The Distortion of Space in Music Production   216
   MATTHEW BARNARD

15 Distorting Jazz Guitar: Distortion as Effect, Creative Tool
   and Extension of the Instrument   228
   TOM WILLIAMS

	Contents	vii
16	'Got a Flaming Heart': Vocal Climax in the Music of Led Zeppelin AARON LIU-ROSENBAUM	245
17	**The Aesthetics of Distortion** TOBY YOUNG	262
	Index	277

Contributors

Matthew Barnard is a composer, researcher, and educator working with music technologies. His applied and theoretical research revolves around the spatial parameters of recording and reproduction and its compositional applications and wider aesthetic implications. This interest is situated within a broader preoccupation with the phonographic spatial tendencies observable in music production practice. He is a lecturer at the University of Hull, UK.

Doug Bielmeier is an associate teaching professor in audio recording and production, at Northeastern University, Department of Music—College of Arts, Media and Design, in Boston, MA, with a doctorate in education and 15 years of experience teaching music, audio engineering, and music technology at the graduate and undergraduate levels. Bielmeier was formerly a freelance engineer in Nashville, TN, and has 15 years' proven success as a studio and live sound engineer. Dr. Bielmeier's live sound work has included working at the Kennedy Center, in Washington D.C., and for then-U.S. Vice President Joe Biden. Dr. Bielmeier was the designer and studio manager of the C.L.E.A.R. Lab recording studio at the Purdue School of Engineering Technology, and has had papers published internationally in the *Audio Engineering Society Journal*, *The Art of Record Production Journal*, and the *Journal for Media Education*. As a researcher, Dr. Bielmeier strives to understand what skills and competencies aspiring engineers need to develop to be successful in the audio industry.

Andrew Bourbon is the Subject Area Leader for Music Technology and Games at the University of Huddersfield, UK. Andrew started his career working in live sound and now works regularly as a producer and recording, mixing, and mastering engineer. Andrew completed his PhD in electroacoustic music, with a focus in multi-channel works. His current research is building on his career as a composer and as an engineer, developing research into approaches to immersive mixing of popular music using Atmos.

Gary Bromham is a composer, artist, music producer and mix engineer. Also a researcher and published author who recently completed a PhD in Computer Science, specifically in music and timbre perception, at Queen Mary, University London. He began his career as a Fairlight programmer and a tape op at Trident Studios in London. He subsequently began his songwriting and music production career in Iceland. During this time, he co-produced the 1988 Icelandic Eurovision entry, made an album with jazz-funk band Mezzoforte, and worked with progressive artist, Bjork. Whilst collaborating with Andrew Ridgeley, of Wham, on his solo album, he was also fortunate enough to work with George Michael on some of his album, Listen Without Prejudice. As a result of his collaboration with George Michael and Andrew Ridgeley he was offered his first publishing contract with Warner Chappell Music in 1990. This proved to be a great success. He has written with many varied and successful artists and songwriters including Sheryl Crow, Bow Wow Wow, The Tubes, Robbie Neville, Eddie Money and The Fixx to name but a few, spending a large period of his time working in New York, Nashville and Los Angeles. Later collaborations include Editors, Graham Coxon from Blur, The Maccabees, Lulu, Delta Goodrem and again Sheryl Crow on her single, 'Soak up the Sun'. He was also signed to EMI Records in 1996 as a recording artist in the band The Big Blue. They released two singles and achieved moderate success. He has just completed his PhD at Queen Mary University London where he is researching the role of retro aesthetics in music production and how some of these features are used in intelligent music production systems. He has several publications in this ever-expanding field. He is currently recording as an artist using the project name Convergence.

Michel Buffa is a professor/researcher at University Côte d'Azur, France, a member of the WIMMICS research group, common to INRIA and to the I3S Laboratory (CNRS). He contributed to the development of the WebAudio research field, since he participated in all WebAudio Conferences, being part of each program committee since its creation. He actively works with the W3C WebAudio working group. With other researchers and developers, he co-created the WebAudio Plugin (WAM) standard and is now working actively on its successor, WebAudio Modules 2 (WAM2).

Niall Coghlan has a career that spans many aspects of the audio production industries, from live sound to interactive media and electronic music production. He currently lectures in Dundalk Institute of Technology, Ireland, with research interests in the relationship between music and emotion, sound system culture, and the musicology and pedagogy of record production. He maintains a creative practice in sensor-based installation works, along with the occasional DJ gig.

Lachlan Goold is a recording engineer, producer, mixer, popular music educator, researcher, and lecturer in contemporary music at the University of the Sunshine Coast, Australia. His research focuses on practice-based music production approaches, theoretical uses of space, cultural geography, and the music industry. In his creative practice, he is better known as Australian music producer Magoo. Since 1990, he has worked on a wide range of albums from some of the country's best-known artists, achieving a multitude of gold and platinum awards.

Jan-Peter Herbst is senior lecturer in music production at the University of Huddersfield, UK, where he is Director of the Research Centre for Music, Culture and Identity (CMCI). His primary research area is popular music culture, particularly rock music and the electric guitar, on which he has published widely. Currently, he is undertaking a funded three-year project that explores how heaviness is created and controlled in metal music production. Herbst's editorial roles include *IASPM Journal* and *Metal Music Studies*, and he currently edits the *Cambridge Companion to Metal Music* and the *Cambridge Companion to the Electric Guitar*.

Ask Kæreby is a Danish composer and sound engineer based in Copenhagen. Kæreby holds a Master's degree in music technology from the Royal Danish Academy of Music and is currently teaching at the Royal Academy of Music, Aalborg (DK). His interdisciplinary and research-based work encompasses experimental composition, soundscapes, and electroacoustic music, and has been presented at Klingt Gut! (DE), Inter Arts Center (SE), SPOR Festival (DK), Nordic Music Days (FO), and 18th Street Arts Center (US).

Jerome Lebrun holds a tenured CNRS researcher position, currently heading the Biomedical Signal Processing group at the I3S laboratory of University Côte d'Azur, France. His current fields of research include signal processing and data mining for computer music and the study of animal vocalizations within environmental soundscapes. He is also involved in the development of lossy ciphered schemes for speech communications and medical apparatus for EEG/NIRS/f-MRI multimodal acquisitions in cognitive sciences. He has a special interest in developing new tools to study how the perception of noise and distortion in speech or music is related to intelligibility and emotions.

Aaron Liu-Rosenbaum is a composer and professor of music technology at Laval University, Quebec, where he serves as director of the certificate programme in digital audio production. His research and creative interests focus on how we communicate and navigate our ever-increasingly technologized society, including soundscape composition, community music, and popular musicology. He is a researcher at the Centre for Interdisciplinary Research in Music Media and Technology (CIRMMT) at McGill University and an affiliate researcher at the Laboratory for New Technologies of Image, Sound, and Stage (LANTISS) at Laval University.

Clive Mead holds a BA in music production from the University of Brighton, UK, and has more than 25 years of experience as an artist, composer, and producer. He has a background in electronic dance music and has also written and produced music in numerous styles for film and TV. He is a specialist in re-creating vintage music styles and produces sample packs in various genres for the industry's leading sample publisher. His PhD research at ICCMR is focused on exploring the relationship between the technology available during different time periods and the composition and production process.

David Moffat is an artificial intelligence (AI) and machine learning (ML) research scientist at Plymouth Marine Laboratory, UK. He works on applying artificial intelligence and machine learning techniques to earth observation data to formulate a better understanding of the natural world. Previously, he worked at Queen Mary University of London and University of Plymouth, as a lecturer working on intelligent music production and music semantics. He is the vice-chair of the AES Semantic Audio Analysis Technical Committee and a member of the AES UK committee.

Austin Moore is a senior lecturer and course leader of sound engineering and music production courses at the University of Huddersfield, UK, where he is also the Music and Audio Production (MAP) research group leader. He completed a PhD in music production and technology, which investigated the use of non-linearity in music production, with a focus on dynamic range compression and the 1176 compressor. His research interests include music production sonic signatures, Atmos in music production, and semantic audio. He has a background in the music industry and spent many years producing and remixing various forms of electronic dance music under numerous artist names and working as an engineer in studios.

Tom Rice is an avid guitarist and pedal enthusiast. Having studied popular music at the undergraduate level, this established a foundation for moving into postgraduate research, where he studied the semantics of distortion pedals and their impact on music production. Now working for an international music retailer as a merchandiser and presenter, this continues the trend of being surrounded by guitars, living the dream.

Ciro Scotto is a music theorist and composer. His research in music theory includes creating compositional systems, producing analyses and theoretical models of the music of the 20th and 21st centuries, mathematics and music, and rock music, especially in the area of timbre. He has published articles in *Perspectives of New Music*, *Music Theory Online*, and the *Journal of Music Theory*. His article "The Structural Role of Distortion in Hard Rock and Heavy Metal" appears in *Music Theory Spectrum*. He is also the editor for *The Routledge Companion to Popular Music Analysis: Expanding Approaches*. He is currently working on a chapter about Dream Theater for a volume on progressive rock. His recent compositions include *Between Rock and a Hard Place*, a work in four movements for electric guitar and percussion ensemble, and *Dark Paradise*, a work for piano and percussion ensemble. Both are available on Ravello Records. He is currently associate professor and chair of the music theory department at Ohio University, USA.

Eric Tarr is an associate professor of audio engineering technology. He received a PhD, MS, and BS from The Ohio State University, USA, in electrical and computer engineering. He received a BA in mathematics with a minor in music from Capital University, Ohio, USA. His interests in audio engineering include digital signal processing, acoustic and electronic system modeling, and auditory perception. His research has focused on the perception of speech signals and the development of signal processing methods to improve speech intelligibility for listeners with auditory prostheses. Eric is the author of *Hack Audio: An Introduction to Computer Programming and Digital Signal Processing in MATLAB (Audio Engineering Society Presents)*. He has created software for many audio companies, including Apogee, Art+Logic, Empirical Labs, Gibson/KRK, Output, ReLab, Sennheiser, and Skywalker Sound.

Tom Williams is a jazz guitarist, lecturer, and musicologist specialising in improvisation, cognition, jazz, and pedagogy. His PhD 'Strategy in Contemporary Jazz Improvisation' (University of Surrey, UK, 2017) created a detailed cognitive and contextual model of how expert-level improvisers develop and use their craft. Tom holds lecturing posts at the Academy of Contemporary Music (Guildford, UK) and the University of Surrey, UK.

Toby Young is a Leverhulme Early Career Research Fellow at The Guildhall School of Music and Drama, UK, drawing on composition, philosophy, and cultural sociology to examine the blurred space between classical and popular music. He is currently working on 'Transforming the operatic voice': a practice-based project that investigates the relationship between singing styles in popular music and opera to widen the appeal of opera to a diversified demographic. Before coming to Guildhall, Toby was the Gianturco Junior Research Fellow at the University of Oxford, running a project on classical music's relationship with electronic dance music. As a composer and producer, Toby has collaborated with artists including The Rolling Stones, Chase & Status, Duran Duran, The King's Singers, Tamil songwriter Ilayaraja, and the London Symphony and Philharmonic Orchestras.

Foreword

Ah, *distortion* . . . what a funny word. In its most literal sense (in audio), it is any change to the shape of a waveform during a process, but if we use the word in this way, every single piece of audio equipment, analog or digital (even wire if it's long enough), can be considered to cause distortion. As audio professionals, we used to use the term to describe the unwanted artefacts of analog processes, and now we use it to lovingly describe the wanted, whether intentional or not, artefacts of everything we do. We often find ways to process audio that are solely focused on distortion, as opposed to it just being the side-effect of compression or equalisation or recording itself. It is the 'glue', the 'warmth', the 'character'. It is desirable and necessary. It is what gives recordings personality and interest. The 'proper' use of distortion is what defines some of the most successful audio engineers in the history of modern record making, measuring that success both by the sales of the records they work on, and also the esteem they are held in by their peers. As Martin Mull or Frank Zappa or Elvis Costello once said, "Talking about music is like dancing about architecture", but talk about it we must, and this book goes a very long way toward giving us the framework to talk about possibly the most important element of music production.

Andrew Scheps

Preface

Welcome to *Distortion in Music Production: The Soul of Sonics*, a topic-specific edition in the Perspectives on Music Production (POMP) series. This book explores distortion from various perspectives, including production techniques, semantic audio, musicological analysis, software design, and qualitative and quantitative studies. The title came from a phrase used in an interview with Mark 'Spike' Stent conducted by co-editor Gary Bromham, for part of what became Chapter 7 in this book. Indeed, the subject of distortion can be somewhat esoteric, even otherworldly, in nature. Sometimes it is explicit as in the case of guitar specific distortion, but often it is subtle as in the case of saturation, colouration, or harmonic enhancement as heard in a studio recording, mixing, or production context. In the Foreword to this book, Andrew Scheps also provides an excellent appraisal of the topic area and offers some excellent insights into the uses of this most important effect in the music production chain.

The book's subject has been a professional and academic research interest of the editors for several years. One of the issues we faced when carrying out academic research into distortion, mainly when writing state-of-the-art literature reviews, was the distinct lack of academic content to include in the review. Thus, this book fills a gap in the current literature and provides academics with a much-needed collection of chapters focusing specifically on distortion in the production process. Moreover, the book will provide professionals with academic research and empirical studies into an area of their work that has not been rigorously documented. Any discussion about colouration and saturation will possibly lead back to the debate about analogue vs. digital technologies. Nowhere is this discourse more pertinent than in the field of non-linearity and distortion. It is arguably the most salient factor in this hotly contested argument. If there is such a thing as analogue 'mojo', it manifests itself more in harmonic distortion than in any other field.

At this juncture, it is worth defining non-linearity and distortion. The most obvious form of distortion is associated with the guitar, be that the riff from '(I Can't Get No) Satisfaction' by The Rolling Stones or the guitar part from one of many heavy metal groups. However, the sound of distortion can be found in many other areas within a music production. To illustrate this point, let us imagine the sound of a piece of music. When one recalls recorded music in their mind, a large part of what they might hear will include the musical components: the melody line, the harmonic changes, the rhythms, and if it's a song, the lyrics. But for some people, particularly those involved in music production, the musical components are not the only attributes recalled in their mind's ear. A representation of the production's sonic signature will also be recollected, a sonic depiction that includes production techniques and aesthetics. For example, if one recalls a production that epitomises the 1980s, they will invariably hear a gated reverb on drums, perhaps a Yamaha Dx7 patch or a Lexicon 224 or 480 hall. But at a lower level, if one were to then recall a production from the 1950s or 1960s, even those with no knowledge of the production process will probably recall a sound that is more lo-fi by modern standards, perhaps duller, and maybe even a little warmer (not forgetting in mono)! What they have heard in the mind's ear is a sound that is shaped by non-linearity. They have heard the colouration effect of tape, the loss of quality from bouncing tape tracks, the slight breakup of a valve preamp and valve mixing console, and so on. As an experiment, take a minute to recall the sound of a dub reggae track from the 1970s. What did you hear? You likely heard a distorted delay line. Now imagine you are listening to some drum and bass, particularly from the late 1990s to mid-2000s. What did you hear? Again, it is likely you heard distortion, this time on the growling and snarling bass lines. Finally, let us imagine a garage rock band from the revival in the 2000s. What did you hear? This time you probably imagined a distorted lead vocal and "trashy" sounding drums that come from a room mic that has been distorted with a compressor.

As we have hopefully illustrated, distortion can reveal itself in many sonic areas within a music production. Distortion is not only about distorted guitars, and the authors would argue that the most interesting, pertinent forms of distortion are also the most subtle. In the mid-1980s, when the shift to digital recording started in earnest, many recording engineers realised that something was missing from their work. They noticed that something special was lacking, but they could not immediately put their fingers on it. In a 1987 interview with Hammond [1], Peter Gabriel discussed using the Mitsubushi X-850 digital recorder on his album *So*. Of interest is the fact that they recorded many of the tracks in parallel using both the digital X-850 and an analogue Studer A80. Gabriel noted that the top-end reproduction of the X-850 was better, but he still preferred the low end of the drums and bass guitar from the Studer. He went on to state that, at the time, he thought he was very much a proponent of digital. Still, this experience made him reconsider, pointing out that something is appealing about distortion in the analogue domain. Digital mixing consoles were also developed during the 1980s and were in mass manufacture by the early 1990s. These consoles often had very clean mic preamps installed, which further compounded the reduced non-linearity. Using a fully digital studio, it was possible to track audio material virtually free of all non-linear artefacts, but, as equipment manufacturers shifted towards digital recording solutions and cleaner, more transparent components, music producers noticed that the character and colour from previous recordings were increasingly diminishing. One can argue this issue was partly due to producers and engineers not changing their recording approach in the shift from analogue to digital (for example, still tracking to hit a target of 0 or still boosting the top end as they had done when recording to tape), but the move to digital and cleaner equipment objectively reduced one important component that had been present throughout recording history—distortion. Thus, since the 1990s, there has been a constant movement back toward introducing this distortion into the signal path. This movement has become progressively more rapid through time, and from the mid-2000s, it has probably taken on the shape of an exponential curve. Mic preamps, signal processors, and software plugins are now developed and marketed as tools to add colour and character to recordings and mixes. A modern digital-based studio will often consist of racks and 500-series lunchboxes filled with analogue devices, usually selected by the owner to impart a particular non-linear distortion onto audio material during tracking. Then during the mix, software emulations of analogue devices that have been designed to introduce non-linearity will be used to further distort and shape the recorded tracks into a mix that is subjectively coloured and objectively distorted but in the 'right' way.

Thus, the authors argue that this book is very timely and a much-needed resource to fill a significant gap in the current literature. We trust it will be a rich resource for scholars, students, and industry professionals from all areas of music production. Furthermore, we hope it will encourage more studies into related areas and act as a catalyst for improving our understanding of how non-linearity has shaped the music production process.

The book is divided into what we considered to be four logical subsections:

- Technology of Distortion
- Perception and Semantics of Distortion
- Retrospective Perspectives of Distortion
- Musicology of Distortion

The first part and subsequent chapters look at the 'Technology of Distortion' and how these have shaped the way we think about processing sound with distortion in its various manifestations. Chapter 1, by Clive Mead, Gary Bromham, and David Moffat, provides an overview of the history of distortion and also creates a narrative for the rest of the book. It seeks to provide a historical timeline for distortion and also offers a definition of not only distortion but also the two more subtle manifestations of saturation and colouration. Eric Tarr subsequently evaluates the current state of software plugin technologies in Chapter 2. Michel Buffa and Jerome Lebrun in Chapter 3 expand the theme of distortion to that of guitar amp simulation in a web-based environment. In Chapter 4, Austin Moore, one of the co-editors of this book, looks at the distortion properties of certain iconic dynamic range compressors (DRCs) and looks at how their respective non-linear properties have helped shape the music production sonic landscape. The section is closed by Andrew Bourbon in Chapter 5, where he assesses the sonic impact of low-order distortion properties exhibited by various microphone preamps on both recording and mixing aesthetics.

Part II of the book, 'Perception and Semantics of Distortion', looks at the descriptive language and terminology used to describe the effects of distortion. Chapter 6, 'Understanding the Semantics of Distortion', by the co-editor Gary Bromham, uses thematic analysis as a method for understanding studio language and descriptors used by music professionals when describing sonic attributes associated with distortion. Esteemed interviewees provided invaluable content and context for this work. Lachlan Goold, in Chapter 7, uses subjective perceptual listening tests as a means of evaluating two different mixes with and without distortion, showing that there is a clear preference for the one using such audio effects. Tom Rice and Austin Moore, in Chapter 8, similarly conduct perceptual listening tests, this time in reference to guitar pedals and the respective descriptive language ascribed to their distortion properties.

Part III focuses on 'Retrospective Perspectives of Distortion' and looks at the retro aesthetics and values attached to vintage studio technologies and the marketing effects of selling such nostalgia to an audience, often with little or no context of the original technologies. Niall Coghlan introduces us to the theme in Chapter 9, where he writes about the continued fascination with hardware and even the rejection of digital counterparts in favour of their esteemed forerunners. Much of this is attributed to the distortion properties exhibited by the pieces of iconic hardware. In Chapter 10, 'Even Better than the Real Thing' by Doug Bielmeier, the importance of harmonic distortion in the contemporary workflow is explored by way of surveys with audio professionals. The findings show that harmonic distortion is an essential tool for all users, even if it is harder to define the specific uses and applications of such processes. In Chapter 11, Jan-Peter Herbst explores the question of authenticity from the perspective of traditional guitar amplifier technology and digitally simulated counterparts. He specifically discusses the reliance on established analogue valve amplifiers and the reluctance to embrace new modelled digital technology such as the Kemper Profiling amplifier in this field. Chapter 12 by Ask Kaerby concludes this section fittingly when he discusses the market forces at play when using nostalgia as a device for selling retro digital music technologies. Digital metaphors and even analogue reissues of iconic equipment imply that 'Retromania', as Simon Reynolds terms these phenomena, are huge motivators for producing and indeed rehashing old technologies.

The final section of the book, Part IV, looks at the 'Musicology of Distortion' from differing perspectives. Chapter 13 by Ciro Scotto discusses the role that distortion has played in creating the sound of rock and heavy metal music in the recording studio. It looks at this phenomenon from both a compositional and an aesthetic perspective and assesses the influence of techniques used in the recording studio that have influenced these genres. In Chapter 14, Matthew Barnard looks at the 'Distortion of Space in Music Production' and analyses ways that the studio can be used as a tool for distorting perception of space. Tom Williams, in Chapter 15, looks at how jazz guitar, not traditionally seen as a place for excessive use of distortion, has become an evolving landscape in the context of progressive attitudes toward the use of the effect. Chapter 16, 'Got a Flaming Heart' by Aaron Liu-Rosenbaum, looks at the use of distortion on the voice and uses Led Zeppelin singer Robert Plant as a conduit for demonstrating this phenomenon. Finally, in Chapter 17, Toby Young discusses the 'Aesthetics of Distortion' and how artistic expression in music, as with other arts, mirrors the cultural and societal influences in our lives. Distortion, in this context, is a means of escaping and breaking free from the norms and restrictions that to some extent limit human beings.

Gary Bromham and Austin Moore

Reference

[1] Hammond, R.: *Peter Gabriel—Behind the Mask* (SOS, Jan 1987), www.muzines.co.uk/articles/peter-gabriel-behind-the-mask/1494

Part I
Technology of Distortion

1 A History of Distortion in Music Production

Clive Mead, Gary Bromham, and David Moffat

1.1 Introduction

Distortion exists in many different forms and manifestations, both explicit in the case of distorted guitar and more implicit types in the case of tape, transformer or tube-based circuits more commonly found in a recording studio. This chapter proposes a narrative and a timeline for both forms of distortion.

It should be remembered that the effect of overdriving electronic circuits has not always been a desired result and was instead rather a consequence of the limitations of the technology, particularly in the analogue domain. Indeed, the irony lies in the fact that, since the advent of digital technology, there has been a feeling that something, sonically at least, has been missing. Julian Rogers from website Production Expert articulates this in a fitting manner when he comments, 'To the frustration of many of those designers who spent their careers trying to eradicate imperfections, as soon as we managed to remove them, we wanted them back again!' [1]

For many, the navigation of distortion in both analogue and digital domains is the story of surprise and contradiction. Whereas the motivation of analogue technology was about approximating perfection, the digital goal has become one about perfecting approximation [2]. Indeed, analogue technology is problematic from a technical standpoint, because it is fraught with difficulties, inconsistencies and non-linearities, but to a great extent this is what defines its aura: the interesting part is arguably (subjectively) everything which is wrong with it! Digital technology, on the other hand, is defined by precision, clarity, and focus, and is more predictable and consistent in behaviour. In cognitive processing the brain notices differences, imperfections, and randomness, and this becomes part of its inherent appeal. To use the analogy (reference point) of film, an out-of-focus, grainy image is often more interesting and appealing purely from an aesthetic perspective, where our imagination comes into play. Indeed, motion blur is an unavoidable side effect of using analogue film cameras, and our eyes have become so accustomed to seeing this effect in film that the clinical video footage captured by digital equipment seems to be missing something fundamental that makes it 'feel' like film. Consequently, modern digital video equipment has features which artificially re-create motion blur. There are parallels to be drawn here with audio perception, which Joshua Reiss alludes to in his inaugural professor lecture at Queen Mary University of London [3, 4]. The lecture titled 'Do you hear what I hear? The Science of Sound' includes a discussion on the connection between auditory and visual stimuli.

1.2 What Is Distortion?

In any field outside of music, distortion is usually considered as the altering of something into an unintended shape. In most signal processing applications, distortion is where a signal can be considered to lose some information-bearing qualities, or an alteration where the primary signal is hidden somehow. Most fields consider distortion to be a negative phenomenon, whether a steel beam in a building becomes distorted and results in building collapse, or in telecoms, where distortion can be unwanted noise included in the signal, or even in the concept of an individual distorting facts, representing a mistruth. Music production is one of the few fields where distortion can be considered a positive. Not only is distortion potentially positive, but the creative use of distortion and signal perturbations can enhance and make the entire sonic image more aesthetically pleasing. For the duration of this book, distortion will be defined simply as 'Output amplitude versus input loudness', or as Alex Case proposes, 'When the shape of the output waveform deviates in any way from the input, distortion has occurred' [5]. A history of audio effects, including distortion, is also discussed by Wilmering et al. here [6].

Producer and mix engineer John Leckie describes distortion as follows:

It adds an edge, there's a danger element to adding distortion. And of course, the amount is . . . it's like you take it to the limit, but you don't go over the limit because it would give the game away. You know, it's an edge thing. It's hanging there, you've got to let the listener still hear what's going on. But by adding the distortion you add an excitement level.

[7]

'Distortion', in audio terms, is the altering of the shape of a waveform and creating a different tonality from that of the original waveform. Traditionally these forms were from the analogue circuits, that were the cause of the deformations of the audio signal. This led to 'saturation' and 'colour' being commonly used terms in audio to describe the types and mannerisms of the signal distortion. 'Saturation' is the process of driving a piece of circuitry almost to the point of hard-clipping distortion, where the sonic properties are heavily muddied by the circuit, but not so heavy that the original sonic clarity and meaning is lost. Tape saturation was commonly used 'back in the day', as it added subtle harmonics and warmth to your sound.

'Colour' usually implies a certain change of timbre, normally brought about by distortion (as in 'change to the waveform') either by actually overdriving something or, for example, by passing it through a transformer or other device which subtly affects the sound, similar to a tone modification or adding warmth to a signal.

Distortion can be thought of as a non-linear audio processing tool. This means that the amplitude of the input signals will affect the amount and type of distortion that is created. Unlike an equaliser, where the input signal being made louder will produce the same equalisation change, a distortion with a louder input will change the nature and tone of the output signal. This property is useful to acknowledge, as the ways in which types of distortion can be used and processed can dramatically change the types and tonalities of an audio signal.

An audio distortion effect will produce variations and modifications to an audio signal and cause the addition of harmonics of the audio signal to be added.

In the case of hard-clipping an audio signal, an extreme example of distortion, the audio signal is increased in amplitude, and any part of the signal above or below a given value are set to the maximum value (see Figure 1.1). The effect of this is squaring off the waveform, which adds harmonics of the original signal, specifically, for a sin wave to be transformed to a square wave, the addition of odd harmonics (3rd, 5th, 7th . . .).

Distortion used in music production is generally not as extreme as this example, where only explicit odd harmonics are added, whilst the smoothing of the edges and transients and squaring of sinusoids are all natural effects of distortions of an audio signal.

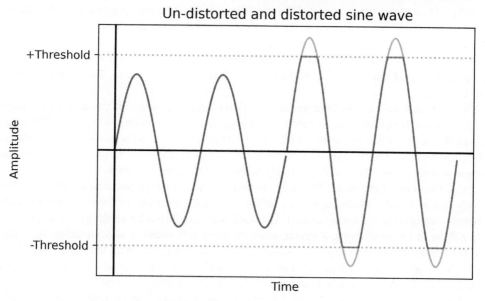

Figure 1.1 Sine wave showing clipped waveform.

Other fields identify these added harmonics or transient smoothing approaches as problematic, while in music production, the added soft harmonics and carefully managed addition of more frequency content to an audio signal can directly lead to the audio signal being perceived as more pleasant and can fill the frequency space more fully than the original unprocessed signal.

It is the distortion's ability to add higher frequency harmonic content, along with smoothing harsh transients that really lends itself to creating a fuller sound, one that can add body to a piece of audio. As such, the distortion effect can really be considered as a spectral audio effect, taking in an audio signal and adding more spectral content, which is heavily defined by the input audio signal, where the variations in the input signal are magnified, making bold and significant changes to the output signal.

1.3 A Timeline of Distortion in Music

> They purposely distorted harmonica. Cos before that with Sunny Terry, it was pure, the harmonica was chirpy and bright, happy and country and rural kind of thing. And when they got to the city, Little Walter mic'ed up the harmonica and it became a whole other instrument really, the electric harmonica. The length of notes, you could add depth if you blew harder, it would distort more.
>
> John Leckie, Producer/Engineer [7]

Over the years, distortion within music production has transitioned from an unwanted side effect of the electronic circuitry into a vital set of audio effects that have shaped the sound and tone of entire genres of music. Here, we outline the evolution of distortion within music, from the first intentional use of distortion, highlighting the key moments in distortion within music.

1951, 'How Many More Years', Howlin' Wolf (Willie Johnson on guitar). Guitar distortion, as created by an overdriven valve amplifier, is recorded commercially for (very probably) the first time. Recorded at Memphis Recording Service (later renamed Sun Studio) by Sam Phillips, the producer and engineer who would later form the massively influential Sun Records label and introduce the world to Elvis Presley, Johnny Cash, Roy Orbison, Jerry Lee Lewis etc.

1951, 'Rocket 88', Jackie Brenston and his Delta Cats. Legend has it that a broken speaker in guitarist and bandleader Ike Turner's guitar amplifier was the culprit, allegedly after it was damaged in a car on the way to the studio (accounts differ as to exactly how). Also recorded at Memphis Recording Service by Sam Phillips, with its up-tempo blues shuffle, saxophones and boogie-woogie piano, 'Rocket 88' is widely considered to be the first rock 'n' roll record.

1958, 'Rumble', Link Wray. The only instrumental song to ever be banned from US radio, mostly due to its title, which capitalised on the '50s media hysteria for teenage gang violence ('rumble' being slang for a mass brawl). The track was born through a serendipitous accident at a live show in Fredericksburg, Virginia, when the band was asked to play an unfamiliar song, and the drummer responded by beating out a rhythm on his drums with the thick ends of his sticks. Wray joined in with a guitar improvisation, and to make the guitar more audible to the audience, a vocal mic was placed in front of the guitar amplifier, unintentionally overdriving the PA amplifier and creating a distorted sound that the audience reacted to favourably. Shortly afterwards the band attempted to recreate the sound in the studio with producer Archie Blayer, and Wray discovered that using a pencil to punch holes in the speaker cone of his guitar amplifier achieved the closest approximation. The menacing feel of the song and Wray's pioneering use of power (5) chords, due to their being comprised of just a root and perfect fifth (with no third to muddy the sound), really accentuated the distortion. Power chords would soon become the basis for rock guitar riffs as we know them.

1961, 'Don't Worry', Marty Robbins. At an otherwise unremarkable Nashville recording session in 1960 at Bradley Film and Recording Studio, presided over by studio engineer Glen Snoddy, guitarist Grady Martin plays his baritone guitar at a session for country A-lister Marty Robbins and is (as was the fashion in country music at the time) doubling the upright bass part on his baritone guitar. Martin is monitoring his guitar sound through a small amplifier in the live room and the recording is a routine one, with nothing out of the ordinary to report. It's only after the band has recorded the song and the session players move into the control room to hear the playback that it becomes apparent that something rather extraordinary has happened. It transpires that the studio's recording console has been built with several improperly wound output transformers and that one has malfunctioned just as Grady Martin played his baritone guitar solo through it. The resulting thick and heavily saturated sound (that would later be christened 'fuzz') divides the musicians at the session, with some wanting to re-route the guitar into a properly functioning desk channel and re-record the song and others arguing to keep this fascinating new sound on the record. The decision is made to release the song as is, and history is made [8].

1962, Maestro FZ-1 Fuzz-Tone, Glen Snoddy and Revis T. Hobbs. Demand from musicians to recreate the sound of Grady Martin's guitar through the broken desk channel prompts Snoddy, the engineer from the 'Don't Worry' session, to enlist the help of former colleague Revis T. Hobbs, to recreate the effect. Although the malfunctioning recording console was valve-driven, they are able to simulate the effect with solid-state electronics and create a design based on germanium transistors. They license their circuit design to Gibson, who manufacture and market it as the Maestro FZ-1 Fuzz-Tone, although it wouldn't be a success until its use by Keith Richards on The Rolling Stones' hit 'Satisfaction' some three years later. The Fuzz-Tone is sometimes incorrectly referred to as the first effects pedal, although it was actually preceded by the DeArmond Trem Trol 800 tremolo pedal in 1941—more than two decades earlier).

1962, '200 Pound Bee', The Ventures. Considered the first notable use of a Maestro Fuzz-Tone pedal on a recording, although some accounts suggest it was a custom-made 'Red-Rhodes' device, also intended to recreate Grady Martin's 'Don't Worry' fuzz sound.

1962–63, Marshall JTM 45 guitar amplifier. Fender and other American-made amplifiers were highly desirable in Britain in the early 1960s, due to their sounds being heard on the highly influential American records of the time. These amplifiers were also rather difficult to find, mostly due to a UK embargo on the importation of American goods, that had lasted from 1951 to 1959; even after the embargo had ended, it was still difficult and expensive to import American instruments and equipment. At the time Jim Marshall ran a musical instrument shop in West London which sold British-made Vox and Selmer amps, as well as American imports that were specially ordered for customers. Marshall's guitarist son Terry had particularly favoured the Fender Bassman, as it was the loudest amp available at the time, and when other musicians who frequented the shop, including The Who's Pete Townshend, began asking for amps with more power than the Bassman, Marshall saw a business opportunity [9].

He asked his shop technical team of Ken Bran, Dudley Craven and Ken Underwood to reverse-engineer a 1959 Fender Bassman (5F6A circuit) so they could build and sell their own modified clones of the design. The difference in available parts between the USA and UK necessitated some changes in the amplifier's design, with one important variation being the use of an ECC83 (12AX7) as the first preamp valve in the circuit, as the ECC83 produces considerably more gain than the 12AY7 used in the copied Fender model. The US-made 6L6 power valves found in the Bassman were soon swapped for British-made 5881, KT66 or EL34s. Additionally, a modified negative-feedback circuit in the Marshall design creates a different harmonic profile, which also increases the level of possible distortion.

1964, 'You Really Got Me', The Kinks. Recorded at IBC Studios and produced by Shel Talmy and engineered by Glyn Johns, the song is widely accepted to feature the first modern rock guitar riff. Guitarist Dave Davies made razor blade cuts in the speaker cone of an Elpico AC55 amplifier, which was then layered with a Vox AC30 to create the distorted guitar sound. Further saturation was added by pushing the valve input stage of the studio's custom-built recording console [10].

1965, Dallas 'Rangemaster' treble booster introduced. An OC44 germanium transistor-based signal booster, it was originally intended to replace high-end loss in 'darker'-voiced British-made guitar amplifiers when fed by humbucker-equipped guitars. Guitarists soon realise that it can be used not just to noticeably increase the output signal from their instruments, thus overdriving the input valves of their amplifiers, but that the Rangemaster adds some distortion to the signal.

1965, 'Satisfaction', The Rolling Stones. Keith Richards uses a Maestro Fuzz-Tone on a demo guitar part that is intended as a placeholder, to be replaced by a horn section. The band's manager Andrew Loog Oldham decides to keep the fuzz part as recorded, and Richards only becomes aware of this when he hears the song on the radio. The record's enormous success creates huge demand for fuzz pedals [11].

1965, Sola Sound Tone Bender. The Maestro Fuzz-Tone is re-engineered by Gary Hurst at Macari's Musical Instruments in London, as the Sola Sound 'Tone Bender'. The circuit was modified at the request of guitarists to provide more sustain (gain) and to be more stable at the colder temperatures inherent in the British climate. The germanium transistors used are very temperature-sensitive at lower voltages, so Hurst increases the voltage from the 3 volts that the Fuzz-Tone used to 9 volts for the Tone Bender.

1966, Arbiter (or Dallas/Arbiter) Fuzz Face pedal introduced. This was essentially a redesigned Tone Bender that was biased colder, to try to overcome some of the Tone Bender's unpredictability at different temperatures. The earlier models used germanium transistors, changing quickly to silicon transistors, which sounded different but were not as temperature-sensitive, resulting in more predictable performance. The Fuzz Face's regular usage by Jimi Hendrix would provide a significant boost to its sales.

1966, John Mayall's Bluesbreakers album (officially untitled, but generally known as the 'Beano' album) is recorded in London. Eric Clapton is playing guitar with Mayall, after his exit from The Yardbirds, shortly before he teams up with Jack Bruce and Ginger Baker to form Cream. Clapton uses a Dallas Rangemaster (this is contested by some, but is highly likely) to overdrive the input of a Marshall JTM 45 amplifier, with a 1959/1960 Gibson Les Paul, fitted with PAF (Patent Applied For) Humbuckers. Some of the sound can be attributed to the valve-based console saturating at the input stages, but it should be noted that Clapton played at what was (considered at the time) incredibly loud volumes. Thus, preamp valve distortion (ECC83), power valve distortion (KT66) and speaker breakup from the low-wattage Celestion alnico speakers in the amplifier were all contributors to the sound. Although he was just using the equipment that was readily available at the time of recording, Clapton's use of a humbucker-equipped guitar into a heavily overdriven Marshall amplifier would create the blueprint for rock guitar sounds for decades to come.

1968, The Beatles, 'Revolution'. The fuzz guitar sound was achieved, not by using an effects pedal, but by recording engineer Geoff Emerick feeding the guitar directly into two overdriving channels (in series) on the valve-based REDD console at Abbey Road Studio. Such innovation would've previously been considered unthinkable by the deeply conservative management at the EMI-owned facility, but by this point The Beatles had been hugely successful for several years and were allowed an unusual amount of leeway by the EMI management, not to mention unlimited studio time and the best recording engineers and technicians at their disposal.

1968, Jon Lord joins Deep Purple and begins using Marshall guitar amplifiers on his Hammond C3 organ, resulting in a far more distorted sound than would be achievable by overdriving the amplifiers into the organ's Leslie speaker. From 1972 Lord adds a Maestro RM-1A ring modulator effect to the output of his Hammond organ, which could be used to provide an extra overdrive stage before the signal reaches the amplifiers [12].

1969, Electro Harmonix Big Muff Pi is designed by Mike Matthews and Bob Myer. Although technically a fuzz pedal, this pedal uses four silicon transistors (two used as input and output amplifiers and two as cascading gain stages) to produce an exceptionally large amount of gain. Soft-clipping diodes then roll off some of the undesirable harmonics created by the distortion process, and this is followed by high-pass and low-pass filters arranged as a sophisticated (for the time) one-knob tone control. Interestingly, advertising literature for this pedal intentionally used the word 'distortion' as a marketing term for the first time. The Big Muff Pi would become hugely successful, spawning innumerable variations, and is still in production (in several different versions) over half a century later.

1969, Randall Smith begins building his 'Boogie' guitar amplifiers. The design begins as a heavily modified Fender Princeton, and these amplifiers will eventually be sold under the Mesa Boogie brand. Smith's main innovation is to include a master volume control, which allows guitarists to overdrive the 12AX7 preamp valves on the input stage of the amplifier to create distorted tones at relatively low volume, with the master volume allowing attenuation of the final output volume. It should be noted though, that whilst this innovation enables preamp distortion at low levels, it precludes the power amp valve distortion and speaker breakup that are also contributors to the desired distorted guitar sound. Although master volume controls had been included on earlier guitar amplifiers (such as the 1966 Guild Thunderbird), these hadn't been designed with the specific intention of allowing preamp saturation. In 1972 Fender introduce the master volume control on their 'Silverface' Twin Reverb guitar amplifier, although Marshall wouldn't introduce a master volume–equipped amplifier until 1975, with their 2203.

Additionally, some guitarists had been using PA amplifiers specifically for their master volume controls, with one notable example being Mountain's Leslie West. West's signature guitar sound was created by using a massively overdriven Sunn Coliseum PA head, fitted with KT88 valves, and driving 4X12 speaker cabinets. This can be heard on their 1969 hit 'Mississippi Queen', which features arguably one of the most iconic rock guitar tones.

1977–78 IC (integrated circuit) Op-amp-based distortion is used for the first time in mass-produced guitar pedals. The MXR Distortion+, ProCo Rat and Boss DS-1 are all introduced in a short space of time. Now guitarists could achieve heavily saturated sounds, without having to play at the extremely high volumes required to induce preamp and power amp valve distortion into their amplifiers. The distortion produced by these pedals would sound rather harsh when fed into a clean amplifier, so guitarists would generally use them to drive an already overdriven amp. The first Boss distortion pedal to be produced in the now familiar compact pedal format was actually the OD-1 in 1977, but this would soon be eclipsed by the DS-1, which would be phenomenally successful, with over one million units sold, and is still in production more than four decades later.

1.4 Types of Distortion

Chris Jenkins, Solid State Logic: Designer of Several Products, Including the 4000 Series Console:

> The limiting factor with distortion was always the tape machine. In those days people weren't particularly looking at ways of minimizing distortion. Because you were going through a tape recorder, the performance of the console was always masked by the performance of the tape recorder. The tape distortion was always far greater than the electronic distortion. So always, the limiting factor was tape. Hence all the things about the levels you align tape machines to and how hard you hit the tape. When they brought in new that could take more level, people cranked the level. But then I was always amazed when I went into the commercial studios at how conservative they were in terms of level on tape compared to what we used to do in my bit of the BBC, where we put way more level on. Tape is frequency dependent distortion. It's not linear distortion. A circuit that distorts linear, it doesn't matter what the frequency is. You reach the clipping point, it will clip. But it is definitely a frequency-dependent non-linear characteristic with the speed. The distortion characteristics at 15ips were very different to that at 30ips. [13]

Distortion can take many different shapes and forms, and there is a range of different terminology which can be used to describe distortion effects:

- **Clipping** (Soft-/Hard-Clipping) is caused by running a signal into an amplifier (often in a mixing console) past the 'analogue maximum'. The result is a 'rounding off' of the waveform, causing distortion. In the digital domain, the waveform reacts more drastically and 'squares off', creating a very intense 'hard clip' sound, as opposed to an analogue 'soft clip'. Here, we're talking about the harder form of clipping that produces a brighter form of distortion, usually due to both the circuit/software and the amount of gain used to boost the signal.

- **Overdrive** as a term generally refers to feeding more signal into an input than its inherent headroom can handle cleanly, but for our purposes it is a type of analogue soft-clipping that can add a warm, yet transparent form of distortion. Usually associated with guitars (and synthesisers), it's used to add a bit of character, but not to drastically alter the sound.
- **Fuzz** is traditionally achieved by clipping transistors hard in a way that makes them sound broken and is a more extreme form of distortion than overdrive. Fuzz distortion has a very noisy and bright characteristic sound at high gain levels. It sounds 'broken'; indeed, the early fuzz distortion devices were discovered by using malfunctioning and less-than-ideal hardware parts.
- **Downsampling** or sample rate reduction simulates the lower fidelity achieved by playing audio through digital to analogue converters (DACs) with a low sample rate (below 20Khz). This emulates the sound of early digital equipment and provides a grainy and potentially harsh sound that can make sounds stand out in a mix without drastically changing their character.
- **Bit-crushing**, similar to sample rate reduction, is achieved by lowering the bit depth of the audio. In extreme cases, the effect makes the audio extremely loud due to the reduction in dynamic range and can dramatically increase the noise floor. Such retro aesthetics, as embraced by artists such as Portishead and Massive Attack, have been discussed by Brovig-Hanssen and Danielsen in *Digital Signatures* [14]. The process for creating such lo-fi sounds and the language associated with this creative process have also been discussed by Bromham et al. [15].
- **Valve** or **Tube** is perhaps the oldest form of distortion to be heard in recorded music. Valves distort in a musically pleasing way, that has many guitarists still favouring them in amplifiers more than half a century after they fell out of use in other applications.
- **Tape saturation** refers to the subtle warming, thickening and transient softening that happens when a sound is recorded to and played back from magnetic tape. This begins as a gentle compression effect, and as the level is increased, the magnetic particles of the tape start to become exhausted, so the audio becomes increasingly saturated. Tape saturation is one of the most musically pleasing forms of distortion, and as tape machines are complex devices, the resulting distortion is non-linear and computationally expensive to emulate.

1.5   Harmonic Distortion, Saturation, and Colouration in the Studio

There is a less overt type of distortion, that manifests itself in more subtle forms, where music equipment distorts, saturates or colours sound in aesthetically pleasing ways. It is important to distinguish between the terms 'distortion', 'saturation' and colouration', as during interviews conducted with producers and mix engineers, it became apparent that there is a definite sense of dissemination among these three states. The perception of such processes is often one of warmth, air, weight or solidity. This type often manifests itself in tape-based formats and in classic mixing consoles such as the Putnam 610 valve console, the EMI TG console at Abbey Road and the Helios designed by Dick Swettenham at Olympic. Later incarnations from Neve, Trident and Solid State Logic all impart a sonic signature which can largely be attributed to the non-linear properties exhibited by the mixing desk. In the case of tape, the saturation and indeed compression characteristics are also part of the appeal and charm.

A similar aura exists for ubiquitous pieces of studio outboard. With the move towards a more 'in-the-box' or hybrid approach, devices such as the Thermionic Culture Vulture, SSL Fusion and Black Box HG2 have gained notoriety for imparting a subtle harmonic enhancement of the sound.

Distortion can also be used for increasing perceived loudness and for shaping of sound and envelope and transient control. By rounding off peaks in the audio, it is possible to increase perceived loudness of audio material. Paul Frindle is the designer of the Sony Oxford Inflator, a tool many mix engineers use for this purpose. The discussion was shaped by a problem a mix engineer faced when trying to increase and match perceived loudness of other mixes. He talked about the idea which was inspired by the valve amplifier design he built in the late 60s/early 70s. He worked out that he could never get the same perceived loudness with solid-state amplifier design due to the way they added harmonic content and shaped the sound. Valve technology envelopes and shapes transient material in interesting ways. This can also be seen in the case of many iconic dynamic range compressors which exhibit similar pleasing distortion properties [16]. The Fairchild 670 Vari-mu compressor is far more appreciated for its ability to colour with harmonic distortion than compress audio. The concepts of misappropriation and context in music production have also been discussed by Lefford et al. with reference to their roles in intelligent mixing systems [17].

Harmonic distortion can be easily created through dynamic range compression, particularly for low frequencies and when the release time is set too fast (too fast an attack time can cause a similar effect). The reason is that the compressor only acts on the waveform peaks, changing the waveform shape and thus introducing odd harmonic distortion artefacts.

Hugh Robjohns, Technical Editor for *Sound on Sound* magazine [18]

1.6   Conclusion

The history of distortion is one of subversion and misappropriation, particularly when one considers, and takes into account, the transition from analogue to digital technology, where the nature of its use has undergone significant change. However, when one chooses to interpret distortion, we acknowledge that it means different things to different people depending on the context and user's perspective. Because of its esoteric nature, it is notoriously difficult to define and articulate the true essence of its impact. Its effect can be edgy, brittle and aggressive or it can be subtle, warm and give body—or indeed all of the things in between.

References

[1] Rogers, J. Can You Really Get the Analogue Experience from Software? *Protools Expert*, www.pro-tools-expert.com/production-expert-1/can-you-really-get-the-analogue-experience-from-software [Accessed 17/04/2022].
[2] Milner, G. (2010). *Perfecting Sound Forever*. London: Granta Publications.
[3] Reiss, J.D. (2018). *Do You Hear What I Hear? The Science of Everyday Sounds*, Inauural Lecture. London: Queen Mary University.
[4] Moffat, D. (2021). AI Music Mixing Systems. In *Handbook of Artificial Intelligence for Music* (pp. 345–375). Cham: Springer.
[5] Case, A. (2012). *Sound FX: Unlocking the Creative Potential of Recording Studio Effects*. New York: Routledge.
[6] Wilmering, T., Moffat, D., Milo, A., & Sandler, M.B. (2020). A History of Audio Effects. *Applied Sciences*, 10(3), 791.
[7] Leckie, J. (2020, February 24). Interviewee, Interview with Gary Bromham.
[8] Scott, J. (2021, September). Fuzz Was the Future: How a Happy Accident Sparked the Rock Revolution of the 60s. *Guitar* (October 2021), 69–70.

[9] Motter, P., & Schu, P. *A History of Marshall Amps: The Early Years*, https://reverb.com/uk/news/a-history-of-marshall-amps-the-early-years [Accessed 28/09/2021].

[10] Massey, H. (2015). *The Great British Recording Studios*. Milwaukee, WI: Hal Leonard.

[11] Dregni, M. (2013). *Maestro Fuzz Tone*, www.vintageguitar.com/17397/maestro-fuzz-tone/ [Accessed 27/09/2021].

[12] The Highway Star. (1995). *Jon Lord Equipment*, thehighwaystar.com [Accessed 27/09/2021].

[13] Jenkins, C. (2020, February 18). Interviewee, Interview with Gary Bromham.

[14] Brovig-Hanssen, R., & Danielsen, A. (2016). *Digital Signatures: The Impact of Digitization on Popular Music Sound*. Cambridge, MA: MIT Press.

[15] Bromham, G., Moffat, D., Barthet, M., Danielsen, A., & Fazekas, G. (2019, September). The Impact of Audio Effects Processing on the Perception of Brightness and Warmth. In *Proceedings of the 14th International Audio Mostly Conference: A Journey in Sound* (pp. 183–190).

[16] Moore, A., Till, R., & Wakefield, J.P. (2016). *An Investigation into the Sonic Signature of Three Classic Dynamic Range Compressors*.

[17] Lefford, M.N., Bromham, G., Fazekas, G., & Moffat, D. (2021, March). *Context Aware Intelligent Mixing Systems*. Audio Engineering Society.

[18] Robjohns, H. (2020, March 9). Interviewee, Interview with Gary Bromham.

2 The Development of Audio Software with Distortion

Eric Tarr

2.1 Introduction

The digital distortion tools used by music producers and mixing engineers are the product of a development process by signal processing engineers and software developers. This chapter aims to provide music producers and mixing engineers with some insight into the functionality of distortion software plug-ins in digital audio workstations (DAWs). This information can be used to better appreciate, understand, and compare different options when making production and mixing decisions.

The various types of distortion, from subtle to extreme, are the consequence of various algorithms implemented during the creation of the tools. These algorithms can be relatively simple and also very complex from a mathematical and computational standpoint. They can be standard and common algorithms, or also highly customized and unique. The development process may involve research and the close study of specific systems for the purpose of algorithmic modelling. Some software is meant to recreate the distortion that occurs in the analog domain, and other software creates uniquely digital distortion.

The creation of software typically follows a timeline with development and testing cycles. Signal processing engineers start by prototyping an algorithm. Initially, this might not look like a fully functioning plug-in. Rather, it may just be a script of computer code that synthesizes and processes test signals. After the proof-of-concept for the algorithm is complete, software engineers create the plug-in that provides a user interface and parameterized control of the algorithm. If possible during this stage, engineers look for opportunities to make the algorithm efficient from a CPU perspective. After the plug-in is functioning, quality assurance engineers perform internal testing and look for bugs in the software. Sometimes this testing finds something that needs to be fixed, and the plug-in is sent back to the software or signal processing engineers to revise. When the engineers believe the software is performing satisfactorily at all of these stages, the plug-in is sent to external beta testers to review and provide feedback. This cycle may be repeated multiple times until the software is ready for public release. It should be mentioned that some software is created with multiple engineers on teams for each discrete stage of development. In other cases, one person might do all of these aspects of development on their own.

2.2 Types of Software with Distortion

Many different types of distortion software are available to use. This software is usually categorized in a DAW as either "Harmonic" software or simply "Distortion." Things like clippers, saturators, rectifiers, and bit crushers are common examples to find categorized this way. Then there are specialized tools for guitar and bass that simulate the entire signal chain from the output of the instrument through pedals, amplifiers, cabinets, and post effects. Distortion is an essential aspect of the signal chain, including things like overdrive and fuzz pedals, as well as vacuum tube pre-amplifiers and power amplifiers. Additionally, other software tools have distortion as a secondary component. Examples include digital models of analog equalizers and compressors, as well as recreations of vintage digital reverberation (reverb) with a low bit depth.

2.2.1 Harmonic Distortion

Conventional types of distortion effects are used to add new harmonics to a signal. Many different types of distortion are intended for this purpose, including Saturation, Rectification, Lo-Fi, Exciter, Enhancer, Hard Clipping, and Soft Clipping. In Figure 2.1, the harmonics produced by a hard clipper for a sine wave input signal of 2500 Hz are shown. One distinguishing factor among these different types is which harmonics are produced. Some distortion effects produce even and odd harmonics, whereas others produce only even or only odd harmonics. From Figure 2.1, it can be seen that the hard clipper primarily produces odd harmonics (e.g. 7500 Hz, 12500 Hz). Another distinguishing factor is the relative amplitude of each harmonic. Subtle types of distortion typically have a relatively low amplitude for the new harmonics, and vice versa for

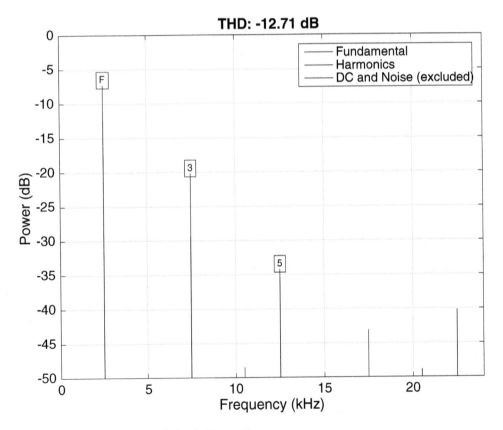

Figure 2.1 Harmonic distortion of a hard clipper effect.

extreme types of distortion. All of these differences in harmonic frequency and amplitude play an important part in the perceptual qualities of distortion. As a side note, Lo-Fi distortion that simulates the process of reducing the sampling rate of a signal can produce aliased frequencies. These inharmonic tones are, more often than not, at frequencies other than even or odd multiples of the source signal.

2.2.2 Simulations of Hardware

Many mixing engineers have a preference for analog hardware effects compared to the sound of basic digital software effects. In contrast to the clean and linear recording and processing of digital technology, the primary difference in analog technology is the nonlinear distortion (arguably). This is true not only for analog distortion effects but also for equalizers, compressors, tape recording, and vacuum tube amplifiers. Simulating this distortion not only becomes an essential part of attempts to accurately recreate hardware effects in software form, but it also serves as a general guide to software developers aiming to create digital effects that are musical and usable.

Considering the entire signal chain of analog recording, many stages of subtle distortion can be simulated in software. Combining multiple distortion effects together by stacking or cascading one type of saturation into another in some ways mimics the flow of a signal that is distorted a small amount by a console's pre-amp and transformers, followed by additional clipping when printed onto tape, and again by other hardware in the signal path.

Another place in the signal chain of a traditional hardware recording studio where distortion could be found is in artificial reverb units. Although it may not be as expected, many classic hardware digital reverbs sometimes had distortion in the signal path. More recently, this is used in software reverb plug-ins that aim to recreate the processing of hardware reverb from the 1970s and 1980s. Two examples are the EMT 250 and Lexicon 224 hardware reverbs. In general, the early hardware products that implemented artificial reverb digitally had a limited amount of processing power and relied on the relatively primitive A/D and D/A converters of the time period. In order to maximize the capabilities of the hardware, audio was quantized to 10 or 12 bits during the A/D conversion prior to the process of adding reverb. Compared to 16- or 24-bit audio, which are now standard for audio recording, the lower bit depth can be considered as a distortion and is audible in some cases. When companies that created digital hardware reverb transitioned to creating software reverb, some of the same algorithms were re-factored to run identically on a personal computer. The sound of the early hardware reverb units became synonymous with the musical aesthetics from the same time period. This spanned a range of genres including pop, hip-hop, rock, and reggae. For music producers attempting to achieve a retro sound from this era, using reverb with a low bit depth is one important part.

In addition to the previous examples, software that simulates a guitar amplifier, or more generally the entire signal chain of processing for a guitar, can be considered a specialized type of distortion software. Bass amp simulation software can also be grouped together here, along with things like a rotary speaker cabinet usually associated with an electric organ. Certainly in the case of the tools used with the electric guitar, many of the modules have a primary purpose of adding some type of distortion.

This could include things like overdrive, fuzz, and even boost pedals. Simulation of these types of pedals in software typically involves simulating the effect of circuit components like diodes [1], bi-polar junction transistors (BJTs) [2], field-effect transistors (FETs), or operational amplifiers (op-amps) [3].

The sound of an actual guitar amplifier is as much about the harmonic distortion it imparts on a signal compared to linear things like the equalizer tone stack. Distortion from the guitar amplifier can come from both the pre-amplifier section [4] and the power amplifier section [5]. Simulations of guitar amplifiers involve modelling the distortion from tubes [6], transistors [7], and transformers [8] in some cases.

2.2.3 **Bit Reduction**

Extreme bit reduction is sometimes used as part of a Lo-Fi algorithm. The process of bit reduction adds quantization noise to the signal. By reducing the bits of a periodic signal, quantization noise is manifested as harmonic distortion because the noise is correlated with the original signal. By adding dither noise prior to bit reduction, quantization noise is decorrelated with the original signal and does not produce the same harmonic relationship.

Dither noise has an important perceptual benefit when applied to a signal. Listeners are much more adept at perceiving things that are harmonically related. This works to the advantage of a listener when they are trying to do tasks like isolate a speaker's voice in the presence of background noise. However, it poses a problem when the bit reduction process adds harmonic distortion to a signal that would ideally be inaudible. The solution is to decorrelate the quantization distortion by using additive noise, which essentially spreads the quantization error across the spectrum rather than focus it at the harmonics of the input signal.

2.3 Aliasing and Oversampling

There can be undesirable, audible problems when using distortion effects if the software is not designed properly. In particular, the problem is digital aliasing. Distortion effects produce harmonics higher in frequency than the original source signal. Harmonics are related to the frequency of a source signal by even and odd multiples. For a given sample rate, those harmonics can theoretically be higher in frequency than the highest possible sampled frequency (called the Nyquist frequency). Due to the limited bandwidth possible for digital signals, those harmonics will not actually occur above the Nyquist frequency. Instead, those frequencies are aliased down to lower frequencies within the available bandwidth. In many typical cases, the aliased frequencies become in-harmonic frequencies relatively to the original source signal (i.e. not whole number multiples). This can be audible and usually perceived as undesirable.

When distortion occurs in the analog domain, there is not the same limit on the bandwidth of the distorted signal. Therefore, harmonics can extend up the spectrum without being aliased to lower frequencies in the same way as digital. Listeners are not able to perceive the harmonics outside the range of human hearing. Nonetheless, this type of behaviour in the analog domain is desirable for digital distortion. It can be approximated by temporarily oversampling a signal as part of the effect.

Oversampling is a multi-step process to increase the sampling rate of a signal prior to distortion and then decrease the sampling rate after distortion. The process of increasing the sampling rate is called up-sampling. The process of decreasing a sampling rate is called down-sampling. By up-sampling, the maximum bandwidth for a signal is increased. By doing this prior to distortion, this process essentially makes space in the spectrum for new harmonics to fit before they are aliased. After distortion is applied to the signal, a low-pass filter (called an anti-aliasing filter) is applied to the distorted signal to decrease the amplitude of harmonics above the original Nyquist frequency. Lastly, down-sampling is applied to a signal to produce a distorted output signal at the original sampling rate, but with reduced aliasing compared to using distortion without oversampling.

The following is an example to clarify some of the details of oversampling. Suppose a signal is sampled at a rate of 48000 samples per second. Therefore, it has a bandwidth that can represent signals up to 24000 Hz. If the signal has a frequency at 10 kHz, it would theoretically have harmonics at 20 kHz, 30 kHz, etc. after it is distorted. Any harmonics that would have a frequency above 24 kHz will be aliased to lower frequencies (e.g. 30 kHz becomes 18 kHz). If this signal is oversampled by a factor of two before distortion, it will have a temporary sample rate of 96 kHz and a Nyquist frequency of 48 kHz. If distortion is performed at this higher sample rate, the 30 kHz harmonic is not aliased to a lower frequency. Instead, it is represented as 30 kHz. At this point, an anti-aliasing filter can be applied to reduce the amplitude of this harmonic below the range of hearing (-144 dB). When the signal is down-sampled to the original rate of 48 kHz, the alias of 30 kHz has effectively been filtered out.

The downside of oversampling is that it requires more CPU processing for a distortion effect. In order to allow for a trade-off between sound quality (higher oversampling) and CPU processing (no/low oversampling), some distortion software will have an option to switch on and off a "high quality" (HD) mode. In other software, the explicit choices of 2x, 4x, 8x, etc. are provided to control exactly how much oversampling is used.

2.4   Waveshaper Algorithms

The algorithms used in these different types of software can be as simple as a single mathematical function (e.g. absolute value) or very complex in order to model the circuitry of an analog system (see Section 5).

A common approach for implementing distortion in software is to use a waveshaper. This general technique applies a mathematical function and/or computer logic to change the amplitude of a signal in a nonlinear way. This type of processing can be visualized using a waveshaper curve, also sometimes called the transfer curve or characteristic curve. This visualization shows a plot of the relationship between the input amplitude to the waveshaper and the output amplitude. The relationship between a signal's waveform and waveshaper curve is shown in Figure 2.2. This type of visualization can be seen in some audio plug-ins including Newfangled Audio Saturate and iZotope Trash 2.

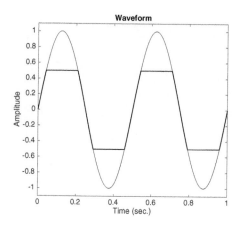
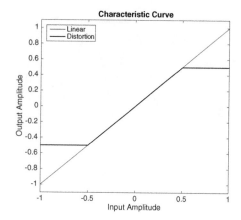

Figure 2.2 Waveform and characteristic curve of a hard clipper effect.

2.4.1 Hard-clipping and Rectification Algorithms

Some of the waveshaping algorithms can be characterized as mathematical functions with discontinuities. In other words, the slope of the function has abrupt changes at one or more points. Therefore, the waveshaper can be considered to have multiple operating regions. Essentially, this means that the amplitude of the input signal is processed one way in one amplitude region and a different way in another amplitude region. Some examples of this type are hard-clipping, full-wave rectification, and half-wave rectification.

A technique like hard-clipping mimics the hard-limiting of converters used during the digitization process. In this type of distortion, the peaks of the signal that go above a certain threshold are clipped to a maximum magnitude value. Otherwise, the signal passes through a linear operating region of the algorithm. This is similar to the behaviour of an analogue-to-digital converter because the hardware has a maximum amplitude it can process before the signal peaks are limited.

Two other examples are full-wave rectification and half-wave rectification. In the case of full-wave rectification, the portions of the signal with negative amplitude are changed to have positive amplitude. In the case of half-wave rectification, the portions of the signal with negative amplitude are changed to have an amplitude of zero. A visual comparison of these two types of distortion is shown in Figure 2.3 and Figure 2.4.

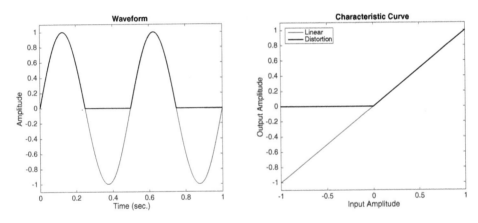

Figure 2.3 Waveform and characteristic curve of a half-wave rectifier effect.

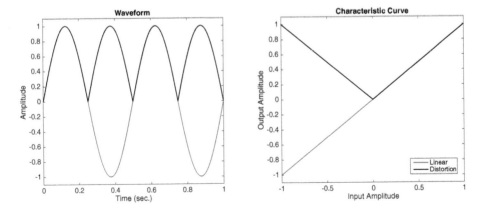

Figure 2.4 Waveform and characteristic curve of a full-wave rectifier effect.

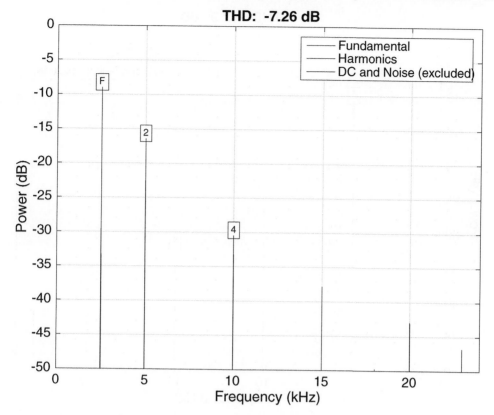

Figure 2.5 Harmonic distortion for half-wave rectifier effect.

For half-wave rectification, the time length of a cycle (repeated pattern) is identical to the time length of a cycle for the input signal. Therefore, the harmonic series for half-wave rectification has the same fundamental frequency as the input signal, shown in Figure 2.5. However, for full-wave rectification, the time length of the repeated pattern is shortened by half. This changes the fundamental frequency of the signal to increase by a factor of two, or an octave. This can be seen in Figure 2.6 of the harmonic series for full-wave rectification, where the lowest harmonic (5000 Hz) is an octave above the fundamental frequency of the input signal (2500 Hz).

2.4.2 Soft-clipping Algorithms

Another category of waveshaping algorithms is soft-clippers. Compared to the aforementioned algorithms, soft-clippers have a smooth curve to map the input amplitude to output amplitude over the entire region of operation. In other words, the function does not have any points of discontinuity. This smooth curve can have many different shapes as long as it is not a straight line, which produces a linear result. Conceptually, these algorithms behave more similarly to analog distortion (e.g. tubes, tape, etc.), in which the amplitude of a signal is gradually distorted as it increases, as opposed to being hard-clipped at some maximum point. However, it should be noted that analog circuits can be used to produce waveshapers very close to the ideal full-wave, half-wave, and hard-clipping algorithms.

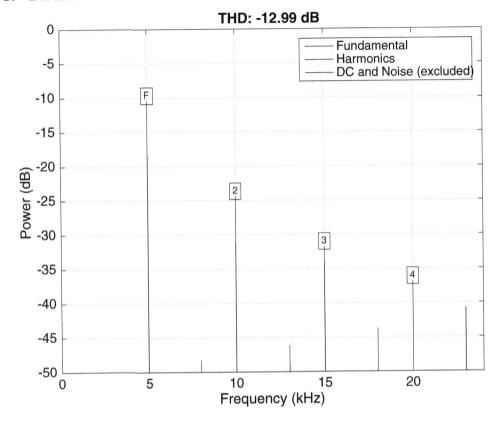

Figure 2.6 Harmonic distortion for full-wave rectifier effect.

One way to create a soft-clipping effect is to use a polynomial function (e.g. cubic or quartic functions) [9]. Figure 2.7 shows the waveform and characteristic curve for cubic distortion. Alternatively, arctangent or hyperbolic tangent functions are also found in audio software as soft-clippers. Figure 2.8 shows the waveform and characteristic curve for arctangent distortion.

Cubic and quartic functions produce a subtler distortion compared to arctangent or hyperbolic tangent functions. Cubic distortion only produces a third harmonic above the fundamental frequency (Figure 2.9), and quartic distortion produces the third and fifth harmonic. Conversely, arctangent and hyperbolic tangent functions theoretically produce an infinite series of harmonics above the fundamental (Figure 2.10). One example of a software effect that appears to use cubic distortion is the Sonnox Oxford Inflator. When the effect slider is increased, the third harmonic above the fundamental is produced. The result is a subtle, musical saturation that can sound pleasing on anything from individual instruments to entire mixes.

2.5 Models Combining Waveshaping with Filtering

On their own, waveshaping algorithms act as an effects block that adds a certain amount and type of distortion, and it is applied the same to all frequencies. In some cases, it is desirable to have a tool that acts consistently across all frequencies. In other cases, it may be desirable to have different frequencies distort in different ways. This can be accomplished by combining

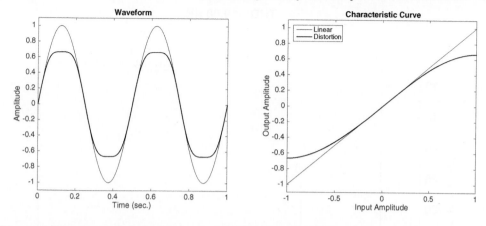

Figure 2.7 Waveform and characteristic curve of cubic distortion effect.

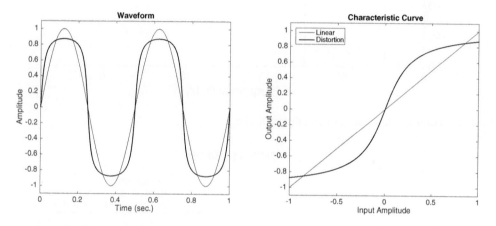

Figure 2.8 Waveform and characteristic curve of arctangent effect.

waveshaping with spectral filtering. There are several options for how this can be accomplished based on combining the effects in series or parallel, as well as in different orders.

As the complexity of a distortion algorithm increases, it is much more difficult to determine what underlying techniques are used in commercial software as this information is kept as a trade secret. The methods described subsequently are general examples found in academic literature on the topic.

2.5.1 Series Combination of Waveshaping with Filtering

Conceptually, the most basic way waveshaping and filtering can be combined is to cascade the effects in series. The various options of combinations are known generally as the Hammerstein-Weiner models. One of the options, described as the Hammerstein model, applies the nonlinear waveshaper first, followed by the spectral filter. Another option, described as the Weiner model, applies the spectral filter first, followed by the waveshaper. In a combined Wiener-Hammerstein model, one filter block is used before the nonlinear waveshaper and another filter block is used

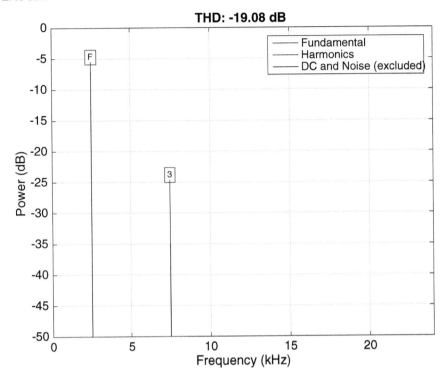

Figure 2.9 Harmonic distortion for cubic distortion effect.

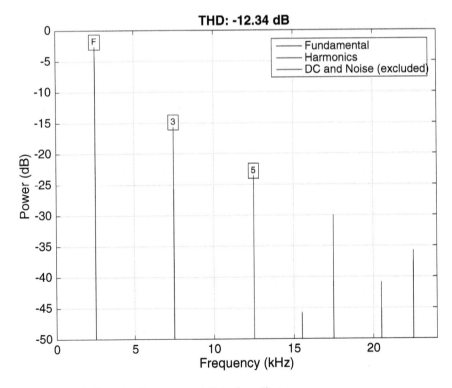

Figure 2.10 Harmonic distortion for arctangent distortion effect.

afterwards. In a combined Hammerstein-Wiener model, the chain begins with a waveshaper, followed by a filter, and then followed by an additional waveshaper.

Here are a few example use cases to discuss the various options. By applying filtering after waveshaping, an engineer could control the tone of the harmonics introduced by the distortion (e.g. reduce amplitude of high harmonics). By applying filtering before distortion, it is possible to change the relative amount of distortion applied to different frequency regions. If a boost is applied to low frequencies prior to waveshaping, this will effectively drive more low frequencies into the distortion, causing them to have a higher amount of distortion than if the filter was removed.

2.5.2 Parallel Processing

Some software algorithms use parallel processing internally in order to achieve other variations of distortion. Multi-band processing can be set up by having a low-pass filter and high-pass filter in parallel with an identical cut-off frequency. Then, the low band and high band can be processed by different waveshapers. As an example, a subtle type of distortion could be used for the high frequencies and an extreme type of distortion could be used for low frequencies. This can be expanded by adding band-pass filters in the middle of the spectrum to increase the number of bands and options for distortion. This general approach can be found in Fab Filter Saturn 2 including user-selectable, multi-band processing.

Another methodology involving parallel processing is called Volterra Series [10]. In this approach, the parallel paths are based on applying different exponential functions to the input signal. One path could use a quadratic function with the input signal and another path could use a cubic function. The highest power of the exponential function used on that path corresponds with the highest harmonic of the fundamental frequency produced (e.g. cubic produces third harmonic). In addition to the nonlinear exponential function on each path, one or more filters are applied before and/or after the distortion function. The processing on each parallel path is sometimes called a *Volterra Kernel*.

A nonlinear autoregressive (NAR) model offers a different approach with interesting characteristics [11]. To understand this model, it is first worth describing the simpler, linear autoregressive (AR) model. The AR model is typically used for basic linear filtering and equalization purposes. It works at the sample level of a signal by calculating the current sample of the output signal by a linear combination of previous samples of the output signal (feedback) and input signal. A linear combination means the various samples from the input and output are added together, which effectively acts as separate parallel feedback paths for the previous samples. To provide different options in the filtering, the samples are first multiplied by a scaling factor before the addition. These multiplication and addition operations are linear operations.

A NAR model is an extension of the linear model. In a NAR, a sample of the output signal is calculated by adding together the linear terms of the previous output and input samples as well as previous samples that have had a nonlinear operation applied to them. This nonlinear operation could be exponential functions like quadratic and cubic functions, as well as discontinuous functions like hard-clipping or half-wave rectification.

The interesting characteristic of the NAR compared to Volterra Series, multi-band processing, or the Hammerstein-Wiener model is that the filtering and distortion are not treated as separate blocks of processing. In the approaches other than NAR, either the filtering happened first on its own followed by a separate block of distortion, or vice versa. In a NAR model, the nonlinear processing is embedded within the feedback paths of the filter. Theoretically, this can offer unique results and is similar conceptually to the interconnected nature of electronic circuits, where filtering and distortion may not always be independent.

2.6 Analog Modelling

Digital software can be created to model the processing of an analog effect. In some cases, this is based on actual measurements of hardware gear. In other cases, the model is based on a diagram of a circuit called a schematic. This diagram describes the components used in a circuit and how they are connected together. A mathematical model of each component is used to calculate how a signal changes while traveling from input to output. There are also ways to model the overall processing of analog effects that are not concerned with the internal components of the system itself.

2.6.1 Black-box Modelling

On one end of the modelling spectrum, black-box techniques treat the inner workings of the audio effect as unknown [12]. Instead, the goal is to learn how input signals change to become output signals after going through the effect. In some ways, it can be of no concern about what happens to a signal while it is inside hardware since only the output is ever used. Therefore, as long as a digital model recreates the overall effect of hardware, it can be very useful.

The general process of performing black-box modelling involves putting various input signals through an effect, followed by measuring, analyzing, and comparing the output signal. The exact input signal for the process typically depends on the type of system to be modelled and also the underlying technique for approximating the system behavior.

Some readers may be familiar with the Kemper Profiling Amplifier. This device uses black-box modelling to analyze the signal chain of guitar effects and amplifiers. In order to capture a profile, users send specific input signals into a guitar amplifier while measuring its output signal. The Kemper hardware then analyzes the result to create a profile or model of the amplifier. Another commercial example (with some publicly available information about it) is the Acoustica Audio Nebula system. The general concept, described on the company's website, is that each piece of hardware is modelled using a "sampling approach" consisting of many different signal measurements for various parameter settings of the device. Furthermore, the company developed a "proprietary audio processing technology, based on Vectorial Volterra Kernels" for its modelling, which is a black-box technique for nonlinear systems [13].

The underlying foundation for this approach is the black-box technique for modelling linear systems [14]. This can be considered the simplified case when the system does not produce distortion. The technique for modelling a linear system is to capture an impulse response (IR). The measurement is based on the assumption that this single snapshot in time captures how the system will always perform at other times in the future. In other words, the system cannot change over time or else the IR is not a valid way to represent it. The term for this requirement on the system is that it is time-invariant. The other requirement for an IR to represent a system is that the system is linear, meaning it has to process samples with a low amplitude in a similar way as samples with a high amplitude. Otherwise, a single IR does not capture the entire region of operation. The main thing an IR does capture is the time delay of an effect. This could be used to model echo effects with individual discrete delays, reverb effect with many closely spaced delays, and filter/spectral effects with very short time delays. An IR does not work for modelling distortion effects, modulation effects, and dynamic range effects as examples.

In order to measure nonlinear systems, other techniques must be used [15]. One approach is to assume the effect is acting as a waveshaper and attempt to measure the output amplitude for various input amplitudes. In order to remove any time delay introduced through spectral filtering by the system, signals with a steady-state amplitude can be sent into the effect. As long as the length of each steady-state step is much longer than the delay of the effect, then it is trivial to recover the output amplitude for each input amplitude.

The parameters of mathematical function can be fit for a line passing through those measurement points. One approach is to fit a polynomial to the curve after selecting an order for the function. Another approach is to fit a trigonometric function (e.g. arctangent or hyperbolic tangent) with several degrees of freedom to optimize. The advantage of the polynomial function is that it can fit a wider range of waveshaping curves, including even-order systems. However, the polynomial function is limited in that the highest harmonic it will produce is the order of the polynomial. The trigonometric functions work well in cases when the measurements appear to already fit their general shape.

The drawback of black-box modelling is that it can be an over-simplification of the complex systems it attempts to recreate. Many hardware effects are made up of individual sub-systems which interact with each other in complicated ways. By treating the entire effect as a waveshaper and filter, those interactions are missing in the model. This has led to the development of other techniques which attempt to model the individual components of a system, along with the way they interact.

2.6.2 Component-level Modelling

Component-level modelling, also called white-box modelling, attempts to create a digital algorithm based on modelling every individual component of an analog circuit. Algorithms use mathematical models for resistors, capacitors, op-amps, diodes, transistors, tubes, etc., and how they behave as interconnected components to recreate the effect of circuitry. One company that is known for using this approach to modelling in their plug-ins is Universal Audio.

There are several benefits of component-level modelling. First, there is the potential for the model to be very accurate because the entire signal path is included discretely. This can be very important for nonlinear circuits because all of the components may influence how the circuit distorts a signal. Second, variable components like potentiometers [16] can be included in the model, allowing for parameterized control of a full range of hardware settings from one model. Conversely, black-box modelling relies on separate measurements for fixed settings, which may or may not have a good way to change from one parameter setting to another.

Generally, the process of creating a component-level model requires turning a circuit schematic into one or more mathematical equations and then turning those equations into computer code for software. A few different methods have emerged in academic literature and commercial software for this general process. These methods include the Laplace Transform and Bilinear Transform for simple circuits [17], Sparse Tableau Analysis [18], Modified Nodal Analysis (MNA), state-space modelling [19], Discrete Kirchhoff (DK) method [20], and Wave Digital Filters (WDF) [21]. All of these methods have their basis in linear circuit analysis and have been extended to also apply to nonlinear circuits.

A mathematical challenge arises when attempting to model circuits with nonlinearities. The challenge is that the nonlinear equations are in a category called implicit equations with no analytic solution. For context, the opposite of an implicit equation is an explicit equation. An explicit equation is one where it is possible to isolate and solve for the variable of interest (called an analytic solution). For implicit equations, it is not possible to isolate and solve for the variable of interest (using the value of the output signal for the circuit). Instead, an iterative process of guessing-and-checking is used to find the answer to the equation. A common technique used in circuit modelling for this iterative process is called the Newton-Raphson method [22].

2.6.3 Gray-box Modelling

There are also techniques for analog modelling that fit between black-box modelling and component-level (white-box) modelling. These techniques are generally called gray-box modelling to indicate that it isn't one of the other categories [23]. Within the category of gray-box modelling there are many variations, and they are sometimes customized for an individual circuit. One noteworthy approach to gray-box modelling aims to preserve the general topology (structure) of a circuit without modelling every individual component.

Some circuits are extremely complicated to model at the component level, and for some circuits modelling each component is unnecessary. For both of these cases, it is realistic to model some but not all of the components. Signal processing engineers might choose to model some of the simpler components of the circuit like resistors and capacitors, but choose to treat other components and sections of a circuit as a black-box (e.g. transformer output stage) due to their complexity. As a particular example, when modelling a compressor, there is the audio path of the signal going from input to output as well as the detection path which a listener will never hear. Engineers may choose to model the audio path at the component level but not the detection path. If an engineer knows a sub-circuit is responsible for converting a signal's amplitude from the linear scale to the decibel scale, that conversion could be done easily as a digital calculation rather than modelling each of the components necessary to do this process in the analog domain.

One approach for gray-box modelling with a general, mature theoretical foundation is Virtual Analog (VA) filter design [24], developed and published by an engineer at the company Native Instruments. This approach includes some aspects of component-level modelling initially as its basis for modelling simple filters. For circuits with greater complexity, this approach relies on modelling the overall topology of a circuit, rather than every individual component. This is accomplished using a Topology Preserving Transform (TPT). This is applied to second-order state-variable and fourth-order ladder filters by separating the topology of the circuit into filter blocks based on an integrator circuit. This allows for circuits like a Moog Ladder filter to be modelled conceptually without needing to model each of the transistor components separately. Additional examples for Korg [25] and Oberheim synthesizer filters have also been shown to be successfully modelled using this approach.

2.7 Discussion

As has been described, many different types of digital distortion are used in audio software. Implementations can range from relatively simple to very complex, each with different purposes and uses. There are several well-established approaches for creating effects with sophisticated distortion algorithms. There also continues to be research to develop new approaches within the field of audio signal processing and software development. In regards to circuit modelling, the trend appears to be that new digital plug-ins will have improved accuracy at replicating their analog hardware counterparts as the technology advances. All things considered, music producers and mixing engineers will almost certainly have new and unique distortion tools to use in the future.

References

[1] Yeh, D. T. Simulation of the diode limiter in guitar distortion circuits by numerical solution of ordinary differential equations, *Conf. on Digital Audio Effects* (2007), pp. 197–204. Accessed from https://dafx.de/paper-archive/2007/Papers/p197.pdf

[2] Yeh, D. T. Automated physical modeling of nonlinear audio circuits for real-tie audio effects—part II: BJT and vacuum tube examples, *IEEE Transactions on Speech and Audio Processing*, vol. 20 (2012), pp. 1207–1216.

[3] Muller, R., Helie, T. A minimal passive model of the operational amplifier: application to Sallen-Key analog filters, *Conf. on Digital Audio Effects* (2019). Accessed from https://dafx.de/paper-archive/2019/DAFx2019_paper_6.pdf

[4] Macak, J. Guitar preamp simulation using connection currents, *Conf. on Digital Audio Effects* (2013). Accessed from https://dafx.de/paper-archive/2013/papers/27.dafx2013_submission_10.pdf

[5] Cohen, I., Helie, T. Real-time simulation of a guitar power amplifier, *Conf. on Digital Audio Effects* (2010). Accessed from https://dafx.de/paper-archive/2010/DAFx10/CohenHelie_DAFx10_P45.pdf

[6] Dempwolf, K., Zolzer, U. A physically-motivated triode model for circuit simulations, *Conf. on Digital Audio Effects* (2011). Accessed from https://dafx.de/paper-archive/2011/Papers/76_e.pdf

[7] Holmes, B., Holters, M., van Walstijn, M. Comparison of germanium bipolar junction transistor models for real-time circuit simulation, *Conf. on Digital Audio Effects* (2017). Accessed from https://dafx.de/paper-archive/2017/papers/DAFx17_paper_28.pdf

[8] Holters, M., Zolzer, U. Circuit simulation with inductors and transformers based on the Jiles-Atherton model of magnetization, *Conf. on Digital Audio Effects* (2016). Accessed from https://dafx.de/paper-archive/2016/dafxpapers/08-DAFx-16_paper_10-PN.pdf

[9] Sullivan, C. R. Extending the Karplus-strong algorithm to synthesize electric guitar timbres with distortion and feedback, *Computer Music Journal*, vol. 14 (1990), pp. 26–37.

[10] Tronchin, L. The emulation of nonlinear time-invariant audio systems with memory by means of Volterra Series, *Journal of the Audio Engineering Society*, vol. 60 (2012), pp. 984–996.

[11] Dobrucki, A., Pruchnicki, P. Application of the NARMAX-method for modelling of the nonlinearity of dynamic loudspeakers, *106th AES Convention* (1990), p. 4868. Accessed from https://www.aes.org/e-lib/browse.cfm?elib=8312

[12] Eichas, F., Zolzer, U. Black-box modeling of distortion circuits with block-oriented models, *Conf. on Digital Audio Effects* (2016). Accessed from https://dafx.de/paper-archive/2016/dafxpapers/06-DAFx-16_paper_16-PN.pdf

[13] Acoustica Audio Nebula. Accessed August 2021 from www.acoustica-audio.com/store/t/nebula

[14] Farina, A. Advancements in impulse response measurements by sine sweeps, *122nd AES Convention* (2007). Accessed from https://www.aes.org/e-lib/browse.cfm?elib=14106

[15] Eichas, F., Moller, S., Zolzer, U. Block-oriented modeling of distortion audio effects using iterative minimization, *Conf. on Digital Audio Effects* (2015). Accessed from https://dafx.de/paper-archive/2015/DAFx-15_submission_21.pdf

[16] Holmes, B., van Walstijn, M. Potentiometer law modelling and identification for application in physics-based virtual analogue circuits, *Conf. on Digital Audio Effects* (2019). Accessed from https://dafx.de/paper-archive/2019/DAFx2019_paper_35.pdf

[17] Yeh, D. T., Smith, J. O. Discretization of the '59 Fender Bassman tone stack, *Conf. on Digital Audio Effects* (2006). Accessed from https://dafx.de/paper-archive/2006/papers/p_001.pdf

[18] Najm, F. N. *Circuit Simulation*. John Wiley & Sons, Inc., Hoboken, NJ (2010), p. 28.

[19] Holters, M., Zolzer, U. A generalized method for the derivation of non-linear state-space models from circuit schematics, *23rd European Signal Processing Conference* (2015). Accessed from https://www.hsu-hh.de/ant/wp-content/uploads/sites/699/2017/10/Holters-Z%C3%B6lzer-2015-A-Generalized-Method-for-the-Derivation-of-Non-linear-State-space-Models-from-Circuit-Schematics.pdf; https://ieeexplore.ieee.org/document/7362548/

[20] Dempwolf, K., Holters, M., Zolzer, U. Discretization of parametric analog circuits for real-time simulations, *Conf. on Digital Audio Effects* (2010). Accessed from https://dafx.de/paper-archive/2010/DAFx10/DempwolfHoltersZoelzer_DAFx10_P7.pdf

[21] Fettweis, A. Wave digital filters: theory and practice, *Proceedings of the IEEE*, vol. 74, no. 2 (1986), pp. 269–327.

[22] Holmes, B., van Walstijn, M. Improving the robustness of the iterative solver in state-space modelling of guitar distortion circuity, *Conf. on Digital Audio Effects* (2015). Accessed from https://dafx.de/paper-archive/2015/DAFx-15_submission_18.pdf

[23] Eichas, F., Moller, S., Zolzer, U. Block-oriented gray box modeling of guitar amplifiers, *Conf. on Digital Audio Effects* (2017). Accessed from https://dafx.de/paper-archive/2017/papers/DAFx17_paper_35.pdf

[24] Zavalishin, V. *The Art of VA Filter Design*. Accessed August 2021 from www.native-instruments.com/fileadmin/ni_media/downloads/pdf/VAFilterDesign_2.1.0.pdf

[25] Pirkle, W. Modeling the Korg35 lowpass and highpass filters, *135th AES Convention* (2013). Accessed from https://www.aes.org/e-lib/browse.cfm?elib=16938

3 A Browser-based WebAudio Ecosystem to Dynamically Play with Real-time Simulations of Historic Guitar Tube Amps and Their Typical Distortions

Michel Buffa and Jerome Lebrun

3.1 Introduction

Since the early 2000s, guitar amplifier digital models gained popularity with devices such as the Pod series by Line 6 or the Axe FX-II amp modeler by Fractal Audio Systems: an all-in-one preamp/effects digital processor that contains hundreds of virtualized vintage and modern guitar amps. In 2002, Amplitube, an audio plugin from IK multimedia, was the first iconic all-software amp simulation on the market, followed soon after by Guitar Rig by Native Instruments. Today we can find hundreds of native plugins (commercial or freeware, with only a few being open source) for digital audio workstations that simulate existing guitar amps or introduce novel designs by their authors.

Software amp simulations are quite convenient for on-the-run situations or when the production budget for recording is low; they are cheaper and more flexible than their digital hardware equivalents, even though some purists may object to using them when compared to their analog original counterpart.

However, so far most of these software recreations of guitar amplifiers relied on standalone, computationally expensive, OS-dependent, proprietary solutions with limited interoperability. In 2012, Google Chrome proposed for the first time a low-latency opportunity to get live audio from a microphone or other audio inputs on Mac OS, followed by a Windows implementation (however with longer latency). Soon, Opera, Firefox and Microsoft Edge also implemented this feature, relying on the Media Capture and Streams API from W3C.[1] Chris Wilson's "Input Effects" demo[2] was one of the first to show real-time sound processing effects written with WebAudio, and proposed implementations of famous effects such as delay, distortion, wah, etc. However, this impressive demo did not allow the chaining of effects, but proved that low-latency processing was achievable. Getting close to the real sound of an analog guitar amplifier remained nevertheless quite a challenge that Chris Wilson's examples did not fully address.

Many papers have been written about vacuum-tube guitar amplifiers modeling and about the characteristics of linear and non-linear distortion effects suited for guitars [1–4]. Two main approaches are usually considered for the simulation of the different parts of a guitar amplifier: one is called the technique of virtual analog aka physical modeling. It consists in processing the electronic schematics using tools like the industry standard SPICE analog circuit simulator to translate the circuit into equations to be solved hopefully in real-time. These general equations are typically nonlinear differential algebraic equations and may be solved using integration methods, roots solver algorithms and sparse matrix techniques. SPICE can even produce C++ code ready to be executed. However, it is often necessary to make huge simplifications and optimizations to achieve real-time processing. This is particularly the case with the modeling of the vacuum tubes used in guitar amplifiers and their interactions with other parts of the circuitry (see [5] and [6] for a review of common techniques, and [7] for the method used by authors of the Guitarix.org project).

Another technique consists in a higher-level emulation, in which "logical" parts are identified (filters, tubes, etc.) and may be emulated manually or by machine learning [8–12] (as with the Kemper profiler/Guitar Rig 6 from Native Instruments,[3] with Neural DSP's gears or Deepmind's WaveNet music generators) using separate or global, explicit or hidden models to achieve perceptual equivalence. This is in theory less accurate as some effects and interactions, such as the subtle current feedback effect of overloaded tubes or the action of the speaker impedance on the power amp/sound tone, may be trickier to consider.

When we started to tackle this problem five years ago, the separate, explicit approach was clearly the simpler and more adapted one to the WebAudio ecosystem, especially its limitations at the time (e.g. custom processing on audio samples was not usable without introducing latency or glitches). Furthermore, the WebAudio API provides high-level nodes (such as the waveshaper node and the biquad filter node) that can be used for an easy, quite accurate modelization of tubes and filters. Namely, when properly used, waveshaping techniques associated with oversampling and appropriate filtering give quite good results [13]. The famous pod XT effect processor by Line 6 relies on such techniques [3].

We followed this "perceptual" approach consisting in emulating the different parts of the electronic circuit of this amplifier using WebAudio, implementing the necessary signal processing algorithms using the available API, and finding adequate solutions to circumvent some limitations specific to the Web browser environment (i.e. thread priority, latency, JavaScript API limitations). Finally, we extensively compared (quantitatively and qualitatively) our realization with the state-of-the-art, i.e. native simulations, mostly commercial, written in C++, and not having the constraints of webapps. These results exceeded our expectations. Meanwhile, we went on refining the models used in the simulation and designed a new framework (Figure 3.1) to reproduce different electronic architectures present in various tube amplifiers found in many musicians' equipment [14, 15]. We can now simulate for example a Fender, a Vox or a Mesa Boogie amplifier, etc. or even create new original designs. These customizable simulations have been tested by professional guitarists, are being used by music schools on an experimental basis and are the subject of a marketing contract by the CNRS in order to be included as plugins [16, 17] in an online commercial digital audio workstation (AmpedStudio.com, see Figure 3.2). Online examples of real-time playing, dry and processed sounds by our plugins can be heard online.[4]

3.2 The WebAudio API Ecosystem

The main specificity of our approach lies in the fact that all our simulations are hardware agnostic in the sense that all you need is a Web browser to run them online in real-time. WebAudio is a W3C-standard JavaScript API[5] that relies on building an "audio graph" by connecting processing nodes one to another. The signal is sequentially processed through the block diagram, where each node crossed can modify the signal (e.g. a filter node may be used to remove high-pitched sounds, etc.). Some particular nodes can also be used as a sound source (audio file, wave generator, link to an HTML5 element <audio> or <video>).

In Figure 3.3, we give some details on this approach; we precisely analyzed the electronic schematics of an amplifier: the different stages, the tubes, filter structures, power supply and output transformers are identified and simulated part by part by JavaScript code relying on the WebAudio API. One big advantage of JavaScript/WebAudio API is the flexible and dynamic way you can manipulate the audiograph (by changing its topology, even while playing guitar) or the parameters of the different nodes; in particular, you can reshape in real-time the transfer functions

30 Michel Buffa and Jerome Lebrun

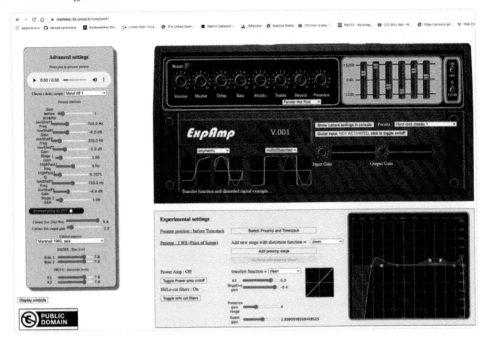

Figure 3.1 Our designer tool (https://mainline.i3s.unice.fr/AmpSim5/): an online framework used to generate the different WebAudio plugins from the next figure.

Figure 3.2 WebAudio amp sim plugins running in the online DAW AmpedStudio.com.

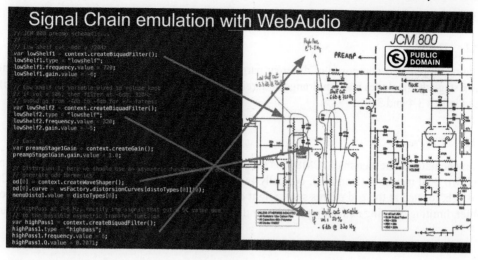

Figure 3.3 A typical high-level simulation: the different stages are identified, and the tubes and filters are simulated part by part using the WebAudio API.

used by the waveshaper nodes. Another advantage of using WebAudio high-level nodes is that most of the audio processing is done in their C++ implementation that lies in the Web browser internals, so performance is generally not an issue. Our simulation has a very low CPU impact and runs audio processing threads with the highest priority. On a MacBook Pro from 2016, we could run up to 15 amp simulations in parallel (all stages: preamp, tonestack, power amp, speaker simulation), in a DAW, without noticing any glitch and with a CPU stress level of less than 50%.

However, one main limitation of the WebAudio API design was that default signal processing is constrained by its inherent block-processing of chunks of 128 samples at a time (3ms at 44.1Khz), and until recently it was not possible to do stream processing at the sample-wise level without introducing glitches and latency. Not being able to perform processing with time-granularity below this 3ms limit caused a lot of issues, mostly in the implementation of the PowerAmp part of the amplifier schematic. Indeed, in its classic push/pull configuration, the power stage includes a crucial negative feedback loop that could not be faithfully simulated without introducing customized solutions to mitigate this major limitation.

This issue was partially solved with the arrival of the new AudioWorklet node and WebAssembly standard[6] in 2018 (a portable binary-code format for executable programs, first to be used on the Web, but also on native environments). Writing custom DSP code in JavaScript, or coding in WebAssembly by hand, is nevertheless quite tedious. Fortunately, some Domain Specific Languages for DSP programming, such as FAUST,[7] quickly proposed WebAudio/WebAssembly as a compilation target [18] and proved to be an ideal framework for developing powerful custom code (we even did some personal contributions to FAUST's WebAudio support) [17].

In the next section, we will detail the different steps used to recreate a guitar amp in the Browser, with some focus on how the PowerAmp stage works, introduce our initial solution to mitigate the block-processing limitations of WebAudio, and devote the rest of this section to detail some dynamic subtleties of tube-based guitar amplifiers that we were able to properly simulate using our new FAUST and WebAssembly-based faithful low-latency approach. We will conclude with a detailed comparison of the different approaches, with special care on the performance measurements (latency, CPU usage, etc.).

3.3 Understanding Tube Guitar Amplifier Simulations

3.3.1 Overview

A guitar amplifier is composed of different parts: usually, a preamplifier stage, a so-called tone stack stage that includes bass, midrange and treble controls, and a power stage (the power amp). See Figures 3.4 and 3.5.

3.3.2 The Preamp

The schematic of the Marshall JCM800 is shown in Figure 3.6. It is composed of several filters and two dual triodes (V1 and V2, typically two 12AX7). The second dual triode, named v2a and v2b, located at the end of this stage (right of the figure), is a DC coupled cathode follower buffer, limiting clipping and acting as a linear driver of the tone stack. The most interesting part

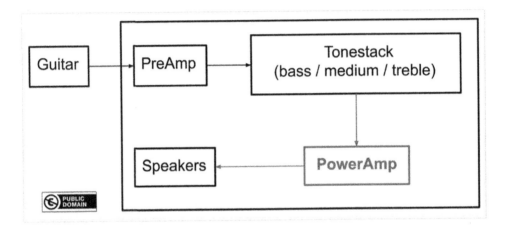

Figure 3.4 The typical stages of guitar amplifiers and their associated signal paths. Some brands (e.g. Fender) may switch the preamp and the tone stack.

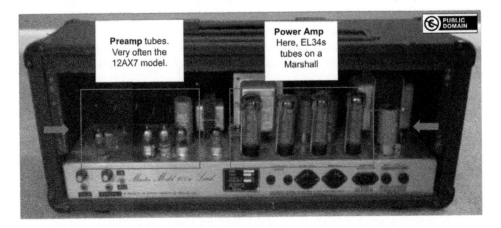

Figure 3.5 Rear-view of a Marshall JCM800 power head with its two amplification stages.

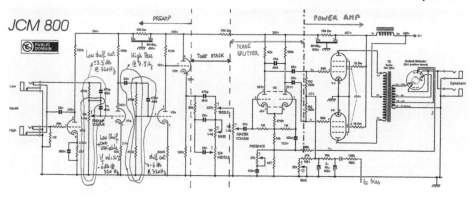

Figure 3.6 Annotated schematic of the iconic Marshall JCM800.

here consists of the filters and the first dual triode v1a and v1b. From the left of the signal chain to the right, we have a low shelf filter with a -3.3dB gain at 720Hz, then another low shelf at -6dB 320Hz, and the first triode stage named v1b, followed by a high-pass filter at 6–7Hz. This is the first part of the preamp. The low shelf filters act as a noise gate by cutting the annoying frequency generated by the guitar, and the V1 part generates odd and even harmonics (v1a is mainly a gain, v2a amplifies and introduces harmonics). Even harmonics are important to perceive the sound as "warm and bluesy", and odd harmonics yield a more "harsh and gritty" sound. When nicely distributed, the mix of both even and odd harmonics is known to be the secret of the "warm and punchy sound" that you can find in "classic rock" and blues music. To model the two parts of this first tube, we used a WebAudio waveshaper node with the asymmetric transfer function described in Pakarinen-Yeh's article [19] (with origins from a patent held by Doidec et al. 1998), shown in the left of Figure 3.7. Some bluesy sounds played real-time can be heard in an online video.[8]

The waveshaper function clips differently the negative and positive portions of the wave ("harder clipping" on the negative portion, while the positive portion of the wave is softly clipped). This asymmetry approximates the duty-cycle modulation seen in overdriven tube amplifiers. This asymmetry in clipping adds both even and odd harmonics, resulting in a richer tone that is characteristic of vintage tube amps. The output signal is no more centered as the asymmetry adds some DC component. The high-pass filter at 6–7Hz rectifies this signal and removes the hum noise that could have been amplified. The second part of the preamp is made of another low shelf filter at 720Hz with a gain of -6dB. This time, for the second triode stage V2, we used a symmetric transfer function (based on a tanh function) for generating more even harmonics. In our implementation, we included about 20 different transfer functions that can be set to the different preamp tubes (V1 or V2), giving more versatility to the original JCM800 preamp design. The ones shown in Figure 3.5 produce a sound rich in both even and odd harmonics, with a high level of distortion/overdrive. This can be heard in the videos of the guitar players who tested our simulated amp. The filters were implemented using standard WebAudio biquad filters.

Special care must be taken during clipping, as these transfer functions may produce harmonics that exceed half of the sample rate and wrap back around to the high frequencies, causing aliasing artefacts in the signal [5]. Some methods such as oversampling are known to allow a larger frequency range before aliasing occurs and thus higher harmonic components in the

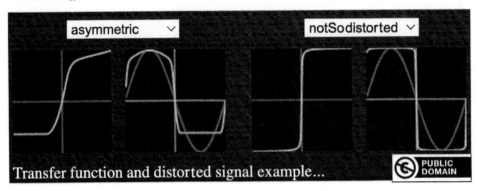

Figure 3.7 Transfer functions used for the two first tubes in the JCM800 preamp.

digital frequency domain. The WebAudio waveshaper node has a property named "oversampling" that unfortunately also increases the latency (depending on its value by "2x" or "4x"). Another approach for limiting aliasing is known as multiband distortion and consists in splitting the signal into several separate frequency bands and applying different amounts or types of distortion in each subband [20]. We implemented this approach in a previous version of the WebAudio amp simulation [21] and achieved very nice, clean sounds that led to an acoustic guitar simulator plugin.[9] That version based on an unusual design, not inspired by any existing amp, can be tried online.[10] The preamp integrates optional oversampling, and we also measured that, in addition, our speaker simulation stage cuts most of the high frequencies that may result from aliasing. These measures have been confirmed during qualitative evaluations, as none of our human testers could hear a difference, with or without oversampling activated.

The preamplifier beefs up the high-impedance, low-level signal coming from the guitar pickup microphones to a lower-impedance, mid-voltage signal that can drive the power stage. The preamplifier also shapes the tone of the signal; high settings of the preamplifier lead to "overload" that creates crunchy/distorted sounds. The power amplifier, with the help of the output transformer, outputs a very low-impedance, high-current signal adequate to efficiently drive a loudspeaker and to produce loud amplified sounds. Another clear aim of guitar amps is of course to introduce desirable distortions too.

3.3.3 The Tone Stack and Reverb

In a reference paper about the modeling of the Fender Bassman amp tone stack [6], reproduced in the JCM800, David Yeh and Julius Smith explained how this circuit (bass, medium, treble) coloring the sound of the electric guitar, presented in Figure 3.8, cannot be simulated accurately by filters in series or in parallel. Though it appears to be simple, this circuit is an intricate network of filters, each influencing the others. Yeh and Julius analyzed the schematics symbolically using the technique of virtual analog, and proposed an exact solution: a filter that responds to user controls in the same way as the analog prototype.

As many other classic amps by Fender, Vox or Marshall share this design, one can find exact implementations of many different tone stacks (only the location on the amp circuit—after the first stage of the preamp on classic Fender amps, or the values of the different resistors and capacitors are different). In our early WebAudio amp, we transpiled the original C++ implementation of Yeh's code from C++ to JavaScript, using emscripten [22], and used a WebAudio

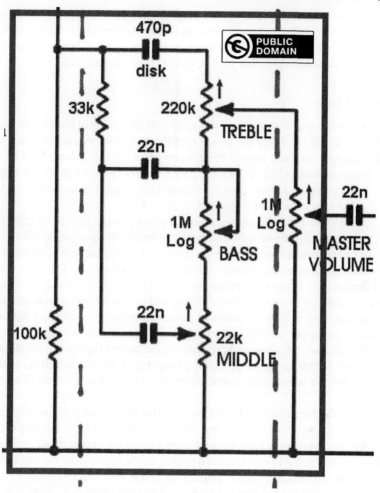

Figure 3.8 The tone stack circuit of the JCM800.

ScriptProcessor node (now obsolete—it was the only way, before 2018, to write custom, sample accurate, DSP code) to reproduce the exact tone stack of the JCM800. However, these experimentations showed that some latency of at least 4ms was introduced. Therefore, we decided to explore further alternatives. In addition, we sometimes encountered buffer glitches when the bass, medium, treble or presence knobs were operated from the amp GUI (as the ScriptProcessor node runs in the GUI thread). This led us to remove this implementation, waiting for the upcoming AudioWorklet node that should have given better performances and no glitches. Meanwhile, we adopted as an alternative solution a set of biquad filters in series: treble filter (high shelf, 6.5 KHz) goes into medium (peaking, 1.7KHz) into bass (low shelf, 100Hz) into presence (peaking, 3.9KHz), and we adjusted the types of filters and their parameters to approximate the frequency response of the real tone stack (we used the tone stack calculator tool for comparing). The result is not a completely faithful replication of the real tone stack circuit, but it "does the job": our testers found it easy to use and managed to shape their sound rapidly. In a real JCM800, the presence filter is in the feedback loop between the input and output of the power amp stage, and we will detail in the following section how to properly simulate that part. We also proposed an optional convolver node with free reverb impulses. We implemented this using a classic wet/dry audio graph to make the "room effect" adjustable. Several impulse responses are included in the online demo[11] (Marshall JCM800 plate reverb, Fender spring reverb, etc.)

3.3.4 The Power Stage

To get a finer understanding of this critical stage, we will further detail how a power amp works, its role in the signal chain and why the power amplifier is so tricky to emulate due to the presence of some intricate feedback loop in the circuitry. Usually solely controlled by the master volume and presence knobs (Figure 3.9), the power amp stage has a profound impact on the sound and overall dynamics. Indeed, the type of sound you can get from the preamp stage alone remains on the lean side in terms of distortions.

Also, compared to the triodes usually used in the preamp stage, the use of tetrodes/pentodes in push-pull topology in the power stage enhances the third harmonics (and generally all the odd ones) [23]. We end up with a spectrum mostly composed of odd harmonics that can be reasonably simulated using symmetrical transfer functions in the waveshaper [5]. Referring to Kuehnel [24] and Denton [25], when you turn up the volume of a tube amp using the master volume knob, the power tubes get more and more distorted and the sound gets thicker and thicker, and also less controlled. And as you get closer to power saturation (i.e. clipping the power tubes and/or saturating the output transformer core, causing the famous grinding effect, a mixture of saturation/oscillations, but also possibly starving the power supply—more about this later in this section), the output dynamics get tighter, resulting in muddy distortions with a heavier sound. These sounds are subjectively characterized as "thick", "heavy", "metallic", "abrasive", "gritty" or "rich".

In power amps, the Negative Feedback Loop (NFB)/presence control (see Figures 3.10 and 3.11) were introduced to extend the usable frequency bandwidth by limiting/reshaping the unwanted distortions originating from the non-linearities of the output transformer at the price of a slightly lower total gain. The NFB takes a portion of the signal from the output transformer secondary winding, dephasing it with a 180-degree phase shift and re-injects it back into the splitter stage (see the highlighted plot in Figure 3.10 and the "feedback" loop in Figure 3.11). Amplifiers without NFBs are usually more powerful (higher gain) with more distortion, but

Figure 3.9 The power amp stage is controlled by the master volume and presence knobs.

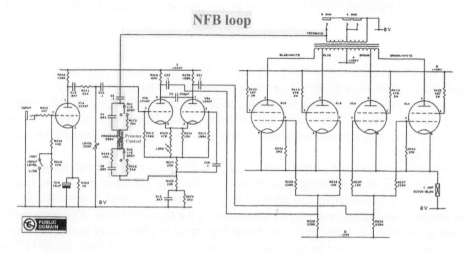

Figure 3.10 The Negative Feedback Loop (NFB)/presence in a MesaBoogie 2:90 with its RC network highlighted. The JCM 800 has similar topology.

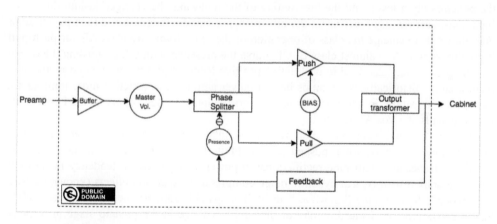

Figure 3.11 Block diagram of a classic push/pull power amp.

their coloration also tends to become rougher and unpredictable when pushed into saturation. The NFB loop smoothes out the sound by reducing the level of distortion in the parts of the spectrum where it is most disturbing (typically mid-range) [24].

Unknown to concert stage or hi-fi power amps, a 'presence' control knob is added to the NFB loop in guitar amps to allow some control on the coloration of the NFB and thus to provide a simple but efficient way of adjusting the brightness and sharpness of the sound at the power amp stage. The presence may be looked at as a global time/frequency control on the brightness of timbre by altering the signal that is fed back from the negative feedback loop. The quantity of negative feedback can be reduced for certain frequencies as the presence knob controls the resistor part of a RC network, hence a tunable filter made of Resistors and Capacitors (RC) in the loop. With less negative feedback in the high frequencies, the sound becomes brighter, louder, more vivid and dynamic. The behavior of this control is very different from those of the tone stack (bass, mid-range, treble), which merely equalizes the output of the preamp. Namely, unlike conventional treble control, which is mainly subtractive, presence control is pseudo-additive in that it limits the fundamental subtractive aspect of the NFB loop. It should be noted also that the NFB/presence has a major influence at the temporal level [4], as the RC networks controlled by the presence knob induce some frequency-dependent group-delay in the NFB loop. This may explain the blurring/sharpening effect that presence has on the attack slopes of the notes played, acting as a "softening/anti-softening" pedal. This clearly motivates our introduction of a curve-based parametric presence control to choose in which frequency band one wants its brightness.

To properly simulate the NFB/presence circuit, lots of parameters are involved, and a lot of care must be taken when adjusting them. In analog tube/transistor-based power amps, manufacturers are very conservative: the allowed range of presence control is restricted so as to avoid unwanted oscillations that may be destructive to the speaker/cabinet. As a consequence, only subtle alterations (mostly upper-mids/lower-high range brightness) are possible, the sound signature being mainly carved through the preamp settings. In our simulations, the curve-based parametric presence control allows us to fully carve the time/frequency coloration at the power amp level to add punch or special effects, like mild oscillators to the output from the preamp and tone stack.

Some other important controls on the power amp are the master volume to adapt gain with the preamp output level, and the bias setting of the tubes may be changed to introduce some light non-linearities/asymmetries in the tube response curves. In real-world tube power amps, bias allows us to change the class of operation of the tubes (typically class AB for push-pull) from almost class A to almost class B. Of course the presence setting for adjusting the overall tone/brightness may also lead to destructive positive feedback if not designed correctly (freq. ranges, etc.). Historically, some amps have a power amp section which is more important in their sound signature; in general, they are the most "vintage" ones—think of the Fenders, the first Marshalls, the Vox AC30, etc.

The most crucial aspect to get more realistic simulation is to properly take care of the real-time dynamics. Namely, when pushed to their limits by large transients, classic guitar tube amplifiers (especially with tube rectifiers-based power supply) have a tendency toward harmonic saturation with typical dynamic range compression/expansion effects (known as "sag" and "bloom", temporal lag (known as "squish") or spatialization artefacts (known as "swirl"). Sag is consecutive to some drooping of the supply voltage in the preamp stage in response to large transients. The recovery from sag as voltage supply returns to normal is coined as "bloom" and is again a well-appreciated effect. Now, squish is linked to some temporal hysteresis induced by some increased time lag in the feedback loop—also in response to large transients. Finally, swirl is linked to saturation of the core in the output transformer from overload, leading to phase inconsistencies and spatial blurring of the chord played (à la Univox Uni-Vibe pedal effect). All those effects proved to be quite tricky to emulate in real time. To tackle this, we introduced a dedicated method to approximate these advanced temporal, nonlinear behaviors of tubes (typically sag/bloom, squish/swirl effects). In the WebAudio API, the slope of the tube transfer function curves can be driven real-time by the power of input signal envelope emulating hysteresis phenomena. Consequently, to mimic this hysteresis and for example the sag in the response of a tube preamp stage, we implemented it with simple dynamic real-time changes in the slopes of the transfer functions in our preamp simulation. By properly adjusting the threshold at which we get the squish, only the higher power transients from the envelope amplitude will trigger controlled change of slope. Figure 3.12 shows a curve animated in real time, with

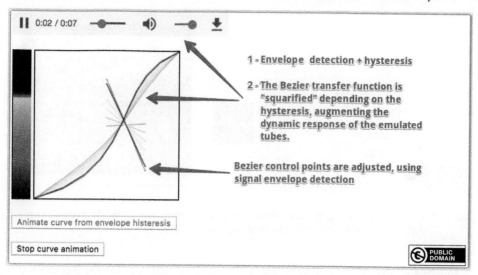

Figure 3.12 A tool developed to adjust in real time the temporal dynamics of the tube simulation based on waveshapers.[13]

hysteresis behavior in the transfer function triggered by the input signal envelope; online videos also propose live examples with real guitar.[12]

Finally, the bias control corresponds to the idle voltage around which the tubes are amplifying signals and thus controls their linearity/asymmetry characteristics. Using bias, one can change the amount of (negative) voltage offset applied to the signal at the grid of the tubes. Colder settings will make the pentodes/tetrodes draw less current, decreasing the overall output volume and potentially introducing cross-over distortion due to the class AB getting closer to class B (which may be what a guitarist looks after to achieve a dirtier/lousier tone). Hotter settings will make the pentodes/tetrodes draw more current, moving closer to class A, with a thicker, louder perceived output level and eventually up to power supply "sagging" (depending on the sagging control setting), adding compression or, in extreme cases, saturation and cleaning up the tone from potential cross-over distortion.

3.4 Implementations

3.4.1 Implementing the Power Amp Stage with WebAudio High-Level Nodes

In our implementation, the presence filter is fully customizable and can be controlled in real time using a graphic equalizer (Figure 3.13) to select the frequency range of the filter. The min and max values of the negative gain in the NFB can also be adjusted using a slider. These tools (NFB and presence) are quite sensitive (at the edge of creating positive feedback with oscillations and Larsen effects, so we provide controls in the GUI to adjust/restrict the admissible range), but this novel presence control provides an utterly powerful and spectacular tool to shape the final sound.

Being able to adjust all the parameters (gains, filter parameters, transfer function of the waveshaper, etc.) is crucial to fine tune this stage using WebAudio nodes. We strongly advise the interested reader to watch an online video,[14] which shows the differences in sound and dynamics with and without this loop (power amp on/off) in our simulation.

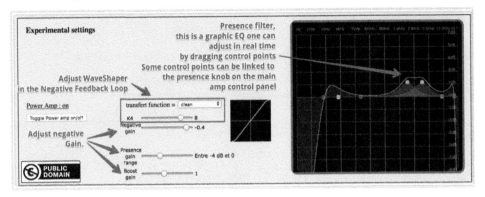

Figure 3.13 The settings of the NFB can be changed/adjusted dynamically on the fly, even while playing, without generating any audio glitches/dropouts.

It sounds and plays very well (we made several user evaluations [14, 15] that showed that guitarists, even professional ones, liked the way the temporal dynamics of a real amp was simulated). We simulate the different tube stages using filters and waveshaper nodes with properly adjusted transfer functions, as shown in Figure 3.11. The green transfer function curve on the left of the graphic EQ (presence bank of filters) is rather linear, meaning that only little distortion is added to the signal, and only at high amplitudes. Being able to adjust this function is also crucial to avoid oscillations. On top of that, the bias control has to be properly adapted to the selected tube transfer functions to guarantee coherent coloration/output volume (making it easier to compare different tubes and choose the right one).

Nevertheless, we encountered a major issue in the implementation of our signal loops in WebAudio. Proper simulations of NFB and presence have been quite difficult to achieve due to some intrinsic limitations of the WebAudio API and divergences/bugs in how browsers generally parse the WebAudio graphs with loops. In the WebAudio API specs, loops in the graph are required to include at least one delay node. Without this delay node, Firefox stops rendering the graph, while Chrome does not complain but adds, behind the scenes, a 3ms delay (the minimal delay from a frame audio buffer of 128 sample-frames as within WebAudio API, the signal is always processed in packets of 128 samples, which means that the minimum value for a delay is 128 / sample rate, i.e. about 128/44100 = 3ms). With the current limitations in WebAudio, and quite strangely, this 3ms delay in the loop to conform to the specs, brings slightly different coloring of the amps between Firefox and Chrome. We have reported these errors and discussed them with the implementers (Raymond Toy from Google and Paul Adenot from Firefox), but so far, nothing has been fixed (as of January 2021). For a good implementation of NFB, a stable delay of fewer samples would be preferable (higher delays increase latency and the softening/smurring of attacks). To faithfully implement loops like the NFB with its RC networks inducing shorter delays, a finer precision at the level of some sample frames is required.

To circumvent these limitations and allow a proper sample-wise accuracy in the processing of loops, we decided to completely re-implement the NFB loop (and other critical parts in general) in FAUST (Functional Audio Stream, which is a functional programming language for sound synthesis and audio processing with a strong focus on the design of synthesizers). These new parts were implemented as AudioWorklets, ending up re-implementing all the processing nodes present in the circuit: filters, gain and wave shaper, in this language.

3.4.2 A FAUST Re-implementation of the Power Amp Stage

With the previously encountered difficulties in simulating the power amp in pure JavaScript/WebAudio, we first tried to reuse the acquired knowledge by recreating as faithfully as possible the signal chain that gave good results in pure Javascript. For example, using the same types of filters with the same parameters, the same transfer function with the core issue of simulating properly a waveshaper in FAUST, the same gain values, etc. And of course, we looked closely at the behavior of the NFB loop, whose delay hopefully should be lower than the 3ms barrier imposed by the previous implementation. This way, we managed to replace the current power amp implementation (made of a dozen high-level WebAudio nodes) with a single FAUST-generated AudioWorklet node.

WebAudio comes with a set of classic biquad filter types: low-pass, high-pass, band-pass, low-shelf, high-shelf, peaking, notch and all-pass. All these filters have a fixed set of parameters: frequency, gain, Q and detune, whereas some of these parameters are not relevant to some type of filters (as Q for low-shelf/high-shelf). FAUST does not come with similar filters out of the box. After trying to adapt existing FAUST filters to behave like WebAudio filter ones, and after some interactions with FAUST and WebAudio implementers, the conclusion was that for a really faithful behavior, it would be better to start from the original C++ implementation of the WebAudio filter API, taken from the Chromium browser source code. The FAUST team did the port and provided us with the so-called webaudio.lib that is now available in the FAUST distribution.[15]

For the power amp tubes, we looked at the way FAUST developers simulated tubes (e.g. in the Guitarix project source repository,[16] in particular in the guitarix.lib file), or waveshapers (as in several distortion plugins such as the ones by Oleg Kapitonov,[17] or by Nick Thompson's Creative Intent Temper Distortion plugin).[18] We found out that most tube simulations relied on C/C++ code and could not be used out of the box (typ. Guitarix tube simulations), as we must be able to run these in a Web browser, and the FAUST toolchain still does not support hybrid FAUST/C source code when the compilation target is set to WebAudio/WebAssembly. The temper distortion simulation, however, used a 100% FAUST-based implementation of a simple waveshaper that produced a warm, adjustable distortion sound that could easily fit our needs. We used it as a starting point. Furthermore, the code contained the definition of a transfer function: *Transfer (x) = tanh (k x) / tanh(k)*, with tanh approximated as $tanh(x) \approx x(27+x^2)/(27+9x^2)$, the parameter k could be adjusted to change the S shape of the curve (i.e. using a more subtle/less aggressive curve) similarly to the one in our pure JavaScript implementation.

Figures 3.14 and 3.15 display the final diagram of our FAUST implementation[19] of the power amp inspired by the temper distortion source code in which we added a presence control (made

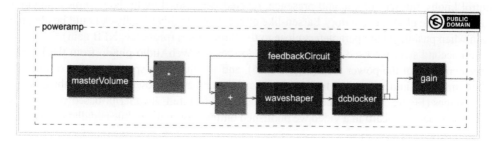

Figure 3.14 Diagram of final implementation.

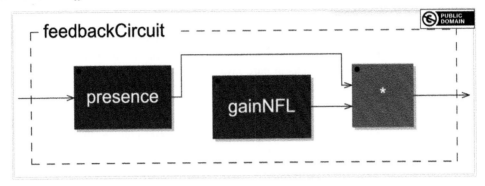

Figure 3.15 The feedback circuit. The presence filter is obtained using a set of peaking filters ported in FAUST from the WebAudio API implementation.

of two peaking filters, at 2kHz and 4kHz) in the feedback loop, and introduced an adjustable negative feedback gain, removing some unnecessary elements (i.e. a resonant low-pass filter at the beginning of our signal chain).

Finally, we also added some GUI elements (knobs) in order to fine tune in real time different parameters, in particular the ones that control the waveshaper (drive, curve, distortion), the NFB gain and the presence filters.

The current implementation of the waveshaper (inspired from the open-source code from the temper distortion) does not rely on pre-calculated point tables, but on a transfer function applied to each sound sample, which leaves some room for speedup/enhancements. The dynamic time response of the tubes is approximated using an amplitude follower placed in the signal chain just before the waveshaper that drives an all-pass filter (and which aims at modifying the DC offset, and thus the slope of the curve). This approach proved to be as efficient or flexible as the pure Javascript method we used previously. Once our FAUST-based power amp re-creation was functional and adjustable, we could proceed to an evaluation phase.

3.5 Evaluations

The first step of the evaluation consisted in listening to the global overall sound when we used the power amp in standalone mode, tweaking the different parameters (master volume, presence, NFB loop negative gain and transfer function parameters), and to check its behavior compared to our previous WebAudio/JavaScript implementation. We did some trials using dry guitar sound samples, but also with a real guitar as inputs. The general feeling is that the two main control knobs, master volume and presence, reacted very closely with both implementations. In addition, the classic oscillatory Larsen-like effects from positive feedback could be obtained again when pushing some parameters close to the limit values (presence, NFB gain).

In a second step, we used the FAUST IDE to create a WebAudio plugin from the FAUST implementation of the power amp (Figure 3.16), and we chained a special version of our tube guitar amplifier simulations in which we bypassed the embedded power amp and cabinet simulation stages (Figure 3.17), and compared with the full-featured, JavaScript-based simulations. Results can be seen/heard in a video we published online,[20] or in the online pedalboard WebAudio application.[21]

The differences in terms of sound/timbre and temporal dynamics are small and subtle. However, we noticed that the FAUST implementation was much more stable and versatile when

A Browser-based WebAudio Ecosystem   43

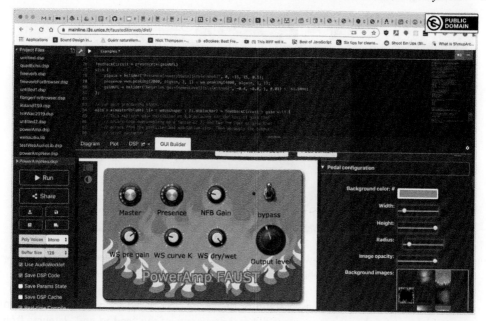

Figure 3.16 For testing purposes, we created a WebAudio plugin from the FAUST code by using the WAP GUI Builder developed and integrated in the FAUST IDE.

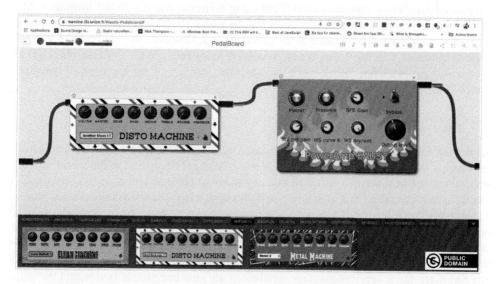

Figure 3.17 The new FAUST power amp plugin in our virtual pedalboard, with the original pure JavaScript power amp bypassed and no cabinet simulator stages.

graphically adjusting the internal parameters of the feedback loop. The processing in our initial pure JavaScript implementation was based on blocks of 128 samples, inducing an undesirable delay of 3ms in the back-fed signal as opposed to a clearly ultra-low latency with the FAUST loop. In the FAUST implementation, the measurement tools (Figure 3.18) validated this sample-wise nature of the processing, with a measured delay of just one sample for the NFB/presence

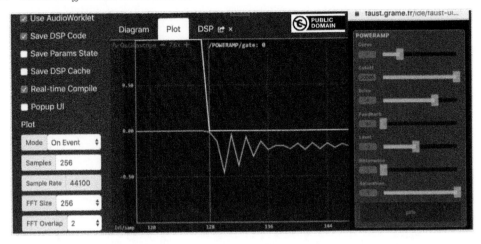

Figure 3.18 The latency with sample-wise accuracy as measured by tools embedded in the IDE. Yellow: a gate signal. Pink: the output from the power amp. X-axis is in samples.

loop. This also explains the increased stability of the FAUST-based loop. Now, in terms of aggregated latency for the power amp, we did measurements of the "end-to-end" latency, from guitar to cabinet, and obtained consistently slightly better values for latency with the new FAUST implementation: around 20–21 ms compared to the 23–24 ms latency of our previous finely tuned JavaScript implementation (both using a Firefox Nightly 75.0a1 browser with an external Focusrite Scarlett and a Macbook Pro 16 under 10.14). All these low values made the latency completely unnoticeable to the guitar players. However, this confirmed a saving of 3ms in accordance with the difference of processing loops between FAUST (sample-wise) and WebAudio API/JavaScript (block-based) and a better stability of the curve-based parametric control of the NFB presence loop.

3.6 Conclusion and Perspectives

Developing real-time simulations of tube guitar amplifiers within a virtual pedalboard ecosystem capable of running in a Web browser was a challenge. Compared to the world of native applications, the browser is a rather hostile environment because you have to share the CPU with the other open tabs. For a long time, we didn't have a low-latency solution, signal processing was limited to blocks of 128 samples, the basic audio nodes provided by the WebAudio API were limited, etc. By designing software at the limits of the WebAudio API possibilities, we participated in the evolution of the standard (one of the authors of this chapter is part of the W3C WebAudio working group), and have shown that the Web platform is capable of running real-time computer music applications. Our simulations are now commercialized within the online DAW Amped Studio, and musical pieces have already been recorded by several guitarists.[22] A major consequence of our work has been the creation of a new standard for WebAudio plugins [16], (aka VSTs for the web) and the release of powerful tools for their development [17] (e.g. online IDE for the FAUST language to produce high-performance WebAssembly code).

Today there are many digital re-creations of analog circuits, but some musicians still claim that few of these re-creations are capable of restoring the warmth and character of their material counterparts. So far we used a white-box approach by constructing approximations of the different elements that make up the circuit inside the amp, such as the preamp, the tubes or the filters for example, and then combining them. While user tests have shown that our solution is competitive with the best commercial products [14, 15], when compared in A/B tests with real hardware, it is sometimes difficult to say if it is "better or worse", but it is clear that the dynamic behavior is sometimes different. Indeed, we don't 100% capture how the different elements inside the amp interact with each other. The internal parts have physical relationships with each other that make an amp unique. These relationships evolve and change continuously, depending on the type of sound being played and the settings of the different parts of the amp. We plan to move to a greybox approach as pioneered by [3–6] that consists in training artificial neural networks to generate algorithms that reproduce the internal workings of analog hardware in detail. For example, when it comes to elements like tubes, for example, mathematical models alone are not always good enough. In our current work, we had to use ad hoc techniques to simulate their complex temporal/dynamic behavior; in addition we had to implement empiric gain compensation between some amp stages, for example, and to fine tune many internal parameters. This modeling technique required a long process of tuning and adjustment to achieve perfect sound. The promise of the algorithms resulting from machine learning systems is that we no longer need to make these lengthy manual adjustments. That's the big difference we're going to explore now.

Acknowledgements

This work was partially supported by the French Research National Agency (ANR) with team WASABI [21] (contract ANR-16-CE23–0017–01).

Notes

1. www.w3.org/TR/mediacapture-streams/
2. https://webaudiodemos.appspot.com/input/index.html
3. https://blog.native-instruments.com/fr/the-making-of-icm/
4. See http://mainline.i3s.unice.fr/distortionBook/sounds
5. www.w3.org/TR/webaudio/
6. https://webassembly.org
7. https://faust.grame.fr/
8. http://mainline.i3s.unice.fr/distortionBook/videos/blues.mp4
9. http://mainline.i3s.unice.fr/distortionBook/videos/acousticSim.mp4
10. https://mainline.i3s.unice.fr/AmpSim/
11. https://mainline.i3s.unice.fr/AmpSim5/
12. http://mainline.i3s.unice.fr/distortionBook/videos/sag.mp4
13. https://jsbin.com/zotaver/edit
14. http://mainline.i3s.unice.fr/distortionBook/videos/designer.mp4
15. https://faustlibraries.grame.fr/libs/webaudio/
16. https://sourceforge.net/projects/guitarix/
17. https://github.com/olegkapitonov/Kapitonov-Plugins-Pack
18. https://github.com/creativeintent/temper
19. https://github.com/micbuffa/FaustPowerAmp
20. http://mainline.i3s.unice.fr/distortionBook/videos/powerampIFC.mp4
21. https://mainline.i3s.unice.fr/Wasabi-Pedalboard/#
22. Online examples: http://mainline.i3s.unice.fr/distortionBook/sounds

References

[1] Chang, C. H. Overdrive/distortion. *DESC9115: Digital Audio Systems*, University of Sydney, Australia (2011). Accessed December 2020 from http://hdl.handle.net/2123/7608
[2] Macak, J. and Schimmel, J. Real-time guitar tube amplifier simulation using an approximation of differential equations. In: *Proc. 13th Int. Conf. on Digital Audio Effects (DAFx-10)*, Graz, Austria (2010).

[3] Yeh, D. T., Abel, J. S., Vladimirescu, A. and Smith, J. O. Numerical methods for simulation of guitar distortion circuits. *Computer Music Journal*, 32(2) (2008).
[4] Difilippo, D. and Greenebaum, K. *Audio Anecdotes: Tools, Tips, and Techniques for Digital Audio*, Vol. 1, Routledge, AK Peters, Natick, MA (2004). ISBN 9781568811048
[5] Pakarinen, J. and Yeh, D. T. A review of digital techniques for modeling vacuum-tube guitar amplifiers. *Computer Music Journal*, 33(2) (2009).
[6] Yeh, D. T. and Smith, J. O. Discretization of the '59 Fender Bassman tone stack. In: *Proc. 9th Int. Conf. on Digital Audio Effects (DAFx-06)*, Montreal, Canada (2006).
[7] Macak, J. *Real-time Digital Simulation of Guitar Amplifiers as Audio Effects*. PhD Thesis, TU Brno, Czech Republic (2012).
[8] Rollo, C. *Black-Box Modelling of Nonlinear Audio Circuits*. MSc Thesis, Aalto University, Finland (2018).
[9] Eichas, F., Möller, S. and Zölzer, U. Block-oriented modeling of distortion audio effects using iterative minimization. In: *Proc. Int. Conf. on Digital Audio Effects (DAFx-15)*, Trondheim, Norway (2015).
[10] Eichas, F., Möller, S. and Zölzer, U. Block-oriented gray box modeling of guitar amplifiers. In: *Proc. Int. Conf. on Digital Audio Effects (DAFx-17)*, Edinburgh, UK (2017).
[11] Schmitz, T. *Nonlinear Modeling of the Guitar Signal Chain Enabling its Real-time Emulation*. PhD Thesis, University of Liege, Belgium (2019).
[12] Wright, A., Damskägg, E.-P., Juvela, L. and Välimäki, V. Real-time guitar amplifier emulation with deep learning. *Applied Sciences*, 10(766) (2020).
[13] *G2 Workshops and Tutorials: Waveshaping and Distortion*. Accessed December 2020 from https://rhordijk.home.xs4all.nl/G2Pages/Distortion.htm
[14] Buffa, M. and Lebrun, J. Real time tube guitar amplifier simulation using WebAudio. In: *Proc. Web Audio Conference (WAC'2017)*, London, UK (2017). ⟨hal-01589229⟩
[15] Buffa, M. and Lebrun, J. WebAudio virtual tube guitar amps and pedal board design. In: *Proc. Web Audio Conference (WAC'2018)*, Berlin, Germany (2018). ⟨hal-01893781⟩
[16] Buffa, M., Lebrun, J., Kleimola, J., Larkin, O., Pellerin, G. and Letz, S. WAP: Ideas for a Web Audio plug-in standard. In: *Proc. Web Audio Conference (WAC'2018)*, Berlin, Germany (2018). ⟨hal-01893660⟩
[17] Buffa, M. and Lebrun, J. Guitarists will be happy: guitar tube amp simulators and FX pedals in a virtual pedal board, and more! In: *Proc. Web Audio Conference (WAC'2018)*, Berlin, Germany (2018, September). ⟨hal-01893681⟩
[18] Letz, S., Orlarey, Y. and Fober, D. Compiling Faust audio DSP code to WebAssembly. In: *Proc. Web Audio Conference (WAC'2017)*, London (2017).
[19] Ren, S., Letz, S., Orlarey, Y., Michon, R., Fober, D., Buffa, M. and Lebrun, J. Using Faust DSL to develop custom, sample accurate DSP code and audio plugins for the Web browser. *Journal of the Audio Engineering Society*, 68(10) (2020), pp. 703–716. ⟨hal-03087763⟩
[20] Clark, J. J. *Advanced Programming Techniques for Modular Synthesizers*, McGill University, Montreal, Canada (2003). Accessed December 2020 from https://tinyurl.com/3xskmv
[21] Buffa, M., Demetrio, M. and Azria, N. Guitar pedal board using Web Audio. In: *Proc. Web Audio Conference (WAC'2016)*, Atlanta (2016).
[22] Zakai, A. Emscripten: an LLVM-to-JavaScript compiler. In: *ACM Int. Conf. Companion on Object Oriented Programming Systems Languages and Applications (OOPSLA '11)*, ACM, New York, NY (2011), pp. 301–312. http://kripken.github.io/emscripten-site
[23] Barbour, E. The cool sound of tubes—Vacuum tube musical applications. *IEEE Spectrum*, 35(8) (1998).
[24] Kuehnel, R. *Circuit Analysis of a Legendary Tube Amplifier: The Fender Bassman 5F6-A*, 2nd edition, Pentode Press, Seattle (2005). ISBN 978–0976982258
[25] Denton, D. *Electronics for Guitarists*, Springer-Verlag, New-York (2013). ISBN 978-1-4614-4087-1

4 Non-linearity and Dynamic Range Compressors

Austin Moore

4.1 Introduction

When one considers studio devices that impart distortion, guitar amplifiers, tape machines and valve preamps are often the first to come to mind. However, dynamic range compressors (DRC), when used outside of their typical working conditions, can introduce various shades of distortion onto audio material. The design of a compressor can have a profound effect on its sound quality and also its distortion profile. The use of FETs, valves, transformers, optical cells, fast-acting time constants and "special modes" such as the famous 1176 all-buttons can both subtly and radically transform the sound quality of program material. This chapter provides an overview of DRC, how non-linearity can manifest at different points in a design and presents a comparison of four classic compressors focusing on their non-linear sonic signature. This work's motivation is to offer the reader a solid foundation on how compressor design affects non-linear sonic signatures. It will create a rich academic source for scholars in this area and provide professionals with a better understanding of using different compressors in their work.

4.2 Why Use Compression?

The first question to address in a study of this kind is why are compressors needed in the first instance? To answer this, we must consider dynamic range. Electronic devices have a finite dynamic range measured between the noise floor at the bottom of the scale and the clipping limit at the top. The amount of dynamic range available in an audio system can vary from device to device. An essential role for the audio engineer is to set an appropriate level that maximizes each piece of equipment's dynamic range and ensures the recording is not compromised with noise or clipping. The process is relatively straightforward if the recorded source is of consistent amplitude. However, during a recording session, most audio material can vary considerably in level, often as a result of the nature of the sound source, musical dynamics and the playing technique of the performer. Thus, audio signals have their inherent dynamic range that can be thought of as the lowest level to the signal's loudest peak. This presents an additional challenge for the recording engineer. They have to maximize the varying amounts of dynamic range offered by the equipment, and they also have to pack the fluctuating dynamics of audio sources into this equipment and avoid distortion (although, in keeping with the nature of this book, it did not take recording engineers long to realize that compressors could be used to actually impart distortion rather than prevent it). To make this process simpler, engineers began to use DRC to automate the tasks. DRC acts upon the dynamic range of audio sources by restricting the loudest peaks and (often but not always) turning up the signal's overall level, thus reducing its dynamic range. The process is achieved by varying the gain of an amplifier and is the reason why compressors

DOI: 10.4324/9780429356841-5

are referred to as variable gain amplifiers. A compressor in its most basic sense is a piece of audio equipment that makes loud levels quieter and quiet levels louder, essentially reducing the dynamic range. In simple terms, a compressor is an audio device in which the output signal does not increase as much as the input signal [1].

4.2.1 Compressor Design

The simplest form of dynamic control to conceptualize is the limiter. As explained so far, all audio equipment has a finite amount of headroom, and signals exceeding this limit result in clipping and non-linearity. Distortion can be used for creative music production effects, but it is not always desirable, and in many other areas of audio work, it can be a cause for concern. For example, overmodulated radio signals can cause damage to transmitters and overloaded PA systems can cause hearing damage. In both these cases, the overloaded audio signal needs to be attenuated as quickly as possible with no further increase in level. This form of abrupt dynamic range control is called limiting, and from a design perspective, the simplest way to limit an audio signal is to use a pair of diodes that hard clip the signal [1].

While this clipping method will accomplish fast level attenuation, it is only utilized as a basic form of dynamic range control for damage prevention and is rarely if ever used in studio production equipment. Due to the hard-clipping process, levels of distortion will be very high and consist of many high-amplitude, high-order harmonics. To reduce this effect, several other elements can be added to the circuit to control the compression performance and preserve the integrity of the input signal. The core components implemented in a basic compressor design are shown in Figure 4.1.

Here, the audio is split as it enters the compressor and is sent to a sidechain block consisting of several components including a level detection section. Various controls specify the threshold level, the ratio amount and regulate how quickly compression starts and stops, often called attack and release or the time constants.

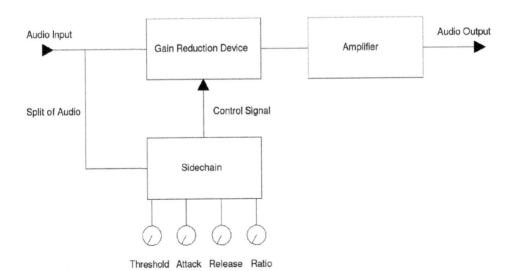

Figure 4.1 Block design of simple compressor.

4.2.2 Attack and Release Controls

Attack and release controls govern the speed at which the compressor starts and stops attenuating the signal once it exceeds the threshold. Depending upon the design of the compressor, this behavior can be executed in a variety of ways, but it is typical for attack and release to be implemented by a smoothing detector filter in the form of a variable resistor capacitor network [2]. In this design, the charging and discharging of capacitors govern the attack and release timings. In the attack portion, the capacitor is charged through a serial resistor while in the release period the capacitor is discharged [3]. This design can be seen in Figure 4.2.

The attack and release controls are often called time constants, a term which is related to how capacitors charge and discharge. The product of resistance and capacitance dictates the length of time a capacitor takes to charge or discharge to a new voltage, and this is called a time constant. Of importance is the fact that after one time constant, the change in voltage is 63% of the total voltage and after five time constants, the voltage change is 99% of its total. More specifically, if the attack time of a compressor is 100ms, then after one time constant (100ms) the gain change of the compressor will be at 63% of its final level. After 500ms (or five time constants) the gain change will be at 99% [4]. As will be shown later, the attack and release characteristics of a compressor can play a role in the generation of distortion. The actual range of time constants offered by a compressor can vary considerably. Some units such as the LA2A allow no control over time constants while VCA-based designs such as the dbx165A allow for a broad range of time constants that can be manually adjusted by the user. It is common for modern VCA-styled compressors to have attack times ranging from 500 microseconds to 100 milliseconds and release times ranging from 100 milliseconds to 3 seconds and beyond. A study by Bromham et al. investigated different compressor attack and release configurations on various musical styles and found the attack control had the most significant effect on the perception of genre [5].

Compressors can be designed to exhibit a degree of program dependency, meaning the speed of compression activity is changed by the acoustic properties of the material being compressed. One method is to vary a compressor's attack and release time depending upon the duration of the signal exceeding the threshold. The motivation behind implementing this behavior is to

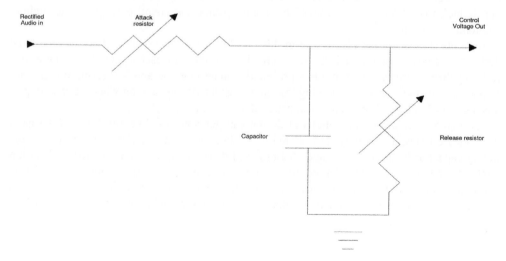

Figure 4.2 RC network of attack and release design.

minimize unwanted artefacts in the time domain. For transients, it is desirable to have a fast release to avoid overlong periods of gain reduction, but the same release time will lead to audible pumping in steady-state signals. Using a program-dependent design will vary the release to prevent this issue [6]. Program dependency is implemented in a hardware compressor by having a combination of time constants in the release section of the circuit. This approach is employed in the buss compressor found in SSL mixing consoles [7].

4.2.2.1 Non-linearity and Attack and Release Controls

Reiss [8] notes a number of audible artefacts are caused by the use of inappropriate time constant settings when using a compressor. These artefacts can be considered user errors rather than flaws in compressor design. They are:

1. Dropouts occurring in the audio where the compressor continues to attenuate the signal after it initially exceeds the threshold. This is due to an excessively long release time.
2. Overshoots occurring where the compressor fails to act quickly enough on the initial transient. This is due to a long attack time.
3. Loss of clarity from the use of fast attack times that attenuate the transient portion of the signal
4. "Pumping" (or fast modulation of the signal) as the compressor releases quickly after gain reduction
5. "Breathing" where the noise floor (or other aspects of the recording that aren't specifically the direct source) are raised in level by the time constants, producing audible amplitude movement of the extraneous audio.

In addition to the previous list, distortion is introduced to a compressed signal by the use of fast attack and release times. Quick time constants lead the compressor to compress within the period of an audio signal and reshape the output. This process can be explained by considering a low-frequency sine wave that has a period length that is longer than the time constant settings. Figure 4.3 shows a 50Hz sine that has been compressed with fast attack and release times; as can be seen, the time constants have changed the shape of the sine wave in a manner that is not unlike soft clipping. This waveshaping becomes more significant as the amount of gain reduction is increased and creates additional harmonic spectra that sound perceptually as audible harmonic distortion.

Both attack and release play a role in this non-linearity, but the attack time has the most destructive effect on the signal due to it affecting the crest of each cycle. The release also reshapes the waveform, albeit in a less profound manner than the attack [1]. But, this has not stopped sound engineers from adopting fast time constant settings in their work, often to impart this distortion on bass, drums and vocal material as a creative effect.

As well as user errors, limitations in design can generate non-linearity that is associated with release times. These flaws are related to modulations that occur in a waveform by the timing capacitor in the release stage charging and discharging and causing ripple distortion artefacts. This time constant behavior affects audio material by creating amplitude modulation that imparts many spurious harmonics to the output signal. This problem is one of the reasons why longer release times, relative to the attack, is implemented in hardware compressors [9].

4.2.3 The Sidechain and Level Detection

Figure 4.1 showed that compressors have a block called the sidechain, and the purpose of this component is to convert the audio signal into a control voltage, which is then sent to the gain

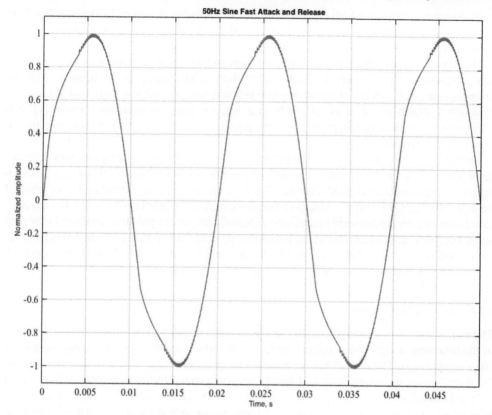

Figure 4.3 Waveshaping due to fast attack and release.

reduction element for attenuation. The sidechain section consists of several stages including a rectifier; a comparator (this is the threshold value where the compressor compares the sidechain signal to the input to decide whether attenuation is required); the time constants; and the ratio control that dictates the amount of the control voltage sent for gain reduction. The ratio and comparator section are thought of collectively as the gain computer section of a DRC [10]. It is at the sidechain block where the control voltage is sent to the gain reduction component for variations in the audio signal to be attenuated in accordance with the ratio setting and time constant speeds.

In a hardware compressor the sidechain can implement either peak or RMS sensing to detect variations in the audio signal. RMS sensing is less obtrusive and more transparent than peak sensing due to RMS sensing correlating better with the perception of loudness [9]. Peak sensing derives its level estimate from a short-term peak in the signal, which is then smoothed using a filter, and the manner in which this sensing occurs is affected by the time constant speed. For example, a short attack time encourages the capacitor to charge almost immediately whereas a longer attack time results in the capacitor charging slowly, smoothing out the peaks used in detection and results in an averaged value. True RMS detection (using the square root of the mean of a measured signal) became accessible with the introduction of VCA-based compressors [1].

There are two ways in which the sidechain block can derive the control voltage, either from the input signal (feedforward) or from the output signal (feedback). These two approaches are shown in Figure 4.4.

Historically these methodologies were implemented because of limitations in the gain reduction component. For example, feedback was utilized in FET compressors because it helped linearize unmusical aspects of the timing law [9], and feedforward was only usable (due to a reduction in distortion) once VCAs had been adopted as the gain reduction element [1]. It is interesting to consider that both of these approaches were implemented to reduce distortion from a design perspective.

Subjectively there are some audible differences between the two approaches, and feedback designs may have a more transparent sound. The reason for this difference is because the feedback style acts gentler on the audio due to it having a smaller dynamic range, which is a result of the signal being fed back to the gain reduction stage multiple times. Feedback also has a technical advantage over feedforward because the sidechain can rectify potential artefacts in the gain stage. Feedback design has a technical limitation, however, that prevents the use of high ratios such as infinity to one (infinite negative feedback is not possible in this design), thus feedback is not used in compressors that necessitate true limiting behavior [11].

4.2.4 Gain Reduction Styles

The majority of dynamic range compressors used in music production can be grouped into one of the following design styles: optical compressors, valve compressors, FET compressors and VCA compressors. There are other variations in compressor design, diode-bridge for example, but these four are the most common. Table 4.1 shows figures identified by Ciletti, Hill and Wolf [12] and should be used as an approximate guide. The amount of available gain reduction will have an effect on sonic signatures and particularly on those that require the use of significant amounts of gain reduction for heavy attenuation. The gain reduction style can have an influence on the nature of distortion, with arguably valve (also known as tube) being the most non-linear.

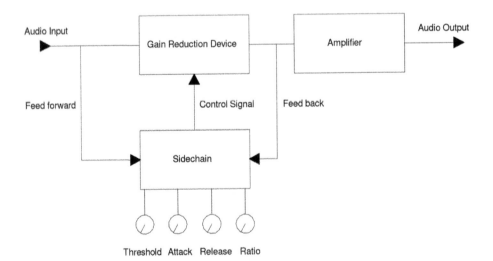

Figure 4.4 Feedback and feedforward compressor designs.

Table 4.1 Gain Reduction Method and Amount

Gain Reduction Style	Amount of Gain Reduction
Optical	12–25dB
Valve	25–30dB
FET	40–50dB
VCA	100dB

However, other aspects of the compressor design can play a more significant role, and this will be discussed in the next chapter.

4.3 Design Analysis of Four Popular Compressors

The following subsections investigate the design of well-known compressors from each of the categories listed previously. This discussion evaluates how compressor design affects a unit's sonic signature and highlights the areas in which distortion can manifest. The analysis will provide the reader with a sound understanding of how each compressor works, which in turn will yield a better, deeper appreciation of the signal analysis conducted in the next chapter.

4.3.1 *Optical Compressor: The LA2A*

The basic design of the LA2A is relatively simple. First the input is passed through an input transformer and into the gain reduction section. Here it is routed through a gain control that drives the signal into 12AX7 and 12BH7 valves before being sent to an output transformer. The use of valves and transformers in its circuit can result in a coloured sound as both valves and transformers can impart distortion when saturated. The LA2A is referred to as an opto-based compressor because it makes use of an electroluminescent panel that shines a light on a photo resistor to change its resistance. This works as a gain control because more light equals more resistance.

The component used to achieve gain reduction in the LA2A is called the T4 cell. It has unique qualities that are not achievable with similar photo resistors and is an integral part of the LA2A's sonic signature. One important aspect of the cell is its program dependency with attack, release and ratio, which all vary depending upon the level, envelope and frequency content of the input signal. For example, short transient material results in a fast release whereas steady-state material results in a longer recovery period. Additionally, the LA2A has a two-stage release, with the first stage taking place over the initial 40–80 milliseconds and the remainder lasting up to 2 seconds. This behavior, which is intrinsic in the T4 cell [13], allows the compressor to work transparently; Ciletti et al. [12] state that its program-dependent response is one of the reasons why it is difficult to make this compressor sound bad.

Aging components can change the sonic signature of analogue equipment over time, and the T4 cell is known to age in a manner that affects attack and release characteristics. Shanks [14] notes that older cells have faster attack times and modulations (or ripples) during the release period.

The LA2A has the simplest user interface of the compressors discussed in this chapter. It features only an input and output control and a VU meter for monitoring levels. The output control is labeled gain and works as an output trim, having no effect on compression activity. The only control the user has to affect compression is the peak reduction control, which when turned clockwise sends a split of the audio signal via an amplifier to the electroluminescent panel. The LA2A is considered a feedback compressor because the signal driving the sidechain is affected by a gain-reduced signal [15]. The sidechain sees the effects of compression on the signal and adjusts the amount of gain reduction accordingly. While the LA2A is often considered to be a transparent compressor, as will be demonstrated later, it does impart a good amount of distortion when compressing signals. However, it has also been used for much more aggressive forms of distortion, and according to engineer Joe Chiccarelli, was largely responsible for the distortion on Jack White's vocals on the album *Icky Thump* [16].

4.3.2 Valve (Tube) Compressor: The Fairchild

The Fairchild is a valve-based compressor and makes use of the variable-mu (also known as vari-mu) form of compression. This method works by sending a voltage through 6386 valves, as the voltage increases beyond a certain point the flow of electrons between the grid, and the gate of the valve is restricted, thus attenuating the signal. The use of 6386 valves is necessary because other valves such as the 12AX7 will not allow for artefact-free gain reduction. Compression happens directly on the audio path rather than being sent to a separate block, and the reason for this design is to make the unit work with high current, low impedance and as little noise and hum as possible. However, the Fairchild does create distortion, and most of it occurs in the valves and the class-A amplifier due to mismatches in the two sides of the amp. This non-linearity is most significant under heavy use, and distortion artefacts increase when using higher levels of gain reduction. The coloration from this compressor, and other valve compressors, is one of the reasons why recording engineers select this style of compressor.

The Fairchild has two large controls on the front panel. One is an input gain stepped in roughly 1dB increments that allow for a maximum of 18dB of gain. The second is a threshold control, which is continuously variable rather than stepped, and turning it clockwise yields more gain reduction. The Fairchild offers more control over compression speed than the LA2A but is limited to six options via a stepped control labelled "time constants". The positions range from 200 microseconds attack and 300 milliseconds release to 800 microseconds attack and 5 seconds release. There are also two program-dependent options at positions five and six. Positions one to three are probably the most used for mixing. Murphy [17] suggests these three time constant positions are suited to the coloured and aggressive styles of compression used in rock music production, presumably because of their fast speeds, which can result in a more aggressive sound due to the rate of compression and its tendency to introduce distortion. Ken Scott, engineer of The Beatles, noted in previous work by the author [18] that part of John Lennon's vocal sound came from compressing the signal hard with a Fairchild to the point of distortion.

The Fairchild does not allow for direct user control over the ratio. Instead, it is changed as the input level increases. The fundamental working principle behind the Fairchild is to set the threshold with the threshold control and then drive it into higher ratios with the input. The factory default setting for this variable ratio is set at +2dBm, and signals that exceed this level are compressed with ratios ranging from 1:1 to 20:1 [19].

4.3.3 FET Compressor: The 1176

The 1176 is called an FET compressor because it makes use of a BF245A field effect transistor (FET) in its gain reduction stage. The FET works as a voltage-dependent resistor whose resistance is altered by a control voltage applied to its gate. In the 1176 gain reduction is created by shunting the audio signal to ground once it starts to exceed a threshold point that is fixed depending upon the ratio setting. The user can alter this threshold point by changing the ratio controls on the front panel, and it rises commensurate with ratio, meaning that higher ratios shift the threshold point higher and vice versa. The knee of compression changes when the ratio is altered, and the curve of the knee gradually turns from soft to hard as the ratio is increased. Internally, the audio signal has to be kept to a small level to stop the FET from distorting during gain reduction activity. Consequently, the 1176 employs a powerful output amplifier to boost the output signal to make up for this initial loss in gain. The 1176 makes use of a 1:1 balanced output transformer that is used to reduce non-linearity and provide impedance matching.

Time constants are implemented by means of resistors and capacitors, which are placed between diodes and the gate of the FET. This design configuration alters the speed of gain reduction, which ranges from attack times between 200 and 800 microseconds and release times between 50 milliseconds and 1.1 seconds. In use the range of the attack time is very limited, and this can be seen in other work by the author [20].

One particularly important aspect of the 1176's design is the bias control, which is used to ensure the FET is never idling at 0 volts but instead at a voltage just under and slightly into conduction. This bias point results in a smoother transition into gain reduction because the FET is not audibly jumping in and out of attenuation as would be the case if biased exactly at 0 volts. An incorrectly set bias creates significant distortion during gain reduction. Therefore, care needs to be taken to ensure it is correctly calibrated and maintained. Variations in bias settings between 1176s may be one of the reasons for differences in their sonic signature and the degree of non-linearity they produce.

The input control of the 1176 increases the input and also sends more of the control voltage to the FET. As the input is driven, the control voltage is increased and creates more attenuation as it is raised towards the threshold point. The ratio is selected via switches on the front panel that can be depressed in some unorthodox configurations to yield unpredictable compression behavior. Nonetheless, this erratic performance has not stopped producers from using these ratio settings in their work. The most well-known of these settings is called the all-buttons mode and is popular to use on drum busses and room mics. What is not quite so well known is that this effect can be achieved by pressing the outer two ratio switches. This configuration works because the ratio resistors are stacked in series. As well as affecting the resistance in the ratio block, pressing multiple buttons alters the bias of the FET, and this affects the manner in which it applies gain reduction. This change in bias results in a significant increase in distortion and time constant behavior and is one of the reasons why engineers use this mode for distortion and colouration effects.

Another interesting aspect of the all-buttons mode is the change in meter behaviour. In all-buttons the meter shifts from 0VU and into the red portion of the VU meter, meaning all the way to the right. This change in the meter has no effect on the audio and occurs from a second FET in the circuit, which controls the VU meter, having its bias altered, and ballistics set out of calibration. The effect is purely cosmetic and plays no role in the sonic signature, but it may affect how a producer interfaces with the compressor. A comprehensive analysis of this compressor was carried out in previous work by the author and covers its non-linear sonic signature in significant detail [20, 21].

4.3.4 VCA Compressor: The dbx165A

The dbx165A is the most modern device discussed and tested in this chapter, and this recent design is reflected in its wider range of controls and the LED that supplements the traditional VU meter. The dbx165A utilizes a voltage-controlled amplifier (VCA) in the circuit for gain reduction, and fully original units make use of a dbx202. A VCA is a special type of amplifier that has been designed specifically to attenuate or accentuate audio levels over a broad dynamic range. VCAs have their gain adjusted by an external DC voltage, and this is an important aspect of their design. Traditionally amplifiers are intended to increase the level of a signal (apply gain) and do not readily lend themselves to the attenuation of signals without potentially becoming unstable and oscillating; a VCA does not suffer from these issues. As well as having little noise, a VCA has a wide dynamic range and, as shown in Table 4.1, yields large amounts of gain reduction, considerably more than the other compressors discussed so far.

The time constants on the dbx165A are expressed in a dB per millisecond scale for the attack and dB per second scale for the release. This specification means an attack time of 1dB per millisecond takes 5 milliseconds to change attenuation by an additional 5dB. While this is a precise way of stipulating attack times it is not intuitive, and the maximum attack time figure of 400dB per millisecond is not going to mean a great deal during a vocal tracking session other than it is the fastest attack time offered by the unit.

The dbx165A offers a wide range of attack and release speeds, and this makes it more suitable for creative envelope shaping than the other compressors. The timing law of the time constants is logarithmic, and the VCA design creates precise and consistent compression speed and timing behaviour. This performance is unlike the other compressors mentioned so far, which have less precise control over their time constants.

The most significant differences between the dbx165A and the other compressors mentioned here are it uses a feedforward design, allows for higher ratios (up to infinity to one) and has an RMS detector that under certain circumstances provide it with a musical and transparent sound. In addition, it features a peak stop setting that can be employed to catch peaks that have been missed by the RMS detector. Therefore, the dbx165A has the option to work rapidly on the detection of program material if need be. The dbx165A allows the user to control the compression knee for settings between soft and hard, although instead of soft knee a gentler curve is called "over easy" on the dbx165A. Soft knee is implemented in this compressor by using an open-loop diode that creates variation in impedance over a range of voltages, and this means a higher voltage produces a sharper knee characteristic [20].

Hicks [21] states that VCA compressors are typically clean and less non-linear than other styles. While this applies mostly to the dbx165A, it can create significant non-linearity when used aggressively with fast time constants. There is much online discussion regarding the distortion generated by the dbx165A, and one commenter at the popular repforums states that "manual mode on these devices=distortion box" [22]. It is likely this comment stems from users using fast attack and release settings with heavy gain reduction, because this design affords users with the option of compressing very heavily and at a rapid pace. One other possible source of distortion was pointed out by Ciletti et al. [12] when they noted that distortion could appear in VCAs because of mismatched transistors, adding that even modern designs suffer from spurious distortion with high input signals. Thus, technical and practical reasons can affect the distortion profile of this device.

4.4 Audio Signal Analysis

The next subsection discusses audio measurements made on four dynamic range compressor styles. The compressors are the Urei 1176 Revision D (FET), Teletronix LA2A (Opto), Fairchild 670 (Valve) and dbx165A (VCA). The measurements discussed in this chapter come from a much larger study [20, 21]. They are presented in this chapter to highlight the distortion profile of popular compressors, which have been used on many music productions from the 1960s to the current year. The tests are not meant to be fully comprehensive but should serve to highlight the sonic signature of each device and shed some light on how their non-linearity has helped shaped the production process.

4.4.1 Test Tone Measurement

Test tone measurements were made to get an overview of each compressor's sonic signature and compare and contrast the results. All audio was processed through the compressors in a professional recording studio in London (UK) with assistance and guidance from the resident in-house engineers who had much experience with the devices under test. All audio was sent directly to and from Prism ADA8 converters at 44.1kHz 24bit. No other devices in the studio (mixing consoles for example) were present in the signal path.

To investigate the distortion, the compressors were fed a 1kHz sine wave of three different amplitudes. This tested for differences in non-linearity as a function of input level. The compressors were sent a signal at 0dBu, +9dBu and +16dBu. When making measurements, all the compressors were adjusted to have no gain reduction or set to a compression off mode if the compressor had one. The compressors were tested using this method as it has been often stated anecdotally by recording engineers that running a signal through a particular device will make it "sound better". Therefore, this measurement was conducted to ascertain how much distortion may be playing a role in this process. Audio with the compressor engaged in gain reduction was also measured to assess non-linearity when the compressor was attenuating the signal. The results from the non-compression activity tests at all input levels can be seen in Table 4.2. The measurements are expressed as a total harmonic distortion (THD) percentage, which is achieved by taking the square root of the sum of the squares of each harmonic amplitude, dividing this by the fundamental (test tone) amplitude and multiplying the result by 100 to achieve a percentage figure.

Under this test, the dbx165A is very clean with insignificant amounts of non-linearity. The distortion rises by a small amount as the input level is increased but the artefacts are inaudible, and the compressor is transparent. When calculated as a THD percentage, these small amounts of non-linearity result in 0% values across all input levels. The Fairchild is also clean in this test, and the plot in Figure 4.5 shows the most prominent harmonics are at 2kHz and 3kHz. Additional low-level harmonics can also be observed at 50Hz, 100Hz and 150Hz, and they appear to be low-level artefacts from mains hum and do not rise in level as the input is increased. Other harmonics are visible that cluster around the 1kHz test tone and are spread apart by 50hz. These are most likely sidebands that have been created in the audio as a result of sum and difference components generated as a product of the test tone and mains hum frequencies. These spurious harmonics rise in level as the test tone amplitude is increased and may play a role in this Fairchild's non-linear sonic signature, particularly at more driven input levels. Without the mains hum and the sideband frequencies, the Fairchild is notably clean when driven with the +16dBu input. The THD figure with the +16dBu driven input is in the region of 0.08%.

Table 4.2 THD at Input

Compressor	THD % 125Hz In Comp	THD % 125Hz In Release	THD % 1kHz In Comp	THD % 1kHz In Release
1176 Rev D	0.40	0.27	0.07	0.01
dbx165A	0.27	0.14	0.14	0.12
Fairchild	0.56	0.35	0.07	0.03
LA2A	2.13	1.18	2.64	2.38

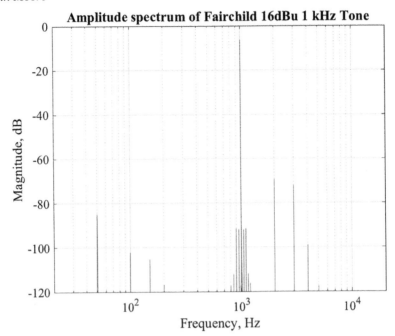

Figure 4.5 Fairchild THD.

As shown in Figure 4.6, the LA2A has a slight amount of mains hum at multiples of 50Hz. The hum artefacts are lower in level than the Fairchild's, and they do not modulate with the test tone frequency to introduce sideband artefacts. Compared to the Fairchild, the LA2A has significantly more non-linearity at all input levels. At the 0dBu input level, the second and third harmonics are -54dB and -85dB down and rise up to -41dB and -57dB by the loudest input level of +16dBu. This distortion results in an audible amount of THD that is rated at 0.5% and 0.9% for the +9 and +16dBu input levels.

The 1176 Rev D has a clean THD result until +16dBu, where the third harmonic increases in level to -52dB down from the test tone. This harmonic is partly responsible for the THD percentage of 0.25%, which is still a low THD figure for a driven input. The distortion from the +16dBu signal is shown in Figure 4.7.

Distortion during compression activity was also measured and reveals a different picture, however. Measurements were made at four frequencies: 125Hz, 250Hz, 500Hz and 1kHz. The most significant differences between the test tone frequencies were between 125Hz and 1kHz, thus they are the results discussed and shown in Table 4.3.

The time constants used in this test made use of "typical" rock vocal settings that were extracted from relevant literature and adapted slightly by the in-house engineers to account for any idiosyncrasies of the units under test. The settings used for the 1176 had the attack at 3, release at 7 (this setting was found to be one of the most common settings in previous work by the author), the Fairchild set at time constant 1, the dbx165A set for a fast release and moderately fast attack. The LA2A has no time constant settings. The 1176 all-buttons mode was not used in this exploration, as it was deemed a "special" mode, used by engineers when they want to overtly distort a signal and introduce amplitude modulation (pumping effects). Moreover, this would have resulted in very biased comparison as the LA2A and Fairchild cannot offer that style

Non-linearity and Dynamic Range Compressors 59

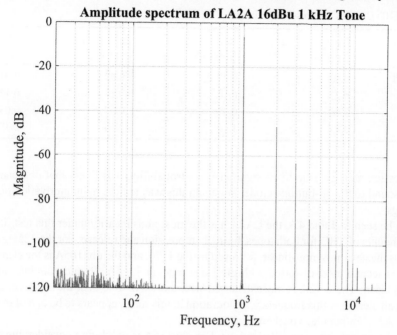

Figure 4.6 LA2A THD.

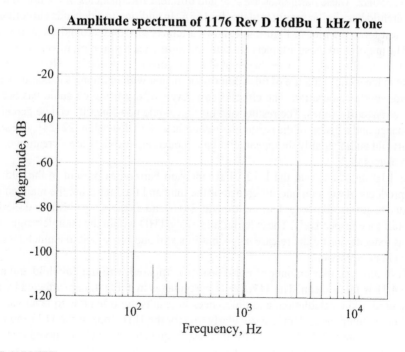

Figure 4.7 1176 THD.

Table 4.3 THD During Compression Activity

Compressor	THD % 125Hz In Comp	THD % 125Hz In Release	THD % 1kHz In Comp	THD % 1kHz In Release
1176 Rev D	0.40	0.27	0.07	0.01
dbx165A	0.27	0.14	0.14	0.12
Fairchild	0.56	0.35	0.07	0.03
LA2A	2.13	1.18	2.64	2.38

of compression activity. Due to its significant controllability, it is likely that dbx can achieve a similar sound to the all-buttons mode, but again this was not the main motivation behind the test settings.

As can be seen in Table 4.3, the LA2A has the most non-linearity under this test. It has significant amounts of THD for both tones during compression and release. For the 125Hz tone, the Fairchild is the next most non-linear, followed by the 1176, and the dbx165A is the cleanest. The dbx165A is second most distorted with the 1kHz tone, and the 1176 and Fairchild are almost equally as clean. The LA2A is more distorted with the 1kHz tone than the other compressors, which are all cleaner at this frequency. Colouration in this range appears to be one of the reasons why the LA2A is a popular vocal compressor.

Figure 4.8 reveals interesting behaviour. The dbx165A is exhibiting considerable non-linearity during compression. As well as prominent second and third harmonics, some additional sidebands cluster around the test tone frequency, spread out in 50Hz increments. Low-level harmonics at the bottom end of the spectrum are evident in this plot and are at 50Hz, 150Hz, 250Hz and 350Hz. These harmonics are sum and difference frequencies of the non-linearity.

The Fairchild has similar artefacts to the dbx165A. In addition to sidebands clustering around the 1kHz test tone, the plot in Figure 4.9 shows artefacts grouping around the non-linear harmonics. The majority of these artefacts are sum and difference harmonics, of very low level, and outside of an audible range where they would be perceived as distortion. However, they may play a role in the sonic signature of this compressor, fusing with harmonics in complex program material for subtle colouration. The effect of low levels of distortion on audio has been investigated in a range of studies. The results suggest that subtle non-linearity can be perceived as a timbral change and a fusion of the harmonics rather than overt distortion [22–24]. The sidebands in the Fairchild audio quickly disappear after gain reduction, thus they are a creation of the gain reduction element.

Figure 4.10 illustrates that the LA2A has sideband harmonics present in the audio (again they are products of the sum and difference of the tone and the hum), but the main colouration effect is due to strong odd-order harmonics, with 3kHz and 5kHz being -32dB and -45dB down from the test tone respectively. These harmonics bring THD within the audible range. The non-linearity is reduced during the release stage, but it is still audible, albeit to a much lesser extent than when in compression.

The 1176 has a similar amount of non-linearity compared with the Fairchild and as shown in Figure 4.11 is fairly clean. The 1176 THD percentages in Table 4.3 are affected by the third harmonic used in the calculation, and differences in results are often due to level variations in this third-order component. This is particularly true for the 1kHz tone of the 1176 and Fairchild, which have the smallest THD results for this test frequency. It should be noted that the 1176 all-buttons mode was not used in this test as the motivation behind this test was to compare the compressors with similar settings. However, previous studies have shown that this setting can

Non-linearity and Dynamic Range Compressors   61

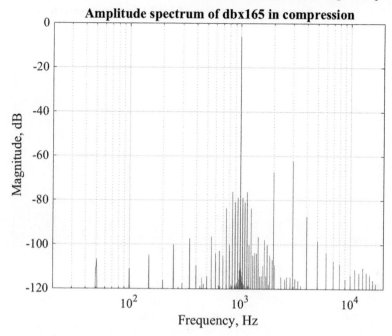

Figure 4.8 dbx165A distortion during compression.

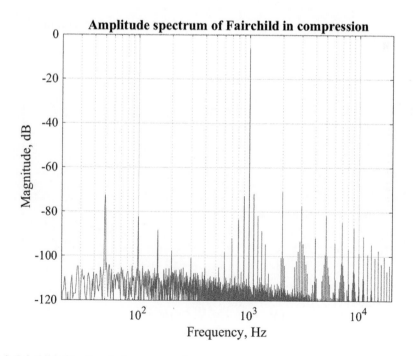

Figure 4.9 Fairchild distortion during compression.

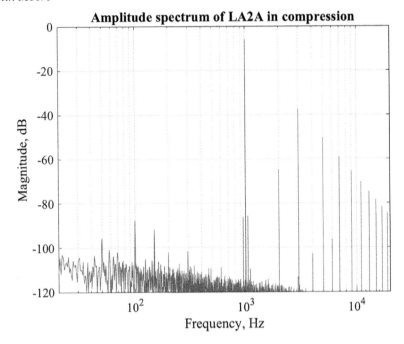

Figure 4.10 LA2A distortion during compression.

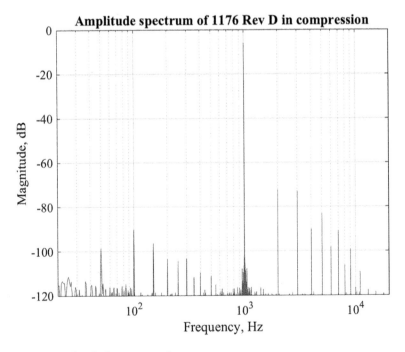

Figure 4.11 1176 distortion during compression.

achieve upwards of 1.5% THD [25]. Thus, the all-buttons should be a suitable choice for non-linear, aggressive vocal timbres.

4.4.2 Vocal Recording Analysis

A vocal recording was processed through the compressors using the settings noted previously and achieving -12dB of gain reduction. The compressors were set to compress hard as this was in adherence with the comments made by hardware designer Paul Wolf, who advises using significant gain reduction when assessing a compressor for its sound quality [12]. To evaluate changes to timbre of the recordings, low-level audio features were extracted from the audio using MirToolbox [26] for Matlab. Previous work by the author found that roughness and features pertaining to perceived brightness (roll off, spectral centroid and brightness) correlated well with distortion artefacts. Thus, only these features are presented in Table 4.4. A thorough review of audio features can be found in the work by Alías et al. [27].

Critical listening highlights that the dbx165A when compressing the voice sounds noticeably distorted, despite showing modest amounts of THD in the previous tests. Presumably the fast attack and release times under heavy compression have distorted the vocal signal in a similar manner to the waveshaping shown in Figure 4.3. Further critical listening also reveals that the compressed signals are brighter than the uncompressed signals and that the Fairchild and LA2A sounds slightly more distorted than the 1176. The reason for this is likely due to the valves used in the Fairchild and LA2A circuits and the opto cell in the LA2A. See the work by Stroe on the non-linear behaviour of optical cells [28]. As can be seen in Table 4.4, the audio features correlate well with what is being heard by the author. Thus, roughness appears to be a good measure of the distortion present in complex program material, or at least vocal recordings.

4.5 Conclusions

This chapter presented a review of how general compression design and the topology of specific compressors can introduce distortion to audio signals. It was shown that the device used for gain reduction, the amplification stages and the time constant speeds could affect how distortion manifests in a compressor's circuit. It was demonstrated that each compressor has a subtle yet unique sonic signature, which is shaped in part by its non-linear characteristics. Furthermore, it was illustrated how this non-linearity can change during compression activity and that its response is frequency dependent. Finally signal analysis, critical listening and low-level audio feature extraction were used to profile each compressor's non-linear sonic signature. The author proposes this as a robust methodology for further studies of this nature.

Table 4.4 Low-Level Audio Features

Compressor	Roll-Off	SpectralCentroid	Brightness	Roughness
Fairchild	10361 Hz	4970 Hz	0.60	169
LA2A	10410 Hz	4943 Hz	0.58	182
dbx165A	10500 Hz	5109 Hz	0.61	197
1176 Rev D	10253 Hz	4826 Hz	0.58	145
No Comp	10095 Hz	4655 Hz	0.55	102

References

1. Dove, S.: *Handbook for sound engineers*. Focal Press, Amsterdam; Boston (2008)
2. Giannoulis, D., Massberg, M., Reiss, J.D.: Digital dynamic range compressor design—a tutorial and analysis. *Journal of the Audio Engineering Society*. 60, 399–408 (2012)
3. Talbot-Smith, M.: *Audio engineer's reference book*. Focal Press, Oxford; Boston (1999)
4. Martin, G.: *Introduction to Sound Recording*, www.tonmeister.ca/main/textbook/
5. Bromham, G., Moffat, D., Barthet, M., Fazekas, G.: The impact of compressor ballistics on the perceived style of music. In: *Audio engineering society convention 145*. Audio Engineering Society (2018)
6. Berners, D., Abel, J.: *UA WebZine "Ask the Doctors!"* February 04 | Drs. David P. Berners and Jonathan S. Abel Answer Your Signal Processing Questions, www.uaudio.com/webzine/2004/february/text/content2.html
7. Reiss, J.D., McPherson, A.P.: *Audio effects: Theory, implementation, and application*. CRC Press, Boca Raton (2014)
8. Reiss, J.: Design of a dynamic range compressor. *131st AES Conference*. New York (2011)
9. Self, D.: *Small signal audio design*. Focal Press, Oxford; Burlington, MA (2010)
10. Berners, D.: *UA WebZine "Ask the Doctors!"* July 05 | Peak vs. RMS Detection, www.uaudio.com/webzine/2005/july/text/content2.html
11. Berners, D.: *UA WebZine "Analog Obsession,"* April 06 | Compression Technology and Topology, www.uaudio.com/webzine/2006/april/text/content4.html
12. Ciletti, E., Hill, D., Wolf, P.: *Gain control devices, side chains, audio amplifiers*, www.tangible-technology.com/dynamics/comp_lim_ec_dh_pw2.html
13. Shanks, W.: *Compression obsession: The amazing release character of the LA-2A*, www.uaudio.com/webzine/2003/june/text/content4.html
14. Shanks, W.: *UA WebZine "Compression Obsession,"* July 03 | Long Live the T4 Cell!, www.uaudio.com/webzine/2003/july/text/content4.html
15. *Universal audio: LA2A compressor manual* (2000), https://media.uaudio.com/assetlibrary/l/a/la-2a_manual.pdf
16. Tingen, P.: *Secrets of the mix engineers: Joe Chiccarelli*, www.soundonsound.com/techniques/secrets-mix-engineers-joe-chiccarelli
17. Murphy, R.: *Analoguetube AT-101—(Fairchild 670 recreation)*, Review by Ronan Chris Murphy—YouTube, www.youtube.com/watch?v=dfC-NhHqp9Q
18. Moore, A.: Tracking with processing and coloring as you go. In: *Producing music*, pp. 209–226. Routledge, New York (2019)
19. *Fairchild recording equipment: Fairchild 670 manual* (1959), http://thehistoryofrecording.com/Manuals/Fairchild/Fairchild_670_stereo_limiting_amplifier_instructions_Schematics.pdf
20. Moore, A.: *An investigation into non-linear sonic signatures with a focus on dynamic range compression and the 1176 fet compressor* (2017)
21. Moore, A.: All Buttons in: An investigation into the use of the 1176 FET compressor in popular music production. *Journal on the Art of Record Production* (2012)
22. Gabrielsson, A., Sjögren, H.: Detection of amplitude distortion in flute and clarinet spectra. *The Journal of the Acoustical Society of America*. 52, 471–483 (1972). http://doi.org/10.1121/1.1913124
23. Gottinger, B.: *Rethinking distortion: Towards a theory 'of 'sonic signatures'*, PhD Thesis, New York University (2007)
24. Petri-Larmi, M., Otala, M., Lammasniemi, J.: Threshold of audibility of transient intermodulation distortion. In: *Audio Engineering Society Convention 61*. Audio Engineering Society (1978)
25. Moore, A.: Dynamic range compression and the semantic descriptor aggressive. *Applied Sciences*. 10 (2020). https://doi.org/10.3390/app10072350
26. Lartillot, O.: *MIRtoolbox—file exchange—MATLAB central*, http://uk.mathworks.com/matlabcentral/fileexchange/24583-mirtoolbox
27. Alías, F., Socoró, J.C., Sevillano, X.: A review of physical and perceptual feature extraction techniques for speech, music and environmental sounds. *Applied Sciences*. 6 (2016). https://doi.org/10.3390/app6050143
28. Stroe, O.: *Complete overview and analysis of an electro-luminescent based optical cell* (2018), https://eprints.hud.ac.uk/id/eprint/34532/

5 Low Order Distortion in Creative Recording and Mixing

Andrew Bourbon

5.1 Introduction

Microphone preamps provide recording engineers with a core tool in the recording process, working as the interface between the microphone and the ongoing signal chain. Microphone preamp designs are varied and often characterised by the sonic signature that they impart onto the signal being amplified. The electronic circuit design is often used as part of the standard classification, with consideration of transformer or transformerless designs, as well as amplifier topologies being associated with particular sonic signatures. In the recording process, engineers will often take advantage of the sonic characteristics of these preamps to enhance the musical sounds being captured. Sylvia Massey [1] notes that the sound of various large format consoles is 'instantly recognizable' and discusses the process of 'creaming the mic pres' as a process of driving signals through a microphone preamp in order to add colouration for perceived dramatic effect. There is more happening within the console than low order harmonic distortion through the mic pre, and it is difficult to separate the impact of this colouration with other non-linearities in the signal path at the end of the production process, but the initial perceived impact of this kind of recording practice is clear and indeed measurable. Engineers seek out particular recording tools for their specific affordances [2], with frequency response, phase response and distortion all measurable factors that combine to present the audible response of the preamp. This chapter seeks to explore a range of microphone preamps with clear and recognisable sonic signatures and the impact of the low order distortion provided by these tools to the signal being processed. As well as measuring the response, the preamps have been auditioned, with their impact on the processed audio described alongside the physical measurements. As well as exploring the impact on the recording process, the impact of low order harmonic distortion on the mix process will also be considered, with microphone preamps and other harmonic tools often used by engineers as part of the contemporary mix process.

5.2 Preamps and Low Order Distortion

5.2.1 Affordances of Low Order Harmonic Distortion

Low order distortion has several affordances when considering the perceived impact on a musical gesture or performance. For the purposes of this chapter, low order distortion will be limited to the additional second and third order harmonic distortion added over the fundamental by specific audio processing devices. Practicing engineers tend to use semantic descriptors when exploring the impact of tools that add distortion, using terms such as 'punch' and 'weight' or 'fatness' to describe the sonic affordances of a given recording tool. Lower order harmonic

DOI: 10.4324/9780429356841-6

distortion tends to be more difficult to hear than higher order distortion, and as such tends to lead to descriptors that are more subtle in nature and are arguably often more subjective and can be harder to articulate. Several interrelated factors led to these perception-based responses, some of which can be hard to separate when looking at the circuitry employed in microphone preamps. Transformers for example exhibit multiple characteristics, all of which contribute to the perceived affordances of those transformers. The focus of this chapter is to document the harmonic distortion present in the preamplifier circuit, rather than looking to measure the temporal distortion as waveforms are modified by the components in the preamplifier circuit.

5.2.2 Low Order Distortion

This chapter will focus primarily on second and third order distortion profiles found within a range of selected microphone preamps. The selected preamps tend only to offer significant higher order distortion at the highest gain settings, and can be used for aggressive saturation when required, however in this study the primary concern relates to the analysis of second and third harmonic distortion, as well as considering overall changes in the frequency response provided by the chosen circuit.

Second harmonic distortion takes place at the first multiple of the fundamental frequency, sitting an octave above the fundamental, with for example a second harmonic at 200Hz above a 100Hz stimulus. Normally measured at a level significantly below the fundamental, the impact of this harmonic on musical material tends to result in a sound that has a greater sense of weight and solidity, making the source feel larger and more resolute.

Third harmonic distortion would take place at 300Hz when the same 100Hz stimulus is used, resulting in a harmonic that sits an octave and a fifth above the fundamental. This is still representative of a highly musical interval, associated with power chords on the guitar for example. The impact of this third harmonic distortion is to create a stronger sense of punch in a sound, helping to lift that element forward in more complex music presentation compared with the more centered size and body provided by the second harmonic. This punch also tends to communicate effectively even on smaller reproduction systems [3].

The amplifier circuits employed in contemporary preamp designs feature various topologies, each associated with a different set of associated distortion profiles. These topologies will be explored and measured through this chapter, providing insight into the measured performance of a range of contemporary recording tools.

5.2.3 Transformers

Design engineers take several approaches in the creation of microphone preamps, with many of the most famous manufacturers having relied on the use of transformers in the audio circuit, both for the addition of gain and for balancing and common mode rejection [4]. Several characteristics of transformers impact on the audio performance of the circuit, many of which could be considered as a form of distortion. The first type of distortion to be discussed is in the response time of the transformer itself, with transformers taking time for audio to move though the core. This leads to a rounding of the transient, subject to the tuning of the transformer, which is often perceived as a sense of fatness often desirable in the recording of drums. The initial magnetising of the core of the transformer also sees further distortion added [3], which provides additional colouration potential for transient sources. In addition to this temporal distortion there is also often an addition of harmonic content associated with the largely symmetrical clipping found in audio transformers, with this magnetic distortion normally expected to be third harmonic in nature. This low order harmonic distortion is 'dramatically less audible' [5] than high order distortion associated with devices expected to add extreme colour such as guitar amplifiers, pedals and more aggressive saturation devices. The affordances of this low order distortion can be highly desirable in all stages of music production subject to musical context.

As level increases, transformers also exhibit increasing frequency specific phase shift and non-linearities in distortion, making them dynamic devices that can effectively be 'dialed in' through the end user driving signal into the transformer to match the specific sonic requirements of a record. The harmonic impact of a transformer is defined by the materials found within the transformer core, the size of the transformer and the nature of the transformer windings [5]. The bigger the transformer is, the deeper the low frequency response and the higher the level of signal that can be handled by the transformer before saturation takes place [6]. It is worth noting, however, that increase in inductance in larger transformers does compromise high frequency response, making larger transformers unsuitable for audio usage. Selection and design of audio transformers has historically been undertaken with the aim of providing the optimal frequency response and amplification characteristics at a given impedance. The harmonic content that is now craved by engineers was an unfortunate by-product of the available technology at the time. It is interesting that modern designs have often embraced active integrated operational amplifier (opamp) technology to improve linear behavior and minimise distortion and noise beyond the measured performance offered by the audio transformer. Despite the improved measured performance of an opamp based circuit, the technical flaws of older transformer designs are still incredibly desirable from a musical perspective.

Distortion found in audio transformers is predominantly third order in nature, resulting in for example new material at 300Hz added to an input signal at 100Hz. The percentage of nickel in an audio transformer directly correlates to the level of distortion provided by the core, with higher percentage of nickel resulting in a less distorted output. Three common core materials are 84% nickel, 49% nickel and M6 steel, with increasing distortion as the nickel content reduces and the highest distortion provided by the M6 steel core [5]. The lower the frequency hitting the transformer, the higher the distortion that is added by the transformer, with the high nickel content transformers exhibiting negligible harmonic distortion above 50Hz. The cutoff frequency continues to rise as the percentage of nickel in the core drops; however, all core materials exhibit lower distortion as frequency increases.

Hysteresis provides further sonic impact on audio signal passing through a transformer. The previously discussed harmonic distortion is normally caused by high levels of audio in the transformer, resulting in greater saturation and audible distortion as the signal level is pushed, with the transformer eventually behaving as an audio clipping device. Hysteresis is the result of the inability of the transformer to react quickly enough to changes in signal, resulting in energy loss as the magnetic state of the transformer is changing. This is particularly apparent with smaller signal levels in the transformer, leading to signal distortion in both the frequency and time domains subject to the design in the transformer and the materials found in the core of the transformer.

As discussed, there is also a relationship between level and distortion in the core of a transformer. As the signal level increases, the core of the transformer moves towards saturation, to a point where the output level of the transformer increases at a slower rate than the input level, resulting in a reduction of crest factor. This transformer compression can also be desirable but will see increased non-linearity in the distortion profile at different frequencies. Pushing a transformer into saturation will cause a much higher percentage of distortion and will be sonically less subtle. Many contemporary designs and software plugins provide an output fader after the preamp in order to allow users to drive the preamp whilst still controlling the output level into the converter or further analogue processing to maximise the impact of this processing.

5.2.4 *Tube Preamps*

Solid state transformer preamps made by manufacturers such as Neve and API are traditionally associated with the addition of third order harmonic material to the audio signal path. These transformer preamps are often sought out by engineers; however, other types of gain amplifiers are also regularly cherished by engineers for their specific sonic signatures, most notably those employing tubes as amplifiers. Tubes will often distort asymmetrically, resulting in tube preamps traditionally being associated with the addition of second harmonic distortion, though the presence of transformers in the audio circuitry will also often see the presence of third harmonic distortion in the circuit. As with transformers, tubes are flexible components that can be designed for minimum low order distortion to offer significant levels of high order distortion. This chapter will focus on the implementation of tubes in scenarios where low order distortion is considered desirable.

5.2.5 *Contemporary Low Order Harmonic Distortion Tools*

A number of flexible distortion tools used in contemporary production are capable of providing the user with control over the addition of low order harmonic distortion. Tube distortion tools such as the Thermionic Culture Vulture and SPL Twin Tube provide engineers with tools that are capable of creating a wide range of harmonic distortion, from subtle low order distortion to aggressive high order distortion. Low order harmonic distortion tools are also increasingly popular in contemporary processing chains. The Waves Scheps channel strip for example offers the ability to add odd, even or a mixture of low order harmonics as part of the signal chain; SSL offer the X Saturator plugin that offers controllable low order harmonics; and companies such as Elysia offer tools such as Karacter, which offer even or odd harmonic processing at both low and high orders, targeting producers through to mastering engineers.

5.3 Analysing Preamps—Measurement

For the purposes of this chapter, the sonic impact of a range of microphone preamps will be explored using virtual representations of the associated hardware. In all cases Unison preamps designed by Universal Audio have been employed, with measurement taking place using Plugin Doctor by DDMF. The Universal Audio Unison preamp combines a plugin which emulates the performance of the chosen audio signal path with a hardware microphone preamp, with the impedance, gain staging and non-linearities recreated through the combination of preamp and software [7]. Though traditionally associated with initial capture of audio, microphone preamps are commonly used in the mixing phase of audio production, particularly as plugin designers have been able to refine the emulation process through use of dynamic convolution and component modelling. Many of the most recent channel strips offer full emulation of the complete circuit, with users able to drive the channel in the same way as would previously have only been possible to achieve using consoles and hardware. There is even greater flexibility in a number of designs, with for example Universal Audio providing a headroom control on their emulations to allow users to take advantage of modern gain staging whilst still providing fine control over gain and distortion in emulations of hardware that were not designed for the reference levels often found in contemporary production, allowing a full range of subtle to more aggressive high order distortion to be added using channel strip input stages.

A range of microphone preamp emulations will be tested from the list of mic channel emulations available on the Universal Audio platform.

Table 5.1 List of Tested Microphone Preamps

Mic Pre	Description
Avalon VT737	Tube
Neve 1073	Class A amplifier, transformer
Neve 1084	Class A amplifier, transformer (same amplifier as the 1073, EQ changed)
Manley Voxbox	Tube
SSL 4000E	Switchable Jensen transformer
API Vision	API 2520 opamp and transformer
V76	Tube
Neve 88RS	Transformer
Helios Type 69	Transformer
Century Tube	Tube
UA610 A & B	Tube

In this chapter the API Vision Channel Strip, Neve 1073, SSL E Channel, Avalon VT 737 and UAD 610B circuits have been tested with a range of measurements taken for all preamps. The selection process was designed to capture the full spectrum of available preamp designs, with solid state and tube preamps selected. Within those selections transformer and transformerless designs are available to be tested, with class A discrete and integrated circuit-based amplifiers also covered by the selection. Two tube preamps have also been tested, one of which is associated with modern recording and the other offering a more vintage aesthetic. Any other signal processing provided by the channel strips was bypassed where possible, with only the preamp circuitry contributing to the measured distortion.

5.3.1 API Vision

The first preamp to be analysed is the API Vision Channel Strip, which is an emulation of the channel found in the console of the same name. The API preamp is often considered to be selected for its characteristic punch and colour and is particularly popular in the capture of transient rich material. The first measurement undertaken explores the harmonic content added by the preamp circuit and can be seen in Figure 5.1.

The measurement was undertaken with a low gain setting, with the Hammerstein model demonstrating the second and third harmonic contribution to the output signal across the entire frequency range. The Hammerstein model employed in Plugin Doctor is a particularly useful analysis tool when exploring non-linearities in harmonic performance of audio circuits. The generalised Hammerstein model employed in these tests allows the impact of a particular harmonic to be demonstrated across the entire frequency range in a single measurement, providing insight into frequency specific distortion added by the preamplifier circuit. It is clear to see from this diagram that the measured distortion in the API channel is third harmonic in nature, with the second harmonic content at such a low level that it is not considered to be contributing to the overall sound of the preamp. One of the interesting observed characteristics here is the relatively consistent contribution of third harmonic distortion across the entire frequency range of the API. As previously discussed, transformers tend to exhibit distortion characteristics relative to the frequency being sent through the transformer, with higher distortion at lower frequencies. At this relatively low gain setting, interestingly this performance is not replicated in the API

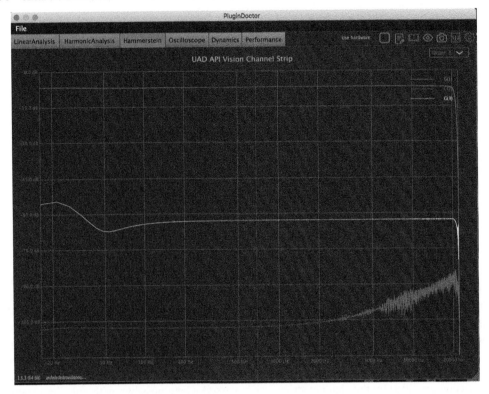

Figure 5.1 API low gain Hammerstein model.

preamp. There is evidence of ripple in the distortion response in the very low frequency range, but the overall distortion characteristic appears to be relatively linear.

As the gain is increased, the same characteristic distribution of third harmonic distortion can be seen, but with an increase in the ripple in the low frequencies, as evidenced in Figure 5.2. The contribution of second harmonic distortion has remained consistent across both measurements.

Engineers will often drive API preamps hard in order to achieve more dramatic colouration, which is made easier to achieve with the addition of an output attenuator at the end of the channel. The API 3124V four channel microphone preamp also has an output attenuator and the ability to change the turns ratio in the transformer to reduce the gain in the transformer as a response to user requirements, showing the importance of controlling the gain staging in the API preamp in order to optimise the sonic affordances of the design. Figure 5.3 demonstrates an extreme gain setting, with significant output attenuation in order to drive the transformer and 2520 opamp into maximum saturation.

When maximum gain is used, there is a much more significant presence of third order harmonic, with the third harmonic actually measuring as the dominant contributor beyond the fundamental in the 20–50Hz range, and offering distortion up to 500Hz. There is significant drop-off in the level of distortion between 60–500Hz, with the level dropping from 45dB below the fundamental and falling into noise floor as the frequency increases towards 500Hz, at which point the level drops to negligible levels. This level of third harmonic distortion is subtle but will have an impact on the listener on the perception of low frequency rich material. There is evidence of greater contribution of second order harmonic content in the lowest frequencies; however, this contribution is still insignificant from a sonic perspective.

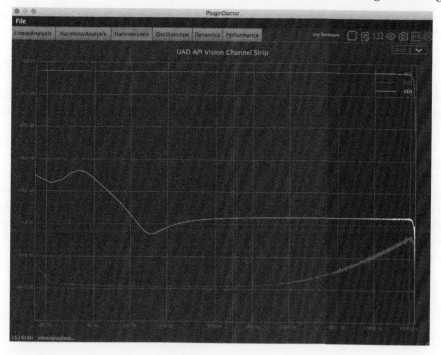

Figure 5.2 API high gain Hammerstein model.

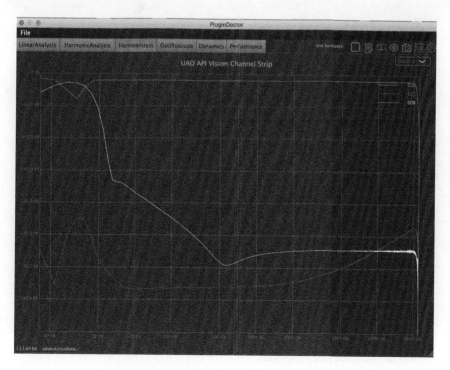

Figure 5.3 API maximum gain Hammerstein model.

As well as contributing harmonic distortion to the input signal, there is also a change in frequency response imparted by the API signal path. Figures 5.4 and 5.5 show the frequency response of the low and high gain settings, with very little impactful difference between the two measurements. Both responses show a largely linear response, with a slight bump in the low end between 20 and 50Hz.

The maximum gain frequency response found in Figure 5.6 demonstrates a change in the frequency response, with the low frequency bump having been smoothed and a gradual increase in level above 5kHz to a peak of approximately 2dB of gain at the top end of the audible spectrum. This change in response coupled with the changes in harmonic content provide significant colouration opportunities to engineers.

As previously discussed, there is also an expectation that there will be frequency specific changes in the phase response of the microphone preamp as a result of the audio transformer. When comparing the phase response graphs generated from the API preamp, there is no significant change in the phase response in relation to the gain added through the circuit. There is a measured phase shift in the API circuit below 50Hz and above 5kHz, contributing to the overall colouration of this microphone preamp, as demonstrated in Figure 5.7.

5.3.2 SSL E Channel Strip

The SSL E Channel Strip is an emulation found in the SSL 4000E console, which is one of the historically most used and indeed most emulated channel strips in contemporary production.

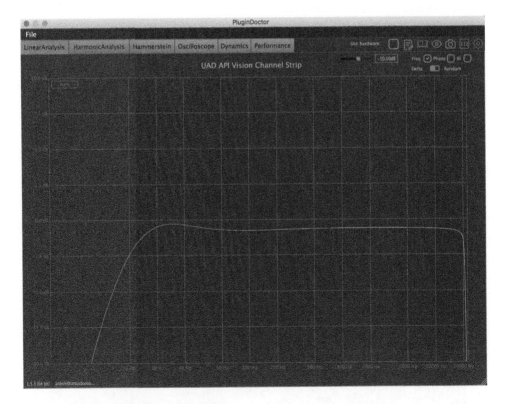

Figure 5.4 API low gain frequency response.

Low Order Distortion in Creative Recording and Mixing 73

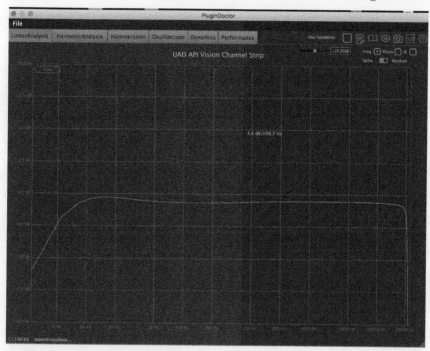

Figure 5.5 API high gain frequency response.

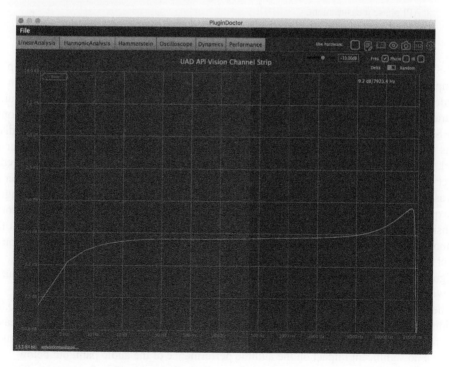

Figure 5.6 API maximum gain frequency response.

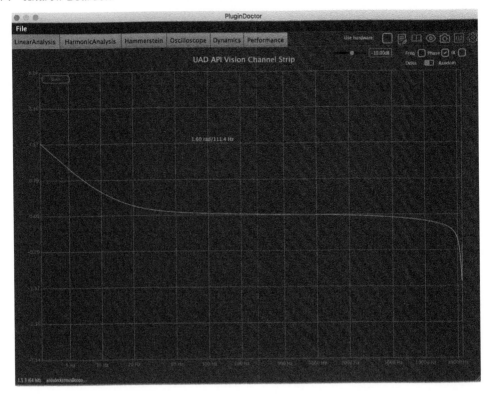

Figure 5.7 API phase measurement.

The SSL E Channel is of particular interest as there is a Jensen transformer in the microphone preamp that can be completely bypassed from the audio path. This particular emulation also provides the user with the ability to run into the line input, providing multiple configurable colouration profiles to the engineer. Figure 5.8 provides a measurement of the harmonic content in the SSL E Channel mic audio path, with the transformer disengaged. Second harmonic distortion is prevalent, measuring approximately 60dB below the fundamental in the Hammerstein model. Though the second harmonic is the prevalent distortion here, at the levels measured the impact on the listener from the additional harmonic content is likely to be subtle, with other features associated with the transformer likely to be important in judging any subjective changes in audio quality and 'feel'. It is also interesting to note that the level of second harmonic distortion is consistent across the frequency range. There is some evidence of frequency dependent third harmonic distortion present in the transformerless circuit; however, the overall contribution and impact of this is limited to the lowest audible octave before dropping into the noise floor.

Figure 5.9 provides a comparison with the same low level of mic gain in the preamp, but this time with the Jensen transformer engaged.

When the transformer is engaged, there is a small reduction in level of approximately 0.8dB. In addition to this, there is a clear change to the distortion profile, with a significant increase in frequency dependent third harmonic distortion. Above 200Hz there is minimal distortion that will have no significant impact on the audible performance of the audio path, but below 200Hz there is a significant increase in distortion down to the lowest audible frequencies. This type

Low Order Distortion in Creative Recording and Mixing 75

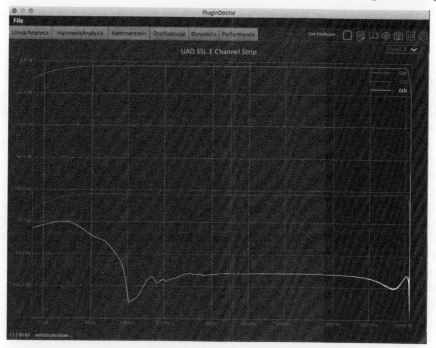

Figure 5.8 SSL low gain transformerless Hammerstein model.

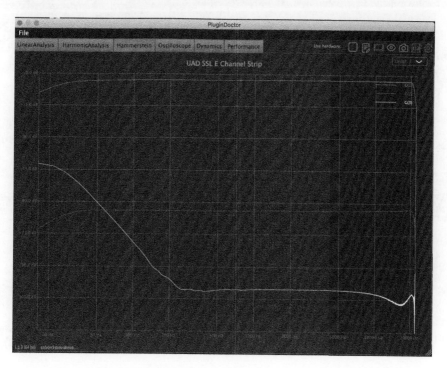

Figure 5.9 SSL low gain transformer Hammerstein model.

of distortion profile is typical of what would be expected from an audio path transformer of this type, with low frequency production elements enhanced through the addition of harmonic content.

The third option available to engineers utilising the SSL E Channel is to select the line input rather than the mic pre input. Figure 5.10 shows the measured response of the line input, with a very similar second harmonic profile to the mic input but with an even lower third harmonic contribution, providing the cleanest input whilst retaining the character provided by the second harmonic distortion.

As the gain is increased, there is little change to the harmonic distortion profile; however, when the 20dB pad is engaged, there is a significant change in the distortion profile, with a marked reduction in third harmonic content, as seen in Figure 5.11.

The change in distortion profile provided by the pad gives another low order dominated distortion colour, which engineers can exploit to manipulate the listener perception of a sound. It is interesting to note that when the circuit is driven into overload the performance of the circuit changes considerably, with the third harmonic becoming dominant and significant addition of higher order distortion into the signal path as demonstrated by the harmonic analysis presented in Figure 5.12. Though this type of aggressive distortion does not fit within the scope of this chapter, it is interesting to note that this distortion profile is available through deliberate overdriving of the preamp circuit.

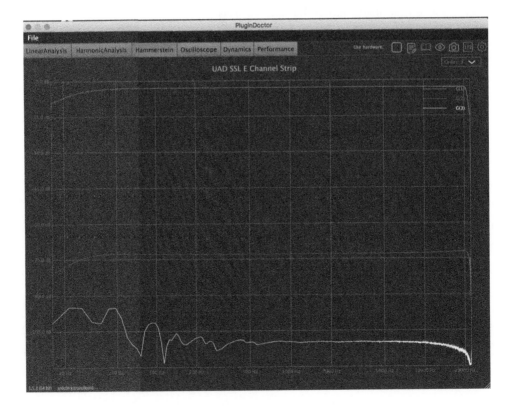

Figure 5.10 SSL low gain line input Hammerstein model.

Low Order Distortion in Creative Recording and Mixing 77

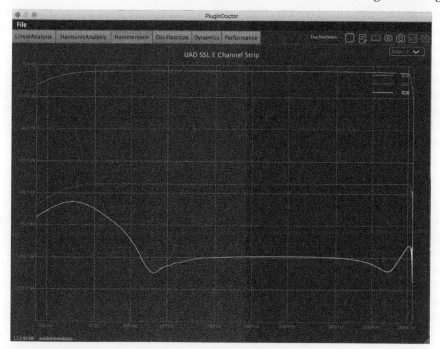

Figure 5.11 SSL high gain with pad and transformer Hammerstein model.

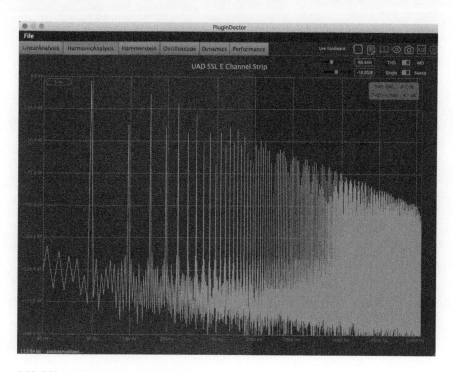

Figure 5.12 SSL overload harmonic analysis.

The frequency response of the SSL E Channel remains consistent across gain levels, with a linear frequency response across the audible range. There is some roll-off at the lowest frequencies, with the filters always present in the signal path, and a change in response when the circuit is driven into overload due to the extremely high order distortion content. The phase response is also largely linear across the frequency range, with some phase rotation in the audio path in the lowest octave of the audible frequency range. The addition of the transformer does not change the measured phase response, with only small changes in the rotation frequency associated with higher gain settings. The SSL E Channel is a highly configurable channel, providing control over third harmonic content through the provided options whilst always featuring second order harmonic enhancement.

5.3.3 Neve 1073

The final solid state preamp to be considered in this chapter is the Neve 1073, which features two gain stages, with mic signals going through two amplifier stages compared with the single amplifier found on the line input. As with the API there are two audio transformers in the signal path, with transformers on both the line and microphone input and the output. The Neve 1073 also features a three band EQ, which has been bypassed for these tests. As with the API and SSL channels tested in this chapter, the Neve 1073 is another preamp that is regularly used by engineers both in recording and for colouration during the production process.

Figure 5.13 illustrates the low order harmonic content in the audio path when using the line input. There is significant frequency dependent low order distortion in the audio path in both

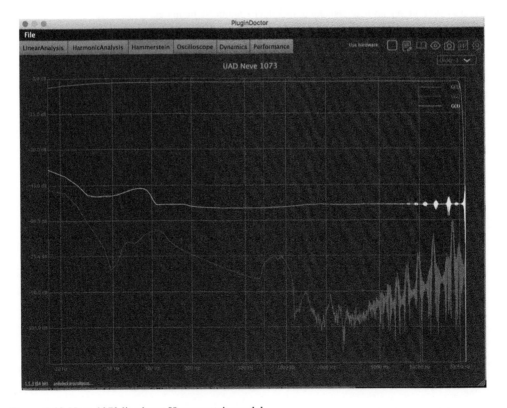

Figure 5.13 Neve 1073 line input Hammerstein model.

second and third order harmonics, but with a greater third harmonic component. The level of both harmonics are significant in relation to the fundamental, with more distortion evident in the Neve than either of the previous preamps analysed. The third harmonic content is exaggerated in the low end but continues throughout the frequency spectrum. At 20Hz the level of third harmonic is within 5dB of the fundamental, with a linear drop-off in level up to 500Hz, where the level of third harmonic drops to 60dB below the fundamental, and therefore has very limited impact on the listener. The second harmonic contribution is more focused in the low end, contributing to the perceived sound of the Neve 1073.

The mic input offers considerably higher third harmonic distortion characteristics to engineers, as demonstrated in Figure 5.14. In this example the circuit is being driven quite hard, resulting in significant harmonic content being added in the signal path. Figure 5.15 by comparison shows the same level of gain, but with a lower input signal and less attenuation at the output. The high signal level plot in Figure 5.14 shows significant levels of third order harmonic distortion, particularly in the lowest octave, gradually dropping in intensity towards 1kHz. Second order harmonic content is present and sonically significant, though to a much lower level than that presented in the third harmonic. Though starting lower, the harmonic distortion follows a similar trend, dropping in level as the frequency increases. By dropping the input level, the signal load in the transformer is reduced, resulting in significantly reduced third harmonic content. The level of second harmonic distortion is not significantly changed, and though there is still a bump in the low frequency region in the third harmonic distortion plot, there is a more even distribution of third harmonic across the frequency range.

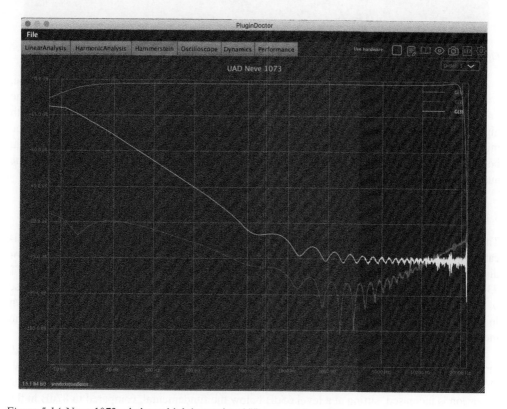

Figure 5.14 Neve 1073 mic input high input signal Hammerstein model.

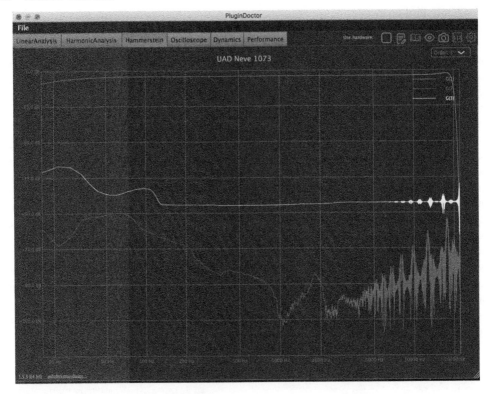

Figure 5.15 Neve 1073 mic input reduced input signal Hammerstein model.

In terms of measured frequency response, the Neve 1073 exhibits a generally linear response, with gentle roll-off below 50Hz and a slight boost in response of 0.2dB above 15kHz through the line input. This response is mirrored in the mic input, with a slightly stronger rise at 15kHz with higher gain levels. There is, however, quite significant phase distortion taking place through the mic and line inputs. The phase plot represented in Figure 5.16 shows significant phase shift in both the low and high ends of the preamp, which is also a core part of the designed sound of the audio path.

5.3.4 Avalon VT 737

The Avalon VT 737 differs from the previously analysed preamps, with the primary amplification device in the microphone preamp being tube based rather than an integrated circuit or discreet amplifier. Traditionally associated with contemporary vocal recording, the Avalon VT 737 also features a custom wound audio transformer in the signal path. The frequency response of the Avalon is extremely linear across the frequency range, with the harmonic content shown in Figure 5.17 also showing extremely low levels of second order harmonic distortion, with no measures of third harmonic content being added by the circuit.

Figure 5.18 provides the Hammerstein model measurement with the mic input engaged, resulting in measured third harmonic content at a level that will not have any significant perceivable impact, sitting well below the likely noise floor of the recording system. Second harmonic distortion is increased, sitting at a level 65dB below the fundamental, compared to 87dB in the line level measurement. Both the measurements show very little audible distortion, resulting in

Low Order Distortion in Creative Recording and Mixing 81

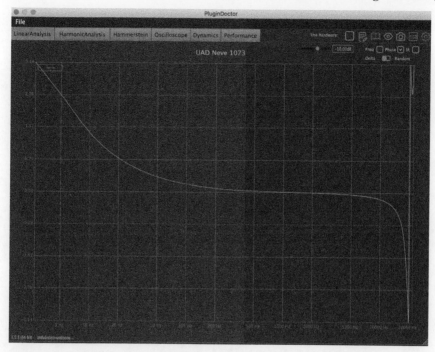

Figure 5.16 Neve 1073 mic input phase response.

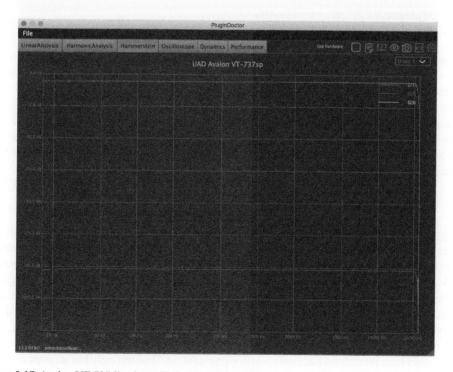

Figure 5.17 Avalon VT 737 line input Hammerstein model.

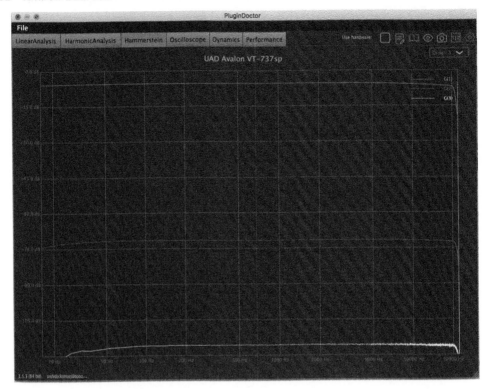

Figure 5.18 Avalon VT 737 mic input Hammerstein model.

what would be expected to be a transparent amplification circuit. There is a small amount of phase shift in the low frequency phase measurement below the audible spectrum, but overall the measured response is almost completely linear.

The Avalon VT 737 is clearly designed as a low distortion unit, and as such offers the most subtle distortion footprint of the preamps that have been analysed in this chapter. The tube design has provided a second harmonic dominance as is to be expected in this topology but has been designed to have minimal impact on the signal path with contextually desirable low level harmonic distortion with minimal phase shift.

5.3.5   UAD 610 B

Along with the Avalon VT 737 the UAD 610 B is also a tube-based design, but with a design that embraces a higher level of distortion than that found in the Avalon. The 610 B is a modern update to the original 1960s designed 610 A tube preamp, featuring a slightly more modern sound and a more flexible two band EQ with multiple frequencies and gain settings compared with the fixed frequency and gain design of the original. Figure 5.19 shows the harmonic content in the audio path based on unity gain in the tube preamplifier. As with the Avalon tested previously, the added harmonic content can be found largely as second harmonic distortion, with an even distribution of harmonic energy across the frequency range. Third harmonic distortion though lower than the second harmonic is still significant, and interestingly has an even distribution of harmonic power across the frequency spectrum. Second harmonic distortion measures

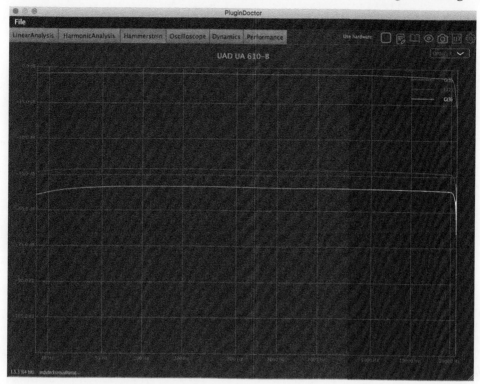

Figure 5.19 UAD 610 B unity gain line input Hammerstein model.

consistently across the frequency range, approximately 40dB below the fundamental, representing a significant and audible contribution to perceived audio.

Figure 5.20 shows a second measurement, with 10dB of gain running through the tube amplifier section. In order to achieve a measurement, the level had to be dropped post initial tube gain, and the output attenuator had to be engaged. The resulting measurement shows an increase in third harmonic distortion in the lower frequencies, likely to be caused by saturation in the audio transformer. Second harmonic distortion has also increased, with a slight low frequency bump the only variance on the liner distribution of harmonics across the spectrum. There is a reduction in harmonic content above 10kHz in both measurements, which reflects the frequency response of the overall unit, with the gentle roll-off above 5kHz visible in Figure 5.21.

5.4 From Measurement to Listening—Evaluating the Impact of Distortion on Musical Reception

5.4.1 Impact on Recording—Impact of the Mic Pre

A set of test recordings were undertaken by the author and have been evaluated to explore the general perceived sonic impact of the mic pre on a recording. For the recording a drum kit was set up with a combination of close mics, overheads and room mics in a large reverberant recording space. No microphone splitters were employed to avoid any issues with impedance loading or tonal contribution from active splitters, and a simple performance was repeated multiple

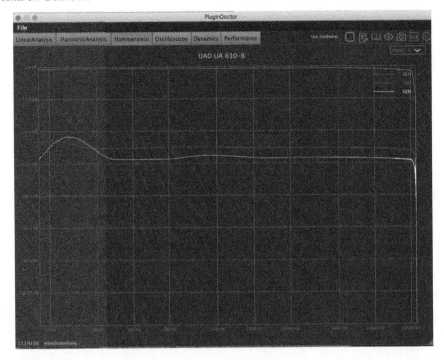

Figure 5.20 UAD 610 B 10dB gain line input Hammerstein model.

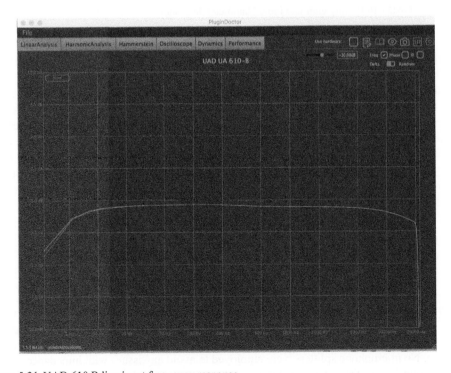

Figure 5.21 UAD 610 B line input frequency response.

times, each through a different emulated preamplifier circuit using the UAD Unison mic pre. In addition to the UAD mic preamps, an Audient ASP880 was also recorded in all three impedance positions for comparative purposes. The list of preamplifiers tested is as follows:

1. API Vision
2. Neve 1073
3. SSL 4000E with transformer
4. SSL 400E without transformer
5. UA610 A
6. UA610B
7. Audient ASP880 Low Z
8. Audient ASP880 Mid Z
9. Audient ASP880 Hi Z

The differences in the presentation of the recorded audio are significant, with changes in stereo image, depth, punch and tone with each of the preamps tested. There are differences in feel between takes; however, the simplicity of the recording and the significant subjective changes to the presented audio from the different recordings still allow valuable conclusions to be made. The observed differences have been described through the analysis of the author in a quality studio environment and are provided as a personal perspective in this specific example. Levels have been matched as closely as possible through the listening, with captured levels kept consistent at the recording stage and loudness normalisation taking place in listening to the recorded examples using a loudness matched playback system. The staging of the recording in terms of levels of contributing room mics and any panning is also maintained across the examples, with the perceived changes in presented image one of the more interesting affordances of the change in preamplifier on the recorded audio. The only changes are minor performance changes and the change of preamp. No additional processing has been added, and where channel strips were employed any processing units beyond the preamp were bypassed where possible.

Harmonic distortion from non-magnetic devices has also been auditioned as part of the testing process, removing the temporal, frequency and level specific impact of the audio circuits, and instead relying on simple low order harmonic distortion. In addition to transient rich, drum based materials, orchestral samples have also been explored to investigate the affordances of harmonic distortion on sample driven orchestral libraries.

5.4.2 Sonic Cartoons and Low Order Harmonic Distortion

Zagorski-Thomas [8] [9] proposes the concept of the sonic cartoon, which provides a useful method for considering the impact of forms of distortion on a musical event. Distortion is one of several tools that allow engineers to manipulate the perception of the original recorded sound [10], resulting in a sense of increased impact, heaviness, energy, and effort subject to the source and the nature of distortion employed. All of the descriptors discussed here are subjective in nature and have been discussed by academics in some detail. Mynett, for example, describes heaviness as representing the weight, size and density of music, and that distortion is a core tool for increasing levels of heaviness in music [11]. Impact is used by the author to focus on the part of a gesture where a body makes connection with another body, so for example the head of a drumstick contacting the skin of a drum. In that moment there are multiple consequences of that impact, with the drumhead exciting the shell of the drum, transferring energy to the resonant head and providing the initial energy release as the stick makes contact. As described in this chapter, transformers add harmonic distortion that can be frequency dependent, is related to the energy already in the core of the transformer and sees temporal distortion taking place. All of these forms of distortion change the relationship between the stick and the drum, resulting in changes in perception in the listener based on their knowledge related to their experience of things being hit, musical genre and associated visual stimuli, all of which impact on the listener's connection with the music that is presented, and all of which can be considered through *sonic cartoons*. As an analytical tool, *sonic cartoons* provide a useful framework for the study of the affordances of distortion in a mixing context.

In considering general recorded drums, the impact of low order distortion is most obvious in the perception of the gesture of hitting a drum. When distortion is engaged on a drum recording there is a clear sense of the drum being hit harder, with increased drive into distortion leading to a more dramatic sense of energy going into the drum being hit. We do not see the same measured response as if the drum itself were hit harder; however, there is an objective sense of the drum becoming hit in a more solid way. When compared with tools that emulate a full transformer circuit rather than simply adding low order distortion, the impact is less dramatic in those tools that do not offer transformer emulation, with the additional transient shaping and increased compression created by the processing also creating the sense of effort associated with distortion and compression. There is, however, a clear sense of change in impact, with the distorted versions offering a much greater sense of conviction in the hit of the drums. Both odd and even distortion afford a sense of increased impact and punch, but with changes in the nature of that impact and the staging of the drums depending on the distortion setting. This ability to shape the sense of performance gesture is a powerful tool for an engineer, resulting in the listener painting a different mental picture subject to the original performance gesture and any processing undertaken.

In the self-created example being analysed by the author as part of ongoing practice, the non-preamp created even harmonic distortion gives a much stronger sense of the drum being hit hard, with a strong feeling of solidity and drive in this example. Kick drum in particular feels tighter and drier, with the lower midrange in the snare feeling considerably more solid. The presentation of the overheads is also significantly manipulated, with a real sense of detail in the stick hitting the cymbals and an overall more coherent presentation of the image of the drums. This contribution of a sense of glue through the addition of subtle harmonic distortion has also been observed by Moore, exploring the affordances of distortion in the mastering process [12]. The sense of a single drum kit, rather than a collection of drums that is provided by this distortion processing is significant, creating a very different picture of the drummer and the performance to the listener.

When exploring the balance of the example featuring odd harmonic distortion, there is again a sense of increased impact. Compared to that of the second harmonic distortion example, however, there is a very different perception of groove. Whilst the second harmonic provides a real sense of hard hitting and coherent drive, there is a sense of lift in the third harmonic example, creating a sense of lightness and transient detail in the bounce of the stick against the source. There is an increased sense of air around the cymbals, and overall a greater sense of the space in which the drums exist. The second harmonic distortion example has a greater sense of focus, whilst the third offers coherency but through air and transient bounce. Placed in context with a track, the impact of the distortion changes again, showing the importance of evaluating distortion in context when undertaking a mix. If we were to compare a different drummer playing a different part, we would see a different result again; the aim here is not to create a typology for distortion, but merely to recognize that the impact of distortion is to change the perceived presentation of the element being distorted. The decision to provide a more focused, driven sound or a lighter, more open groove is entirely dependent on the context of the recording, with choices made through the production and mixing process in response to a musical language.

A similar impact can be heard when exploring the same harmonic distortion process on music created with orchestral sample parts. When engaging the distortion on programmed strings, it is possible to manipulate the playing gesture, increasing the sense of the bow dragging on the string. Adding depth creates a sense of friction, bringing out detail and a sense of reality in the performance audible in sample. By moving from second to third harmonic it is possible to move the focus, with the second harmonic creating a sense of richness and warmth, and the third harmonic offering a sense of edge and friction. The third harmonic distortion created a sense of the bow physically dragging on the strings, which results in a significant change in emotional intensity in the presentation of the music. Again, the appropriate choice is entirely dependent on the context: in this case the composer was keen to add intensity and a sense of aggression, which was supported by the choice to embrace third harmonic distortion on the string sections. The brass parts also received third harmonic distortion with significant drive, adding not only a sense of the instrument being played harder, but also allowing the instrument to cut through the mix.

5.4.3 *Preamp Comparison—Sonic Overview*

The following section includes a subjective analysis of the test recordings described in section 4.1. The API preamp provides a solid and consistent stereo image, with solid transient detail and strong punch, particularly in the snare. Transients are rounded but punchy, giving a solid sense of being hit with conviction. The room sound presents as a continuation of the close mics, particularly in response to the snare. In terms of frequency response, the entire kit feels balanced across the frequency range, providing an honest but punchy representation of the room sound across the kit. Cymbals feel wide, with good image detail; however, there is a degree of smear in the presentation of the cymbals compared with some of the transformerless preamps tested.

There is a significant change in perspective when comparing the Neve 1073 to the API Vision channel strip. The Neve 1073 feels slower in the low end to the API, resulting in a slightly leaner kick drum with less sub. Slowness in this context is perhaps a controversial descriptor and would benefit from further definition. In this case slowness is being used to describe the way in which the preamp responds to the onset of a transient. In this case the transient and the sustain and low end of the preamp feel somewhat separated, with the sustain of the low end not feeling as closely consequential to the onset of the transient as in other examples. This response may be desirable or may be a problem for a given engineer subject to the desired feel of the recording, relationship with other elements and overall presentation. The snare feels lighter than in the API and presents with less cohesion between kick and snare. The top end of the Neve feels more open but less focused in terms of transient detail, but does offer an increased sense of width and air in ride cymbal. The difference in feel between the top end and low end of the Neve 1073 creates a drum presentation that feels less coherent and slightly boxy in nature compared with the focus of the API that exists across the entire frequency range. It is interesting how this presentation correlates with the measured responses of these preamps, with more measure frequency dependent distortion and significant phase shift in the Neve manifesting in a slightly less coherent and focused recording in this case. It is important to note that these differing responses could be desirable in different situations, with a change in performance feel provided between the two recordings. It is also important to note that the performance impact would change dramatically based on the source—a vocal for example would be impacted very differently from the drum recording. It may also be that drum overheads are better in a different preamp to kick and snare; all of these observations are extremely subjective and need to be contextualised by engineers in making appropriate selections.

The SSL channel strip has been recorded both with and without the transformer. With the transformer engaged there is a detectable sonic impact, with more of a sense of bite in the snare and an increased sense of player impact to the listener. There is also more of a sense of compression in the presentation of the drums through the transformer, though the room mics are less pronounced, and there is a slight reduction in the sense of detail when the transformer is engaged. The low end is less pronounced as the transient is rounded, with all elements feeling more pushed with the transformer engaged. When the transformer is disengaged the image relaxes, with kick and snare all moving slightly backward in the image and feeling wider and more open. Hi hats and cymbals are not hugely affected by the instantiation of the transformer, which is also reflected in the measured harmonic performance of the SSL 4000E. There is a general increase in detail in the cymbals with the transformer disengaged, with the resultant removal of magnetic distortion and hysteresis from the signal path. The impact on performance feel here is important, though relatively subtle, with the temporal and frequency-based distortion in the transformer creating a different image of the performance to the listener, particularly in kick and snare. In the transformer recording the kick is punchier but offers less open low end, with mid-range punch also enhanced on the snare. The transformerless recording retains the core identifiable qualities of the SSL channel with transformer, but with a more open and faster feel with extra detail and depth but without the drive and cohesive centre punch of the transformer. In comparison with the API and Neve preamps, the SSL feels like it offers less control, with more speed and aggression. There is evidence of mid-range grit and a hardness in the top end that feels fast and could be perceived as exciting depending on what performance manipulation was intended by the engineer.

The UA610 B preamp was explored in this drum recording session, representing the tube mic preamp choice in this test. The measurements of the 610B demonstrated second and third order harmonic distortion present, with more frequency specific distortion from the transformer in the low end and a general roll-off of high frequency content in the frequency response. In listening to this preamp, it is clear that the measured response is indeed reflected in the recording. The 610B presents a different performance picture to that provided by, for example, the API vision channel strip. The drums have a larger-than-life quality, but also a sense of fatness from the presentation of transients. There is an enhanced sense of groove, which combined with the solidity of the second harmonic present gives a subtly different feel to those recordings taken through the solid state preamps tested here. Despite showing a gentle roll-off in frequency response, the recording does not feel closed but does exhibit slightly less width overall. The 610A preamp was also recorded in this test, with the 610B being a more modern development of the 610 circuit, as discussed in this chapter. The 610A does show similarities in tone but felt more vintage in the capture of the performance, with the most dramatically different feel of all the recordings undertaken. Those present in the session all immediately identified this as an older sounding preamp, with the 610B offering similar affordances but with a much cleaner and more modern sound than the 610A. There was evidence of compression and indeed saturation in the low end based on the tested gain staging, resulting in a less subtle change in presentation than has been presented in the other tests.

As a point of reference the recordings were also captured using the Audient ASP880 preamp, which is a high quality transformerless design often selected for its lack of colouration. Unlike the Universal Audio solutions, there is no software hardware hybrid, instead providing a pure hardware solution with variability limited to a high pass filter and three impedance settings. The three impedance settings did provide voicing alternatives, with the lowest impedance offering a slightly looser feel with a wider, more slappy kick drum. The mid-impedance provided an open, balanced recording with fast transients and a general lack of colouration. The highest impedance setting led to a slightly narrower sound and felt more like the drums were bring forced slightly more to the listener. This led to a sense of greater strain in the performance, which in this case felt less relaxed and more inconsistent. Perhaps the biggest takeaway from this recording was how different this transformerless low colour circuit felt from the other tested microphone preamps. The impact of the circuitry on the presentation of sound and performance from the transformer designs led to a considerably more dramatic change in performance gesture and feel than the differences between any two of the selected transformer preamps. The transformerless preamps feel considerably more truthful to the sound that was presented in the room on this occasion, but the sense of musicality provided by the different preamp choices that exhibit non-linearities certainly offered musical benefits akin to channel processing often applied in mixes at the recording stage.

5.5 Conclusion

It is clear from the measurements and the recordings undertaken that the choice of microphone preamp can play a significant role in the presentation of recordings. Unlike higher order distortion devices, the changes afforded by microphone preamps when used to introduce small levels of colouration tend to be at the more subtle end of the processing spectrum; rather than taking over the sound, they manipulate the sense of the sound as a source, adding solidity, punch, cohesion or air as non-exhaustive descriptors that can be associated with this type of processing. As well as being used at the capture stage, contemporary plugins and hardware provide engineers with the opportunity to explore the affordances of these preamps on their productions at multiple stages of the process. Low order harmonic distortion is a key component in the sound of these devices, resulting in a perceived performance gesture change depending on the nature of the device employed. Pure distortion devices provide clear performance gesture perception changes, with more complex interactions provided when using emulations of tube and transformer devices to manipulate audio through combinations of low order harmonic distortion and the various other forms of non-linearities afforded by the components of audio preamplifier circuits discussed in this chapter. It has been fascinating to explore the relationships between the measured performance and the impact on recordings, and further testing to explore measurement of perceived punch [13] and extended listening tests will be undertaken in future to further explore the affordances of low order harmonic distortion and the impact of magnetics on recording.

References

[1] Massey, S. *Recording Unhinged*. Hal Leonard, Milwaukee (2006).
[2] Clarke, E. *Ways of Listening: An Ecological Approach to the Perception of Musical Meaning*. Oxford University Press, Oxford (2005).
[3] Waterman, T. *HMX—The Benefits of Harmonic Distortion*! Accessed August 2021 from https://blog.audient.com/post/31266771685/hmx-distortion
[4] Woods, L. *The Measurement of Audio Transformer Colouration*. Masters Thesis, University of Huddersfield (2019).
[5] Whitlock, B. 'Audio Transformers'. In: Ballou, G. (ed.) *Handbook for Sound Engineers*, 5th ed. Focal, New York (2015).
[6] Winer, E. *The Audio Expert. Everything You Need to Know About Audio*. Elsevier, Oxford (2012).
[7] Unison Basics. Accessed August 2021 from www.uaudio.com/blog/unison-quick-tip/
[8] Zagorski-Thomas, S. *The Musicology of Record Production*. Cambridge University Press, Cambridge (2014).
[9] Zagorski-Thomas, S. 'Sonic Cartoons'. In: Papenburg, J. & Schulze, H. (eds.) *Sound as Popular Culture*. MIT Press, Cambridge, MA (2016).

[10] Bourbon, A. & Zagorski-Thomas, S. Sonic Cartoons and Semantic Audio Processing: Using Invariant Properties to Create Schematic Representations of Acoustic Phenomena. *Proceedings of the 2nd AES Workshop on Intelligent Music Production*, London (2016).

[11] Mynett, M. 'Maximum Sonic Impact: (Authenticity/Commerciality) Fidelity-Dualism in Contemporary Metal Music Production'. In: Bourbon, A. & Zagorski-Thomas, S. (eds.) *The Bloomsbury Handbook of Music Production*. Bloomsbury, New York (2020).

[12] Moore, A. 'Towards a Definition of Compression Glue in Mastering'. In: Braddock, J.P., Hepworth-Sawyer, R., Hodgson, J., Shevlock, M. & Toulson, R. (eds.) *Mastering in Music*. Routledge: Oxford (2020).

[13] Fenton, S. & Lee, H. 'A Perceptual Model of Punch Based on Weighted Transient Loudness'. In: *AES: Journal of the Audio Engineering Society*. 67, 6, pp. 429–439, 11 p (2019).

Part II
Perception and Semantics of Distortion

6 Understanding the Semantics of Distortion

Gary Bromham

6.1 Introduction

When I emailed producer and engineer Mark 'Flood' Ellis to arrange an interview to discuss the subject of distortion, he replied with a particularly amusing but pertinent statement:

> It would be lovely to talk about a subject that is very close to my heart/ears/soul. I think it would be great for Alan (Moulder) and I to do it together as we have different methods, but similar taste/perception/application of one of the most misunderstood tools in the sonic palette. I'm around and ready to spill the beans!

Indeed, the ensuing conversation with Flood and Alan Moulder at their Assault and Battery studios in London proved to be both revelatory and insightful and, in many ways, shaped the narrative and direction for some of this chapter.

The chapter proposes a corpus of descriptors for distortion based on interview comments obtained from producers and mix engineers. The research took the form of semi-structured interviews conducted with several award-winning producers, mix engineers and product designers, which were conducted over a period of two years during the writing of my PhD. Though the primary motivation was to discuss their perspectives on distortion, the comments collected proved invaluable in the discourse on music production values and on the subject of retro and vintage aesthetics. Much work has been done in this area in the last 20 years in research, but it has often been anecdotal and somewhat opinion-based in nature and often written by scholars who are not well versed as practitioners. Some of the concepts, purely by virtue of their esoteric nature, rely on a level of understanding which is not always obvious or evident in this domain. As a practitioner with more than 30 years of experience, I have tried to unpick, understand and articulate some of these insights. The limited view of music production as a purely technical process has in many ways held back our understanding of this relatively new field of research. It is an interdisciplinary area of study where aesthetics and specifically emotional content and value are arguably more important than technical details. For engineers and producers with high levels of tacit knowledge, the understanding of the technical parameter space is in many ways a given and assumed to be the case. Mapping a technical parameter space to a specific emotion or, in the context of this work, a semantic value is the challenge of this work.

6.2 Background and Historical Context

Before discussing comments and reflections that came from the multiple interviews conducted over a period of 18 months, it would be appropriate to say something about the current state of research in the area of semantics in music production as this forms the premise of the chapter. It would also be pertinent to state that the method for analysing the texts primarily used a grounded theory approach where themes emerged from the interviews. This method has been developed and used to great effect by several scholars, including Braun and Clarke [1], Bryan-Kinns [2], Pestana [3], DeMan [4] and Reiss [3], in their work. Grounded theory seeks to discover or construct a theory from data gathered from comparative analysis—in our case from reflections and comments obtained from interviews with professional producers and engineers.

Distortion can be loosely defined as the difference between the input and the output of an audio signal: "When the shape of the output waveform deviates in any way from the input, distortion has occurred" [5]. Quite simply stated, it is output amplitude vs. input loudness.

Many adjectives for describing sound exist, though these are often less well defined for timbre [6]. The study of timbre perception is problematic in the sense that we often assume that it is everything about a sound that isn't pitch or time. It is easier to use antonyms than synonyms when trying to determine a definition. Existing work by McAdams [7], Saitis [8], Seidenburg [8] and Zacharakis [9] have provided a sufficient definition of timbre, and in the case of Saitis and Weinzerl [10], Zacharakis [8] and, more recently, Wallmark [6] its semantic associations. Saitis and Weinzerl's chapter 'The Semantics of Timbre' in the book *Timbre: Acoustics, Perception and Cognition* is particularly pertinent to this research. They comment,

> Comparisons with similarity-based multidimensional models confirm the strong link between perceiving timbre and talking about it.
>
> (p. 119)

For the purposes of this chapter, when analysing the impact of distortion, I shall use the word 'colour' as a loose definition, for it also suggests many attributes including character and/or quality of a sound. Much has been written about timbre perception and, most relevant to this chapter, timbre semantics, where cross modal associations have also come to play a central role, such as 'dark-bright' and 'smooth-rough' for example [6]. Multi-sensory associations also bring some level of integration and connection between tactile or visual elements with auditory ones. For this reason, a level of what is referred to as 'semantic crosstalk' is inevitable and indeed permeates the responses and comments obtained from many of the interviewees for this chapter. Distortion is a challenging proposition when attempting to define its inherent properties for manipulating sound. This is reflected in the fact that there is currently no existing ontology on or framework for distortion and the accompanying terms that are used to describe it. The Audio Commons research and consequent deliverable conducted by Pearce et al. [11] from the University of Surrey goes some way toward offering a list of descriptors, but it is less focused on distortion, and it does not use audio professionals exclusively in the gathering of data. It does, however, provide an ecosystem for describing timbral attributes used for library sound searching and specifically for the creative reuse of audio content. Moore does discuss the distortion properties of certain well-known hardware compressors here [12], and develops this theme in his journal paper on the specific semantic descriptor 'aggressive' in this context [13]. Though there are several works where audio effects have been defined on a technical level, including those from Case [5], Reiss and McPherson [14], and Wilmering et al. [15], little detailed study has been conducted on the aesthetic properties and their meaning, other than those offered anecdotally or in passing reflection. Wilmering, along with co-authors, has come closest to offering a sufficiently technical definition and an accompanying semantic ontology [16]. This chapter builds on existing theory and research in an attempt to unravel some of the mystery associated with the phenomenon.

A supplementary factor, which was mentioned on numerous occasions in the interviews, was the importance of emotion in the music production process. The ability to articulate the emotional effects of distortion and other audio effects, either through sound manipulation or language, is integral to an understanding of the domain. Psychologists such as Juslin [17], Sloboda [18], Gabrielson [17] and Hargreaves and North [19] have all discussed music psychology from differing perspectives, and where possible I have tried to acknowledge and contextualise this in the research. It is, however, beyond the scope of this work to describe their individual contributions in detail but suffice to stress the importance of their work should the reader be interested in further investigation.

6.3 Semantics in Music Production

Distortion takes on many different meanings dependent on the perspective and context of the user. Its perception, and the consequent understanding of the ability to communicate its attributes, is often dependent on respective levels of experience acquired amongst practitioners. Porcello analysed the effects in two case studies where he evaluated the effectiveness and impact of apprenticeship on the ability to communicate and articulate musical ideas in a studio context [20]. He also makes the salient point that shift from traditional apprenticeship to institutional learning has further complicated this issue due to the way that professional competence, linguistic or otherwise, is acquired. Because of the ambiguities attached to distortion specifically, which is an esoteric concept, subtle or more explicit in nature, it might easily be misrepresented when communicating ideas in the recording studio.

Trying to put something into words that is natural and organic and, in some cases, unquantifiable is problematic in music. Academia and science seek to find solutions and articulate explanations, which have to some extent fallen short of understanding the true meaning. You often lose the essence when attempts are made to put something into words. Questions arise in this search: Why should they put it into words? Maybe it's purely artistic. How do you explain the inexplicable? Attempts to demystify the subject have been unsuccessful.

6.3.1 Semantic Crosstalk

Musical timbre is notoriously difficult to define as it invariably uses metaphors, descriptors and adjectives that are borrowed from non-auditory senses [6]. Melara and Marks [21] coined the phrase 'semantic crosstalk' when discussing inter-relationships and cross-modal correspondences between the senses. Porcello, in 'Speaking of Sound' [20], identifies a problem where less experienced engineers find it difficult to describe 'acoustic phenomena' in language. Porcello makes the salient point about the role that relative levels of experience play in the understanding of studio speak. These findings emerged from ethnographic studies conducted with engineers with varying levels of experience in a studio. Though ambiguity in language is common in Neuro-Linguistic Programming (NLP), it is not so common in music. There are differences in language, and we need a perceptual structure for unpicking this to be easier. Some of the misunderstanding which occurs during comparative semantic crosstalk stems from, and is directly related to, the number of hours spent in a studio environment. Indeed, the importance of internship and apprenticeship in the quest to learn such skills cannot be overestimated in this context, arguably something a degree program in music production struggles to teach. This becomes part of the discourse and negotiation in a studio environment. Tchad Blake comments,

> the whole thing about words and being a messenger and being a receiver, everybody hears things differently and hear words differently.
>
> [22]

Is there agreement and consistency among producers, engineers and indeed artists when describing sonic attributes which are associated with distortion, saturation and colouration?

A better understanding of semantic crosstalk will help in our understanding of our emotional response to sound and is very relevant in the context of this field of research. It may also go some way towards explaining the inability on some level to articulate sonic attributes on the part of music professionals. Both Moulder and Dave Bascombe commented on this principle in the interviews when discussing the differences between 'muddiness' and 'warmth' as an example: 'It means different things to different people!' [23].

Wallmark highlights a similar problem but articulates this in a more academic way: 'Metaphors reflect and shape how we perceive, conceptualise and appraise musical sound' [6] (p. 1).

Both Wallmark and Porcello have proposed a lexicon and corpus of descriptors, but this field still requires far more investigation to form a coherent ontology, particularly in the case of terms associated with distortion. Perhaps a better question to ask ourselves would be whether cognitive processing is compromised when timbre and semantics are mismatched?

6.3.2 'Studio Speak'

Discussions about timbre in the recording studio are arguably as much a social, shoptalk phenomenon as they are about sonics. The language used to communicate musical direction, stylistics or aesthetics in a music production context are central to the music-making process. 'Getting a sound' is part of a discourse and negotiation that takes place among artist, producer and engineer. Artistic vision or agency used when shaping sounds is critical to the creative process and important when constructing a sonic signature. As Porcello expounds,

> As attention to sound, for its own sake has become deeply embedded in studio work, so has the need for engineering professionals to be able to discuss it in finely detailed ways. The problem is how to render acoustic phenomena concretely in language.
>
> [20] (p. 734)

These patterns can be seen through the comments made by the interviewees when discussing the themes identified as central to the understanding of how distortion works in music production. The role of distortion, in all its various manifestations and incarnations, in moulding and shaping sound, is both implicit and explicit, subtle and overt, colourful and transparent.

6.4 What Is Distortion?

The meaning and impact of distortion has changed over the course of time. From the speaker ripping (literally!) sound of The Kinks' 'You Really Got Me' to the sound of Sonic Youth and the Jesus and Mary Chain in the 1990s to the subtle warmth heard on *So* by Peter Gabriel or, more recently, *Back to Black* by Amy Winehouse. This chapter focuses on more subtle, less overt forms of distortion than those more traditionally used by guitarists.

In the analogue era, distortion was often an unintended consequence of the technological limitations, but in the digital era where precision and clarity have become the motivation, there is a sense that something is now missing in the audio domain. Indeed, distortion was a side-effect or consequence of the limitations of the technology rather than an intended aesthetic statement. It has been aptly remarked that as soon as designers managed to eradicate some of the imperfections and, pertinently, distortions present in analogue designs, we wanted them back again!

[24]. Producer and engineer John Leckie, who trained at Abbey Road Studios in the early 70s, comments when asked if he was deliberately using distortion,

> No, rarely. No, you were on your guard against it really. If something was distorted, you'd turn it down. You would have to get rid of the distortion.
>
> [25]

Distortion, of the analogue variety, is often considered the secret sauce to providing, and returning, some of this aura in a digital age. If one uses the analogy of film, an 'in-focus' picture is not always the most interesting image for it leaves little to the imagination. When sculpting or manipulating a sound, introducing graininess, some obfuscating of the source material, can often lead to more interesting sonic results. As Andrew Scheps comments, 'Things get less see-through when I add distortion' [26].

Bascombe refers to a similar lo-fi aesthetic when comparing the effect with those produced by vinyl:

> People are convinced that vinyl sounds better. Well, it doesn't technically, we know it sounds worse. But it's that worseness that gives it the quality, the character.
>
> [23]

The concept of 'character' is mentioned several times in the interviews in the context of distortion.

The side effects of misappropriating or abusing pieces of sometimes iconic technology is a factor in the manifestation of characterful aesthetics. Flood comments, 'Technically it may be wrong but aesthetically it is correct' [27].

6.4.1 Distortion, Saturation, and Colouration in the Studio

Distortion is a mysterious, esoteric phenomenon which has the capacity to alter perception, mood and sonics in the most interesting forms. The sub-title of this book, 'The Soul of Sonics', came from a phrase used by Mark 'Spike' Stent during our interview when describing distortion. The phrase also intimates and suggests a sense of mystery and otherworldliness about the phenomenon, which is often hard to articulate with conventional language. Indeed, some of the terms which arose during the interviews, often spoken with some hesitancy and considered response, suggested that it may almost be beyond words. As Trina Shoemaker so aptly articulates,

> Distortion is a harbinger of chance in music production. Distortion ushers in a part of the sound that did exist but needed rediscovering. It manifested so abundantly in analogue technology.
>
> [28]

6.5 Perspectives on Distortion

6.5.1 Critical Listening Skills and the Role of Experience

Much of the work in this chapter pre-supposes the position that experience, competence and, most importantly, highly developed critical listening skills play a huge part in the perception of distortion. Many of the concepts and constructs discussed in the chapter work on the assumption that listeners have sufficiently developed critical listening skills where they can perceive noticeable differences. Unfortunately, in the experience of the author, this is often quite rare and goes some way toward explaining why this work has focused solely on professionals and not on those with less experience. The interdisciplinary approach adopted also draws upon an innate appreciation of aesthetics and a high-level awareness of the role that emotion plays in the responses from the interviewees. A more hermeneutic phenomenological approach to conducting the research is used, where an understanding and a recognition of the value of experience and competence might be seen as determinant factors when evaluating and articulating sonic attributes. Knowing how to communicate these emotions is not always easy for a practitioner who often relies on intuition to sculpt a sound.

In the specific case of distortion, this is arguably more difficult than learning to listen to audio-processing actions such as equalisation, compression or spatialisation for example. Distortion, saturation and colouration, and their accompanying side-effects, are much harder to perceive and indeed quantify without highly developed critical listening skills and subsequent experience of using the effect in its various incarnations and manifestations. Because it is often an unintended consequence of manipulating sound, particularly in the analogue domain where it is also notoriously difficult to control, the effects are not always immediately obvious to the untrained listener. Several of the engineers referred to the effects of adding harmonic distortion as 'EQ for free'. Its impact on both spectral and temporal masking when combining with other sounds also makes it problematic. Now that engineers tend to mix in the box, it tends to place more emphasis on the importance of capturing character, colour and harmonic distortion at the recording stage. This is diametrically opposed to a modern DAW-based methodology which relies more heavily on a 'let's fix it in the mix' philosophy.

The idea of adding colour is an interesting analogy as it also correlates with the idea of synaesthesia. Colour, conventionally, is a visual stimulus which has become closely associated with sound. As described previously, this is a form of cross-sensory correspondence. Trina Shoemaker describes this phenomenon in the context of her working method as follows:

> Distortion isn't really to paint colour, it's a paint thinner. It's used to change colour. I look at everything visually, olfactorily and sonically when I'm mixing. I can go on a whole journey when I'm doing a mix.
> [28]

In the box, this method requires more of an accumulative approach where small amounts are added in several places with multiple instances of plugins to achieve a similar effect to one where a single process may have sufficed in the analogue domain. Tchad Blake makes the salient comment that he often finds himself using three compressors to achieve a similar effect that could have been achieved with one processor in the analogue world,

> I might have to use three compressors in the plug-in world where one would have done the job in the analogue one. In the plug-in world you start getting a squaring of waveforms, it doesn't sound that desirable to my ears. I find I need two or three all doing a third of the job to get what I think is the flavour. Not the actual sound, just the flavour that I'm looking for.
> [22]

Indeed, digital counterparts do not behave in the same way as analogue ones.

Alan Moulder agrees, 'Completely. It never sounds the same. There're a million and one digital distortions but none of them do that drive thing, that little thing that you get with analogue' [29].

Flood expands, 'It's a bit sort of, I don't know, it needs to travel down a wire, there needs to be some electrons involved' [27].

Tchad Blake talks about the importance of viewing digital software emulations as new tools, not the same tools. He goes to great pains to stress that 'I don't want it to sound the same, I just want it to feel the same' [22].

6.5.2 Misappropriation in Distortion and 'Happy Accidents'

Using pieces of iconic studio equipment in unintended ways has often yielded interesting creative results. Subverting the technology and deliberately distorting the sound for creative purposes was a legitimate goal in this context. The now iconic Fairchild 670 Compressor is a perfect example of this where driving the tubes and transformers produced a pleasing harmonic distortion which glued the sound together. This was arguably more interesting than using it purely as a dynamic range compressor, the purpose for which it was designed. There are numerous examples of this repurposing of not only analogue technology but also their digital counterparts to achieve similar aesthetic results. Danton Supple articulates this when he says,

> I think when I first started it was the term that everyone was fearful of because it was seen as a mistake. But even then, when it was a mistake, I remember thinking, 'Hmm, I kind of like what it's doing'. And I think once you realize there are multiple levels of distortion, it's such a huge, wide range of sonics you're talking about, all the different harmonic distortions, you know, electrical distortion, valve distortion, things that happen, then you realize it just adds so much excitement to it. And you don't even know what those harmonics will be till you actually distort it.
>
> [30]

Gareth Jones echoes this sentiment when discussing his work with Depeche Mode. He talks about the surprise factor and excitement when processing audio with boxes which distort the sound in interesting ways and where happy accidents are allowed to happen:

> That was a big part of my early work with Depeche, bringing this to the table. Of course, they loved it and that's why the creative relationship worked so well, sonically I was able to play around in this marvellous playground and the more weird shit like this we did, the more people seemed to like it.
>
> [31]

The idea of allowing happy accidents to happen freely in the context of distortion permeates many of the discussions, and the notion that there is no such thing as doing something the wrong way is mentioned frequently. When talking about working with the Jesus and Mary Chain where nothing was off-limits, Flood comments,

> They just pushed that saturation level to where what we'd probably been brought up on before was incorrect and this suddenly became an open field of nothing's incorrect.
>
> [27]

Alan Moulder talks about the importance of using guitar fuzz and distortion pedals for re-amping in his workflow to achieve an aesthetic where experimentation and creative expression are the route to producing something unique and characterful.

6.5.3 Pseudo Loudness and Perceived Volume

Adding distortion to a sound or mix will generally result in an increase in the perceived loudness of the material, or as Dave Bascombe referred to it, as a corresponding increase in 'pseudo loudness'. During an informal chat with Paul Frindle, the designer of the Sony Oxford R3 mixing console and some of the renowned Sony Oxford plugins, he shared the story behind the Inflator plugin, a weapon of choice for many mix engineers. He told the story of how a friend of his could not match the loudness of a rival engineer's mix, which made Frindle consider the reasons why this might be the case. Using principles from valve amplifier design, he was able to simulate the effects of the tube saturation with software to facilitate both spectral and timbral shifts. A similar effect can also be obtained by saturating a transformer.

Cenzo Townshend also remarked that distortion is often perceived far more easily when listened to at lower volumes. He suggests that there is a need to monitor quietly:

> To hear the effects of distortion you need to turn the sound down which is counter-intuitive as distortion is often perceived as being a loud thing.
>
> [32]

Moulder articulates this in a slightly different but no less interesting way:

> Distortion is the creative alternative to simply pushing up a fader. I use it more for sitting something in lower.
>
> [29]

6.5.4 Mix Bus Magic and Finding the 'Sweet Spot'

'Mix bus magic' is a term we have adopted to describe a saturation effect obtained by driving the stereo mix bus on an analogue mixing console in a specific way to achieve a harmonically rich, punchy and warm saturation effect. 'Finding the sweet spot', as it is referred to, is a skilful process where experience is a key factor in judging how far to go. Several comments were made in reference to this phenomenon. Flood says,

> When you're recording or you're mixing, everything, in my opinion, needs to be just on the edge. Because it adds this low-level harmonic distortion. Which is vital for the sound.
>
> [27]

Moulder echoes these words when talking about his experience when working with legendary engineer Andy Johns:

> he spent a long while at the beginning just getting the gain structure of the board for his mix just right, how hard he was hitting the console and everything, how high the faders were pushed up, then the master fader. And the engineer in the studio said nobody had ever got the board to sound like that. And it was all due to how he was driving each part. He got distortion that isn't perceivable as distortion, not hearing crackle or whatever, he was getting that drive that's just on the edge.
>
> [29]

Moulder continued,

> All these small differences stacked up makes a massive difference.

Spike Stent also talks about the importance of getting this process right:

> Yeah, I think it's just something that I naturally happened upon. I think basically what happened with distortion, with me, and colouration, was how I used to run the SSL. It was always too clean, and I discovered that if I drove it in the red, or the lights were on red, it started to sound good!
>
> [33]

He goes on, 'I found that if I drove the channels more it started gluing together in a nice way.'

Though sharing some of the same sentiment, Andrew Scheps also saw a downside to this way of working due to the inherent inconsistent behaviour of an analogue desk that later persuaded him to move into the box and away from the console,

> what's the point of using a Neve? If you don't push it, you don't get anything, it's a really clean circuit. So, you push it, you get some distortion but then you also get the cross talk, the collapse at certain frequencies of your stereo buss, all the stuff that comes with it.
>
> [26]

It could be said, anecdotally, that small steps make for big changes. Adding distortion has an accumulative effect where small amounts added in several, sometimes subtle places produce a pronounced effect. As with so many elements in a mix, it is the small things that matter.

6.5.5 Taming Transients and the 'Compression Effect'

When we add small amounts of distortion to a sound, we also create a 'compression effect', rounding off the transients in the material similar to a dynamic range compressor. In altering the envelope and shaping the temporal attributes of a sound, we can increase its perceived impact. In a similar fashion to the behaviour of a limiter, we are able to increase the perceived loudness by limiting transient material which might otherwise inhibit the possibility of increasing volume. In the interviews this effect was described in various terms.

Alan Moulder remarked, 'my way of thinking of saturation is compressing and rounding off your transients and mooshing them.'

Mike Exeter speaks in similar terms,

> I want to use saturation to tame the transients, to shape them. I don't want to use things like transient designers. I want to use something that just clips off the ends in a musical way.
>
> [34]

Recording to tape produces a similar effect as Moulder comments,

> It makes a big difference to me cos if you record vocals on tape, it sorts out half your problems. With all the transients being rounded off, it gives it body.
>
> [29]

6.5.6 'EQ for Free'—Adding Harmonics and Colouration

Adding harmonics to the fundamental frequency will change the spectral centroid of a sound, and it will often be perceived as being brighter. Similarly, augmenting an individual sound or mix with harmonic distortion through saturation will also change the frequency content. Dave Bascombe referred to this condition as 'EQ for free'. Indeed, many of the engineers I spoke to commented that they were using harmonic distortion more than EQ, and if they were using equalisation the character and the sonic identity of the unit was made more interesting by the distortion properties exhibited by the circuit. Scheps reflects, 'What makes the API EQ special is the distortion properties' [26].

Harmonics, depending on the type you are adding, can change the perception of the sound entirely. Even order for example, as commonly found in tube circuits, will tend to sound warmer, thicker or add body, whereas odd order harmonics, as found in tape-based circuits, will tend to be more aggressive, raspy or brighter sounding. Flood comments,

> Yeah, if you want something really angular, to stick out and sort of go 'eeee!' you want to go for the odd harmonics.
>
> [27]

He expands, when talking about the effect in relation to colour, alluding to the possibilities of it being synaesthetic in humans,

> Maybe distortion is the one thing that a human being can react to. And it doesn't quite know the parameters. Colouration is really a sort of mental outlook, like 'that was a bit of a boring sound', you just put the volume up and ooh, that's added a nice colour to it. And suddenly you're reacting to something in a positive way rather than it being a sort of negative.

To some, the interesting part of the sound of distortion comes from the non-linear, random behaviour it exhibits, particularly when driven hard. As Dave Bascombe articulates, this is part of the allure and indeed, character and colour of the sound:

> I think a lot of that non-linearity in devices that are distorting is attractive. It's two things—the non-linearity which is attractive to us and then the gimmickiness of it, the ear candiness of odd sounds, distorted sounds.
>
> [23]

Spike Stent also talks about warmth as being an essential ingredient in a mix which he hears when adding subtle distortion,

> Yeah it needs to have some heart to it, otherwise it will sound sterile and brittle. I've definitely been through that period in my early mixes and thankfully I came out the other side. I'm distinctly aware of the warmth that you've got to have and which you get from distortion, otherwise it turns the listener off.
>
> [33]

6.6 The Ecology of Distortion

Some of the comments obtained from the interviews showed interesting associations and correlations between distortion with occurrences and events seen in nature and in our environment and surroundings. Life distortions or ethereal distortions, as Trina Shoemaker referred to them, have parallels in sound. The use of synonyms and antonyms such as 'natural' or 'unnatural', 'harmonious' and 'perilous', or more commonly, 'warmth' and 'brightness' are all used to describe sonic attributes and situate distortion in the sonic palette. Tchad Blake spoke of how distortion can have dangerous connotations,

> distortion to me signifies danger. It's danger. And when you hear it, you're programmed to go 'F***! Look around, it's fight or flight!'
>
> [22]

Eric Clarke's book on perception and musical meaning, *Ways of Listening*, considers the impact of environment and space on sound. This is also pertinent to some of our understanding of the semantics terminology used in music production, specifically on distortion [35]. From an ecological perspective, sound that is distorted, intentionally or unintentionally, can often be related to events that occur in nature. Removing a sound from its natural environment could be considered a distortion in some sense. The sound cannot be considered separate from the space it inhabits.

Distortion of space is an effective tool used to great effect by many mix engineers, whether through re-amping a signal or applying distortion to a reverberant sound. It is a tool that is used to transport the listener from 'realism' to 'escapism', as it was occasionally referred to in the interviews. Tarek Musa talks about sound needing to be unsettled:

> Aggravating the room makes the room sound bigger. Say you're recording drums, really distorting those room mics makes the drums come alive for me.
>
> [36]

The goal of music production is rarely to create realism or embrace authenticity; it is far more often used to facilitate escapism or artificiality. The oddness or unnatural behaviour is part of the aura and charm of distortion and resonates in the context of the ecology of sound. When Spike Stent referred to distortion as being the 'soul of sonics', I believe he was tuning in on that otherworldliness that permeates the psyche and the emotion that comes from the sanctity revealed when processing and manipulating audio material with distortion.

6.6.1 Contrast and Sound Staging in Distortion

To add perspective and depth in a mix, distortion is an important tool in the engineer's sonic armoury. The skill is in knowing when and where to use it and in not being overzealous with the amounts or the places you add it. Tchad Blake says,

> If you want to bring something out of the muck and you have no other tool but to make everything else worse, that's what you gotta do. So, if you had a whole track that was distorted but you wanted the vocal to sound clear, but the vocal wasn't clear, it was kind of distorted, well just distort the whole rest of the track and now the vocal will sound a lot clearer, right? So, I think being able to shift the space in your head around in that way can be really good sometimes.
>
> [22]

Alan Moulder also says,

> All these small differences stacked up make a massive difference. The most effective trick is to not have too much of it. It's that trick of having an environment of sort of cleanliness and having one or two thing that really overstep the boundaries.
>
> [29]

Explicit use of distortion must be implemented sparingly whilst implicit or more subtle harmonic enhancement can be used more liberally. Creating a sonic signature is about bringing out character, introducing different colours and perspectives by using different types of distortion, often in small amounts.

6.6.2 Taste and Stylistics in Distortion

Many of the interviewees referred to distorted sounds as being 'tasteful' or 'distasteful' without specifying what this actually meant to them. The words 'subtle' and 'unsubtle' were also used in a similar vein. Distortion could also be 'explicit' or 'implicit', 'overt' or 'disguised'. Regardless of the descriptor used, it was apparent that the type of saturation or colour added was a matter of taste, which was governed by the stylistics of the individual. Taste is a difficult thing to define in music production, yet it shapes so much of both the technical and artistic process and is inextricably linked to style and aesthetics. Style and preference are crucial factors that separate one creative from another. Engineers and producers are hired for the sonic signature they impart and have spent time harnessing. As Gareth Jones commented in our interview,

> we get hired for our aesthetic if people like what we do. It doesn't mean that X's mix is better that Y's mix. What it means is some people like my aesthetic and some people don't.
>
> [31]

Agency and intention are key factors in the negotiation between artist and producer and engineer. Distortion is a conduit for developing some of these stylistics which separate some of the interviewees from one another. In *The Social and Applied Psychology of Music*, Hargreaves and North discuss factors which contribute to this discussion. Despite the notion that there is no accounting for taste, in music production at least, it seems there might be some determining factors which would suggest otherwise. Distortion is arguably one of the factors where taste is attributable to past events. Tchad Blake talks about avant-garde guitarists shaping preferences in his early life, 'nostalgia proneness' as it is referred to by Hargreaves and North [19]. The early hip-hop records influenced him with the explicit use of noise and distortion in the records, and he talks about the importance of contrast where a dirty backing track juxtaposed with a pristine, clear vocal was used to great effect. He also uses the interesting analogy of film and using different lenses to create perspective—putting sounds out of focus. This is also a good example of semantic crosstalk taking place where visual stimuli has influenced auditory stimuli. He goes on to say,

> It's contrast, isn't it? Humans are built to really notice contrasts. In a sea of stillness, you'll notice the one tiny little movement, or in a sea of movement you'll notice that one tiny stillness.
>
> [22]

Contrast in music production is only one of many tools used to create perspective and depth, but it is a crucial one in the context of bringing out character in a recording. In the quest for creating a sonic signature and bringing out character, Bascombe highlights the importance of identifying the parts in the recording that need to shine and be given the attention they deserve to bring out the character.

6.7 Current and Future Directions

Because of the nature of the way that producers and engineers now work, where they are adopting a predominantly in-the-box approach, several have switched to a hybrid workflow where they use a mixture of analogue and digital technologies to make up for some of the deficiencies experienced. Companies such as Overstayer [37], Thermionic Culture [38] and Malcolm Toft [39] have shifted their marketing focus to reflect this change in attitude. At the time of this writing, Malcolm Toft—designer of the Legendary Trident consoles, a guru in audio design and known for extremely high specification analogue products traditionally with low distortion levels—is about to release the PUNISHR 500 series module [39], which uses the terms 'destruction' and 'obliteration' as motivational marketing material. The resurgence of historical brands such as Helios, Universal Audio and Chandler EMI along with the rejuvenated established ones of API, Neve and SSL reflect a focus on adding colouration, saturation and distortion rather than seeking ways of removing it. Indeed, we have seen a conscious shift by equipment manufacturers to overtly increase levels of harmonic distortion, to put back some of the mojo missing from digital audio recording. Analogue summing mixers which have become ubiquitous in studio setups are less focused on transparency and more on colouration. Acoustica Audio [40], using convolution techniques, have also shown an uncanny ability to capture some of the magic we associate with analogue studio technology in their plugin El Ray, itself a model of the legendary RCA BA6A Compressor, famous for its distortion properties. Austin Moore, in his work on distortion in dynamic range compressors, has researched and documented this phenomenon and continues to be a leader on research in this field. The McDSP [41] Analogue Processing Box (APB) technology takes this principle to its logical conclusion, where it has married the sound of analogue with the convenience of plugin control in a DAW. Digitally controlled analogue sems to be an excellent convergence of the two technologies. Wes Audio [42], Tegeler [43] and Bettermaker [44] pursue a similar technological and marketing direction. Whether hardware or software, it is becoming increasingly common to find semantic descriptors used instead of purely technical terminology. There are many examples of this, but the Scheps Particles plugin by Waves [45] and many of the offerings from both Klevgrand [46] and Goodhertz [47] use equally abstract interface terminology for describing sound.

As the use of artificial intelligence in music production becomes more prevalent and common, there is a greater need to understand many of the concepts outlined in this chapter because they highlight the importance of context in this discussion. Indeed, the importance of subverting technology to make bold aesthetic statements has to some extent been largely ignored in this discourse. The importance of outliers in datasets and exceptions to rules is arguably where creativity finds itself situated in music production. Navigating this domain will become increasingly crucial if we are to avoid making music that largely conforms and adheres to established rules. Many of the producers interviewed commented on the prescriptive, generic and mundane nature of current music production, largely attributable to technologies offering one-stop, preset-focused solutions. Steve Lillywhite articulates this point when he talks about the problem encountered with too much choice and lack of limitation with production tools:

> when it was analogue there was little choice, it goes back to that limitation factor—great art comes from limitation. And the idea that with very limited things, back in the day, we came out with lots of varied art. But now everyone has the option to do anything. Every record sounds the same, mainly because they go through each sound and go ooh, that

sounds good and then okay, now a snare drum. Basically, everyone chooses the sound that individually sounds good. Put it all together and you've just got everything that sounds good just cancelling each other out. You need perspective.

[48]

Several scholars have identified this as a factor that needs addressing, including Lefford, Moffat and the author of this chapter [49]. Nowhere does this quest to understand the idiosyncrasies of music production manifest and permeate itself more than in the use of semantic descriptors in intelligent music systems. Many studies have been conducted, including some by the author, which seek to reveal this phenomenon where rating scales for audio examples are measured against perceived descriptive language.

So how do we connect and map semantic descriptors to a technical parameter space? This is challenging. If the goal is to create a feisty, relaxed or intimate sound, how can we use distortion, or other audio effects for that matter, to create this? The language used to communicate such aesthetics and interactions must also be a prominent factor in this discussion. This is very pertinent and important in the context of developing intelligent mixing systems where descriptive language is arguably as important as the scientific terminology used.

6.8   Conclusion and Final Reflections

We can conclude from the interviews conducted that distortion, in its various manifestations and incarnations, is esoteric, ethereal and difficult to clearly define aesthetically. It can operate on a subtle almost imperceptible level where it excites harmonic content, or at a 'full-on', bombastic, explicit and even chaotic one. We can surmise that it almost certainly provokes an emotional response in the music, which is difficult to articulate and define from a purely technical standpoint. On a semantic level it might be a conduit for conveying danger but at the same time conveying warmth, beauty and even transcendence in the music, or as Trina Schoemaker more light-heartedly comments,

> Hey, distortion may just be the result of panic, the pressure to just get a sound.

[28]

We have proposed a lexicon of descriptors in the body of this chapter which contribute to the discussion. However it manifests itself, or affects our mood or perception, I return to the comment from Mark 'Spike' Stent in our interview that proved the most resonant for me and had the most gravitas. Maybe it is simply,'the soul of sonics.'

References

[1] Braun, V., & Clarke, V. (2012). *Thematic analysis*. Washington, DC: American Psychological Association.
[2] Bryan-Kinns, N., Wang, W., & Wu, Y. (2018, July). Thematic analysis for sonic interaction design. In *Proceedings of the 32nd international BCS human computer interaction conference* 32 (pp. 1–3), Belfast. https://doi.org/10.14236/ewic/HCI2018.214
[3] Pestana, P. D., & Reiss, J. D. (2014). *Intelligent audio production strategies informed by best practices*. London: AES International Conference.
[4] De Man, B. (2017). *Towards a better understanding of mix engineering* (Doctoral dissertation, Queen Mary University of London).
[5] Case, A. (2012). *Sound FX: Unlocking the creative potential of recording studio effects*. New York: Routledge.

[6] Wallmark, Z. (2019). Semantic crosstalk in timbre perception. *Music & Science*, 2, 2059204319846617.
[7] McAdams, S. (2013). Musical timbre perception. *The Psychology of Music*, 35–67.
[8] Siedenburg, K., Saitis, C., McAdams, S., Popper, A. N., & Fay, R. R. (Eds.). (2019). *Timbre: Acoustics, perception, and cognition* (Vol. 69). New York: Springer.
[9] Zacharakis, A. (2013). *Musical timbre: Bridging perception with semantics* (Doctoral dissertation, Queen Mary University of London).
[10] Saitis, C., & Weinzierl, S. (2019). The semantics of timbre. In *Timbre: Acoustics, perception, and cognition* (pp. 119–149). Cham: Springer.
[11] Pearce, A., Brookes, T., & Mason, R. (2017) Audio commons. In *Deliverable D5*, 2, Evaluation report on the first prototypes of the timbral characterization tools. Available: www.audiocommons.org.
[12] Moore, A., Till, R., & Wakefield, J. P. (2016). An investigation into the sonic signature of three classic dynamic range compressors. In *140th international AES convention*, Paris. Available: http://www.aes.org/events/140/
[13] Moore, A. (2020). Dynamic range compression and the semantic descriptor aggressive. *Applied Sciences*, 10(7), 2350.
[14] Reiss, J. D., & McPherson, A. (2014). *Audio effects: Theory, implementation and application*. New York: CRC Press.
[15] Wilmering, T., Fazekas, G., & Sandler, M. B. (2013, November). The audio effects ontology. In *ISMIR conference proceedings* (pp. 215–220), Curitiba. https://doi.org/10.5281/zenodo.1415004
[16] Wilmering, T., Moffat, D., Milo, A., & Sandler, M. B. (2020). A history of audio effects. *Applied Sciences*, 10(3), 791.
[17] Gabrielsson, A., & Juslin, P. N. (2003). *Emotional expression in music*. Oxford: Oxford University Press.
[18] Juslin, P. N., & Sloboda, J. (Eds.). (2011). *Handbook of music and emotion: Theory, research, applications*. Oxford: Oxford University Press.
[19] North, A., & Hargreaves, D. (2008). *The social and applied psychology of music*. Oxford: Oxford University Press.
[20] Porcello, T. (2004). Speaking of sound: Language and the professionalization of sound-recording engineers. *Social Studies of Science*, 34(5), 733–758.
[21] Melara, R. D., & Marks, L. E. (1990). Processes underlying dimensional interactions: Correspondences between linguistic and nonlinguistic dimensions. *Memory & Cognition*, 18(5), 477–495.
[22] Tchad Blake, Interviewee, Interview with Gary Bromham. 26 August 2020.
[23] Dave Bascombe, Interviewee, Interview with Gary Bromham. 7 September 2020.
[24] Rogers, J. Can you really get the analogue experience from software? *Protools Expert*. Available: www.pro-tools-expert.com/production-expert-1/can-you-really-get-the-analogue-experience-from-software [Accessed 17th April 2022].
[25] John Leckie, Interviewee, Interview with Gary Bromham. 24 February 2020.
[26] Andrew Scheps, Interviewee, Interview with Gary Bromham. 7 September 2020.
[27] Mark 'Flood' Ellis, Interviewee, Interview with Gary Bromham. 9 December 2019.
[28] Trina Shoemaker, Interviewee, Interview with Gary Bromham. 4 May 2021.
[29] Alan Moulder, Interviewee, Interview with Gary Bromham. 9 December 2019.
[30] Danton Supple, Interviewee, Interview with Gary Bromham. 5 February 2020.
[31] Gareth Jones, Interviewee, Interview with Gary Bromham. 15 April 2020.
[32] Cenzo Townshend, Interviewee, Interview with Gary Bromham. 27 July 2020.
[33] Mark 'Spike' Stent, Interviewee, Interview with Gary Bromham. 5 May 2020.
[34] Mike Exeter, Interviewee, Interview with Gary Bromham. 23 April 2020.
[35] Clarke, E. (2005). *Ways of listening: An ecological approach to the perception of musical meaning*. Oxford: Oxford University Press.
[36] Tarek Musa, Interviewee, Interview with Gary Bromham. 17 August 2020.
[37] Overstayer. [Online]. Available: www.overstayeraudio.com [Accessed 28th April 2022].
[38] Culture, T. [Online]. Available: www.thermionicculture.com [Accessed 28th April 2022].
[39] Toft, M. Malcolm Toft [Online]. Available: www.malcolmtoft.com/products [Accessed 28th April 2022].
[40] Acustica audio [Online]. Available: www.acustica-audio.com/store/products/elrey [Accessed 21st April 2022].

[41] McDSP [Online]. Available: www.mcdsp.com/apb/ [Accessed 14th May 2022].
[42] Audio, W. [Online]. Available: https://wesaudio.com/ [Accessed 14th May 2022].
[43] Audio, T. Tegeler audio [Online]. Available: www.tegeler-audio-manufaktur.de/Home/Index [Accessed 14th May 2022].
[44] Bettermaker [Online]. Available: www.bettermaker.com/ [Accessed 14th May 2022].
[45] Waves [Online]. Available: www.waves.com/plugins/scheps-parallel-particles#andrew-scheps-mixing-scheps-parallel-particles [Accessed 14th May 2022].
[46] Klevgrand. Klevgrand [Online]. Available: https://klevgrand.com/ [Accessed 14th May 2022].
[47] Goodhertz. Goodhertz [Online]. Available: https://goodhertz.com/ [Accessed 14th May 2022].
[48] Steve Lillywhite, Interviewee, Interview with Gary Bromham. 20 September 2020.
[49] Lefford, M. N., Bromham, G., Fazekas, G., & Moffat, D. (2021, March). Context aware intelligent mixing systems. *Audio Engineering Society*, 69(3), 128–141.

7 An Ecological Approach to Distortion in Mixing Audio

Is Distortion an Expected, Rather than an Unwanted Artefact?

Lachlan Goold

7.1 Introduction

Recording practice reluctantly accepted distortion and associated unwanted artefacts until the development of digital recording technology in the late 1970s. Initially, digital recording systems were costly and only available to the upper echelon of the music industry. Choosing analogue recording equipment was often a compromise between price and linearity metrics, comparing the input to the output signal in which distortion exists as a non-linear occupant in between these signals [1]. Laboratories tested Total Harmonic Distortion (THD), frequency response and signal to noise specifications to market analogue hardware devices to professional recording studios [1]. By the 1990s, digital recording technology dropped dramatically in price, with formats such as the ADAT obtaining much better laboratory performance metrics than analogue recording technology [2, 3]. However, in use, early digital systems were found to impart "cold" recordings lacking analogue warmth [4]. By the time DAW recording became an established medium for both the professional and home studio, digital recording technologies continued to carry a perceived inferiority to the large-format studio, particularly concerning mixing [5]. This deficiency in DAW operation and use led plugin developers to use convolution and emulation technologies to recreate classic analogue equipment [6] and thus re-introduce the non-linearities that were once unwanted. What was once undesirable has changed to "desirable" with positive semantic descriptors such as character and colour [7]. Distortion has an association with a poor-quality recording or playback [1, 8] despite distortion and harmonic saturation commonly used by professional mix engineers to mitigate the perceived coldness of the now ubiquitous digital recording format.

This research will interrogate recording industry professionals and laypeople using blind listening tests of completed mixes. I mix two songs and create two versions of each song—one version mixed in a naturalistic manner with whatever levels of distortion the mix engineer felt appropriate to the program with the artist's approval and one version with all intentionally (or unintentionally) added distortion removed and meticulously gain matched. I analyse the music based on a theoretical model that explores both the perception and cognition of listening. I will then conduct a survey to determine whether a respondent prefers the mix with or without distortion. The analysis of the data from this survey will begin to reveal if the absence (or reduction) of distortion is preferred over the original mix, to help understand if distortion has moved from an unwanted artefact to a legitimate and expected part of digital audio processing. As distortion has an adverse history, this study attempts to approach distortion pragmatically and multimodally using a combination of quantitative and qualitative research methods. I aim to understand the current role of distortion in music production.

DOI: 10.4324/9780429356841-9

7.2 Context

The purpose of an analogue mixing console is to sum together audio signals with the highest transparency possible [9]. Due to the nature of analogue circuitry, some distortion will be present during this process. The mixing console is one part of a complex system that contributes to increased distortion levels in the analogue studio. Many mix engineers use the console's distortion characteristics to their advantage in enabling a "sonic signature" of sorts. In an interview, UK-based mix engineer Spike Stent explains his technique of using distortion in the mixing process, which involves the clipping of individual SSL channels [10]. Producer Tchad Blake is renowned for using liberal amounts of distortion when mixing and has done so since 1992 [11, 12]. So too, recording and mixing engineer Marcella Araica explains how she uses distortion and saturation to subtly achieve depth in the low end of her mixes [13]. McIntyre [14] interviews several producers from the Newcastle area in New South Wales in which most mention how they use various methods of utilising distortion in the recording process. The application of distortion discussed here is in the recording and mixing process with a particular creative goal irrespective of the recording medium or summing process used.

As music production techniques move away from analogue systems to more linear digital systems, the ability to adopt distortion colouration has flowed through to plugin technology. Andrew Scheps' signature Waves plugin, the Omni Channel, includes an individual hardware console type saturation in a dedicated distortion section [15]. In the development of Chris Lord-Alge's Mix Hub plugin, he explains how he did not want to emulate a channel from his console accurately [16]. He adds, 'I wanted it to be able to overdrive and distort with a lot more headroom because it's digital' [16]. The gain structure in digital mixing is different from analogue techniques, but this does display the desire to emulate previous practice, as discussed by Bourbon [6].

A multitude of plugins are available to add distortion in the music recording process, which are often marketed as a panacea for dull and lifeless recordings. De Man and Reiss [17] discuss distortion improving the quality and presence of a recording. Recording engineers from any genre tend to make their equipment choices using 'practice-based manifestations' of discussions on internet forums and tutorial videos [18]. Examples of this (in a software context) are plentiful on software developer websites. For example, the waves.com website claim distortion can achieve 'more musical results overall and without making the track sound clipped' [19]. Distortion is positioned as another tool for manipulating digital audio files with an alternate or extended range to traditional tools with semantics such as, 'more forward, rounder or warmer' used to describe its actions [19].

Distortion has a mixed reputation within music production scholarship, with many research projects focusing on associated unfavourable characteristics of varying types of distortion [20–22]. Zagorski-Thomas [18] asks: '[w]hy do we like some forms of distortion?' in an article addressing the musicology of record production. Distortion is a complicated artefact and can occur in many forms, such as linear and non-linear distortion, intermodulation distortion, and high-order and low-order distortion. One source of distortion generation is the transformer, commonly found within large format consoles at the input and output stages. Simply put, a transformer either steps a voltage up or down depending on the ratio of windings between two isolated coils [23]. Through this process of transforming a voltage, the resultant sound may experience colouration with low-order harmonics that can add musically to the sound [23]. Similarly, valves and tape also contribute to low-amplitude, low-order harmonics, often termed soft clipping. These types of low-order distortion are considerably less perceivable than high-order distortions [24]. Whitlock [24] explains that the 'awfulness of distortion' is most apparent in high-order harmonics, such as the '7th or 13th for example'. Additionally, intermodulation (IM) distortion is also very audible, but a transformer reduces IM significantly [24]. Whitlock [24] discusses the inadequacies of THD measurements, as low-order harmonics are much less audible than higher-order harmonics. In summary, it is the low-order harmonic distortion generally produced by soft clipping (valves, tape, console channel emulation, mix buss saturation and transformers) that plugin developers emulate due to the more subtle results. The distortion types discussed in this chapter are artefacts (either intended or unintended) generated by plugin application processes emulating soft clipping.

Due to the complexities involved, some academic research analysing distortion focuses on one single track within a song, rather than the entire song [7]. Additionally, singular processes, such as mix buss compression, are analysed under laboratory conditions [25]. De Man [26] considers the difficulties in examining professional mix practice due to an artist's unwillingness to share potentially inferior material. This research looks to expand on previous approaches and use "real audio material" to determine if a listener perceives distortion within the context of a contemporary music mix, as suggested by Toulson, Paterson and Campbell [21]. As a practising mix engineer, I have unique access to release quality recordings with artist consent. Worthy of further research would be to investigate the large number of experienced producers who have moved into purely mix-based practice.

My research has shown the recording studio to exist in three temporal phases: the laboratory era, the factory era, and currently, the domestic (or DIY) era [27]. All eras have experienced technological innovation and the progression of recording processes [1, 28]. The DIY era is marked with portability and predominance of software-based digital manipulation of audio [29]. The trend toward software-only recording continues to increase, resulting in plugin design adaptation in emulating the colouration stages of various types of rare and expensive analogue hardware. Whether this trend is good for the recording practitioner or not is the subject of much debate [28–30]. I argue that technological innovation in the recording studio is constantly tempered with the needs of the consumer more than the recording practitioner. New playback mediums and methods of consumer reproduction have always changed the nature of the recording studio in that an artefact needs to meet the limitations of that medium. A current example would be using distortion to add harmonic material to low frequencies so that a lower-pitched instrument is still perceivable on small playback systems even though the playback system may not be capable of replaying the fundamental frequencies. The human brain hears the harmonic and interprets the fundamental frequency as being there.

Additionally, another characteristic of distortion is to increase gain due to this added harmonic material, and therefore loudness and dynamic range require discussion; however, a detailed analysis goes beyond the scope of this chapter. Enderby and Baracskai [31] discuss how distortion adds to the perceived volume and presence of a signal, while production blogs describe distortion as a "secret" tool [32]. Perceived loudness is not a quantifiable term and sits in the discipline of psychoacoustics [63]. Colloquially known as a descriptor for how loud a signal may be anticipated upon playback, perceived loudness is generally associated with increased Root Mean Square (RMS) levels [63]. Early analysis of loudness by Chalupper [33] showed that despite the non-linearities introduced by psychoacoustic processing, there is no deterioration to the sound. Concomitantly, Taylor [63] discusses the 'deleterious effects of hypercompression' and various mythologies surrounding mastering loudness. Discussions on overall final mastering levels cause passionate debate among professional audio engineers, producers and artists with the introduction of every new reproduction format, most recently concerning digital technologies of CD and streaming. Digital recording distribution platforms such as CD that allowed for much greater dynamic range moved antithetically to these frameworks, and techniques to reduce dynamic range led to louder and louder masters. Fenton, Wakefield and Fazenda [1], Milner [5], and Toulson, Paterson and Campbell [21] discuss the predominant negative attitude toward a reduced dynamic range, dubbed the 'loudness war' [34], among professionals in-depth. While these discussions have a largely scientific base, they have ignored the consumer, and the methods and place of music consumption. While music listening trends have moved to less-than-ideal circumstances such as laptop speakers, earbuds and smartphones [35], music streaming services attempted to normalise audio as long as the preference is enabled in the playback software. Regardless, a reduced dynamic range suits the portable styled playback mediums (particularly in noisy environments), and the loudness war debate continues among conservative practitioners. Throughout this research, I take on a consumer-led approach to playback, and a similar overall volume and dynamic range has attempted to be maintained between comparative mixes.

Mixes completed from software-only process is referred to as "in-the-box" mixing. These processes allow for a mixing engineer to apply pre-prepared session templates to a multi-track complete with an array of plugins predetermining a starting point for mixing. In many ways, a template emulates the environment of an analogue large-format studio but with much more adaptability. Analogue mixing approaches impart subtle distortion from the very moment the first track is auditioned and summed through the mixing console to the master fader. Using a DAW, a mixing engineer can build their perfect virtual recording studio and apply it to every mix they make with the added ability to drastically (or subtly) alter the virtual studio according to the program material [36]. Using this approach allows a mix engineer to work quickly from a place where an already established sound is applied to the source material, much like the analogue studio did as part of its natural system [36]. A mixing template thereby reduces the amount of time required to start building a "sound" around the song. Using these software processes emulates the practice many mix engineers developed in an analogue environment out of necessity. Subtle distortion is invariably a part of these templates and is applied subjectively by the mixing engineer. This chapter analyses my mix template by replacing all plugins in the template that introduce distortion with an alternative plugin free of harmonic distortion.

7.3 Theory

The application of ecological tropes in academia, and specifically music, is not a novel concept [37]. Many diverse fields have used an ecological approach as a metaphor for natural environments, all using the parameters of ecology to denote different things in differing circumstances [37]. As an example, in music, Archer [38] used ecology to look at the relationship between the semiotics of music makers and music consumers. I will focus my research on an ecological approach to perception developed by Gibson [39] in the field of visual perception and further adapted to music by Clarke [40]. Scholars such as Schafer [41] use the frameworks of visual perception and apply them to aural perception. The ecological approach looks to examine individual perception within the confines of an environment [42]. Lefford [42] posits that music recordings are mimetic of an ecological world. In this chapter, I view the environment of a music recording as a representation of ecology to examine the role that distortion plays in a contemporary music mix. The remainder of this section examines my approach to this methodology.

The dictionary defines *ecology* as 'the set of relationships existing between organisms and their environment' [43]. Schafer [41] describes an acoustic ecology as the relationship between life and sound that cannot be studied in a laboratory. Gibson [39] posits an environment is made up of objects, energy and a medium. Lefford [42] states that '[e]nergy such as light or sound is perceived when it interacts with the environment's medium and objects' [42]. In ecological theory, Clarke [40] posits that when listening to music, the construction of meaning is almost exclusively an experience of perception. In order to perceive and make meaning of an event, a participant needs to engage in the environment of that event [40]. Schafer [41] elaborates on the finer details of an environment in relation to the habits, mood and familiarity with the environment. He demonstrates this with the example of industrial noise being unobtrusive until the significance of social circumstances highlights the apparent intrusion [41]. Similarly, the same can be said of small sounds when a listener is focused [41]. The focus of this research is to determine whether an interruption in the environment, such as removing added distortion, affects the meaning being made in the artefact.

Ecological theory proposes how an individual perceives sound differently to the previously dominant methods of music perception. Previously, perceptual psychology and auditory perception used an "information-processing" approach to music cognition whereby a listener makes sense of what they are hearing by using a 'variety of "processing" differences largely based on mental representations or memory processes of one sort or another' [40]. Simply put, the information-processing approach conceives music perception as a series of cogitative levels moving from stimulus-derived physical properties through to more complex and abstract methods of cognition derived from cultural perspectives [40]. The flow of information between these levels is not unidirectional as the processing of any stimulus will be affected by 'generalised preconceptions and expectations derived from previous experience' [40]. Individuals may experience a different interpretation due to a different mindset when encountering a stimulus [40].

The ecological approach found significant problems with the information-processing approach. Rather than the perceiver constructing meaning from the perception of information, more significance is placed on the structure of the environment and how that environment affects how information is perceived [40, 42]. Clarke [40] posits that perception is an active action. 'Perception is essentially exploratory, seeking out sources of stimulation in order to discover more about the environment' [40]. An example of active listening would be when the listener turns their head to locate the provenance of a sound in order to seek a better balance of direct to reverberant sound, therefore improving perception [44]. A recording will block many of our natural perception-action tendencies and prevent us from actively exploring its source [40]. Many physical embodied reactions to music, such as foot tapping or dancing, are a result of transforming these natural responses [40]. Despite this, listening to a recording involves a feedback loop of actively and reactively seeking clarity and meaning from a representational environment. With pre-recorded stereo material, we can cognitively determine that the environment is a representational system fixed to the playback system [44].

Bourbon and Zagorski-Thomas' [45] notion of sonic cartoons is useful as a way of interpreting the representations of sound in the recorded format. Additionally, we can use Gibson's theories of visual perception, with the approaches I have described previously, and apply those to a completed mix as an ecology unto itself. Gibson proposed that knowledge in the mind is established on previous experience (potential affordance) of a set of encounters reduced down to a given encounter (invariant properties) [45]. Perception is continually evolving and adapting in so much as a listener will become increasingly conscious of differences in an environment that always existed but were previously undiscovered [40]. This indicates that if a completed work of music is representational of an ecology, and an element of that ecology is removed, the listener may be able to reductively detect that element even though it was previously undiscovered.

7.4 Methodology

My research methodology uses a mixed-methods approach [46]. Conventional techniques in measuring the performance of audio equipment often encompass objective and subjective measurement techniques [47]. I will consider multiple versions of a completed artefact with a similar approach. To gather objective empirical data, we need to consider the fundamental components that make up music. Every track that makes up a mix includes a complex array of frequencies (or pitch), harmonics and timbres [1]. A full music production blends these tracks at different amplitudes to create a mosaic with subsequent sonic texture and timbres [1]. The complexity of this signal is difficult to measure in a useful manner using laboratory techniques, such as frequency response or THD measurements [47]. However, using a combination of data gathered empirically from the recordings, autoethnographic observations and subjective listening tests alongside the ecological theory will begin to elucidate the role of distortion in contemporary music production. Using distortion in the mixing process is often subtle, not measured (or truly measurable) and applied uniquely in each application. The best way to approach its presence would be in using a reductive technique.

When evaluating a piece of audio equipment, subjective listening tests allow judgement on what is heard [47]. Moore [47] adds that empirical evaluations of audio equipment is valuable but is of little use to the music producer. Libshiftz and Vanderkooy [48] suggest that blind A/B tests are a suitable testing mechanism for subjective evaluation, and they propose that only a highly controlled laboratory-style investigation is of use. I am testing whether an element is observable in the ecology of a completed artefact. Additionally, I am concerned whether a layperson or professional prefers the presence or absence of distortion in that artefact. Blind A/B subjective listening tests in a naturalistic environment will allow me to determine not only if a listener subjectively prefers the presence or absence of distortion, but also the efficacy of more prevalent consumer listening devices. As stated, distortion is often used to compensate for the lack of bandwidth in compromised listening conditions.

A simple A/B online survey will allow me to draw upon as many respondents as possible and will give the qualitative part of my research a quantitative result [46]. Additionally, the respondents will answer some simple questions related to their choices. I anticipate an expectation bias due to mood, the participant's response to the text (likes or dislikes the genre style or song) and listening conditions. This research is a pilot work that I hope to expand into a more extensive test using more music genres to reduce some of these biases.

Survey responses will be thematically analysed to identify patterns (themes) within the data [49]. The thematic analysis will aim to gain an understanding of participants' perception of distortion within the sample material provided. The research design is as naturalistic as possible, in that the initial mixes are realised creative artefacts. Naturalistic design attempts to research conditions that are 'committed to the primacy of natural context' [50]. Additionally, my research aims to 'have direct relevance for the theoretical questions' [51]. Using these approaches will allow me to analyse the data and derive relevance to the theory.

My objective analysis will use Youlean Loudness Meter software [52] to generate a histogram profile of each mix and determine a consistent loudness for each substituted plugin. This visual will display any overall loudness differences in each mix. I will use Matlab to generate a comparative waveform analysis to compare loudness discrepancies between the different mixes. These data will then be discussed. This approach captures a multimodal approach to the research in using survey response raw data, images created by the artefacts, an analysis of the words from the survey responses and details of the actions undertaken by me [53].

7.4.1 Method

The content for the A/B tests will include an original mix intended for public release (all mixes are unmastered). The second mix will be a version of the artist-approved mix with any distortion added during the mixing process removed with the goal of maintaining the original balance and overall volume as much as practicable. The basis of the initial survey is on two alternative styled pop/rock songs, both including a band performance of acoustic drums, electric bass guitar, acoustic and electric guitars and vocals. I was not involved in creating the multi-track for these recordings in any way. I will also ask the participants in the survey why they made the choice they did and what playback system they used to make their choice. In using the respondent's own playback systems, I will address the naturalistic and pragmatic approach to this research. The presence of distortion will most likely be perceivable by a professional under laboratory conditions. I am most concerned whether a consumer can detect distortion in a situation where they would typically consume music. I am looking to see if the playback medium affects the result. Many approaches to using distortion in a mix are to compensate for poor playback situations.

I started my investigation by analysing the mix template I would apply to the supplied multi-track. I approached the mix template in stages, starting with the mix buss. Initially, I duplicated the mix buss plugin path and bypassed all plugins. I then introduced a 220 Hz sine wave through the mix buss and monitored the output on a spectrum analyser (Fab Filter ProQ2 [54]). If I observed added harmonic frequencies to the fundamental 220 Hz tone as I activated a plugin, I found a suitable substitution to remove this distortion. For example, if an equaliser (such as a Pultec emulation) imparted distortion, I replaced it with a Fab Filter ProQ2. Compression creating distortions were replaced with the Klangheim DC8C 3 [55]. I replaced any channel emulation or circuit modelled distortion with a Waves L1 limiter [56] (The actual version I used did not include the Dither and Quantize features, and on the mix buss I used a Sonnox Oxford Inflator [57]).

I matched equalisation curves using the technical specifications provided from the plugin manual, to best emulate the curves initially produced. As compression is a more complicated process, all settings were checked with a signal generator playing a 220 Hz tone at -10 dB for 20 seconds, pink noise played at -15 dB for 20 seconds (-10 induced distortion in the standard six buss), and an example mix song to ensure that overall loudness was consistent. The Pro Tools peak meter and the Youlean Loudness Meter were used to ensure Peak Level (PL), Dynamic Range (DR) and Loudness Units Full Scale (LUFS) matched between the original plugin path and the newly created reduced distortion alternative.

Distortion proved to be the most challenging plugin to replace adequately. As part of the distortion saturation process is to increase perceived volume, I used a limiter to compensate. Comparing the gain (LUFS, via the Youlean plugin) between the mix buss saturation and the Waves L1 limiter using Pink Noise, the settings obtained are shown in Figure 7.1.

However, to match the LUFS meter setting with an actual mix, the threshold was reduced, as shown in Figure 7.2.

On the mix buss, no other plugins affected a difference in the LUFS meter, producing identical results to the Pink Noise test. This example demonstrates the shortcomings of using a signal generator in the place of actual program material. Using a tone burst method, as discussed by Moore [7], may have helped, but in this instance program material enabled the creation of a practical model for use.

Earlier I explained the general process of in-the-box template mixing. After loading in my mix template, I separate the multi-track into groups that are then routed to various summing stages. At these summing stages, I use many analogue channel emulation plugins, compressors and EQs to emulate the analogue studios where I learnt to mix. Grammy award-winning engineers such as Andrew Scheps guide you through their mix template, providing you subscribe to the aggregator's website [58]. The plugin replacement process I described for the mix buss is then applied

Figure 7.1 Pink noise settings on distortion replacement.

Figure 7.2 Program settings on distortion replacement.

throughout the various summing stages of my mix template. Repeating this style of plugin substitution can be applied to any mix template, and although it will not wholly eradicate distortion, it will significantly reduce the harmonic content of the completed mix, while maintaining the gain structure of the original mix. No plugins on individual tracks were changed so that the individual character of the performance was not altered. The two completed versions of the two songs were then deidentified, uploaded to Soundcloud [59] and links added to a Microsoft Forms [60] with a simple survey asking the participant to choose which mix they preferred, A or B.

Additional details included if the participants identified as a professional audio industry practitioner, the participants' playback system and a quick response to describe why they made their choice. Allowing the participants to choose their own playback system is in line with this experiment's intentions of applying a consumer-led approach to the ecology of a completed mix. Additionally, the participants were not limited to one playback source. Many approaches in applying distortion within mixing practice are to compensate for poor playback systems, such as earbuds over full-range professional monitors. The design of this experiment hopes to go some way toward understanding if this compensation is necessary.

7.4.2 Limitations in the Method

While using A/B testing software such as De Man and Reiss's [61] would be desirable, I used Soundcloud in order to meet the demands of publication. While Soundcloud converted 24-bit 48k masters to 320 kps AAC files, all files were played back in this format, applying any additional artefacts created in this process evenly to all files.

The method employed in this research does not exactly compensate for the "sound" of distortion, but creating a more robust and efficient method goes beyond the scope of this current study. The extra gain created in the distortion process is fluid, spectrally complex and program-dependent whereas a limiter is fixed, and when used in place of heavy distortion, may produce undesirable pumping effects potentially audible to an experienced respondent. Generally, the use of distortion in the mix template is subtle; any instance of more aggressive distortion is generally applied sparingly in the final mix. Additionally, the timing elements of compression were difficult to emulate, but using technical specifications from the manual and fastidious checking of consistent LUFS levels provided a suitable process for replacement. Mixing using a predefined set of open-source code plugins, such as SAFE plugins, would obviate many of these issues. Open-source plugins would allow the user to toggle between distorted and clean operation easily and create a more repeatable outcome. However, these processes would go beyond a naturalistic inquiry, in that they would alter the regular tools of my creative practice.

The Microsoft Forms survey link included some information regarding the research aims—such as my goal to evaluate the role of distortion in a contemporary music mix—as is standard practice suggested by the ethics department at my university. This would alert professional practitioners to attempt to listen for distortion, rather than simply answer my question of which mix is better and why and not knowing what the differences in the mixed versions may be. This would undermine my research question of: is distortion an expected artefact? Non-professional listeners may not have the playback equipment or the experience to evaluate subtle amounts of distortion. In future iterations of the test, I would seek to include less research information on the project information section, thereby creating a double-blind subjective test. Additionally, using Soundcloud and Microsoft Forms made it unable to randomize the order of playback in the A/B test.

When the artist-approved mix was complete, I then applied the new "clean" template to the mix, ensuring to match the average levels and general mix balance. At times, I noticed

118 *Lachlan Goold*

the compensating limiting adversely affecting the clean mix. I did not attempt to correct this, and I assume those listening on professional monitors or headphones may discover these artefacts.

7.5 Results Data

In total, I replaced 30 plugins that generated some sort of harmonic content with 37 equivalent plugins (at times, one plugin required two plugins for accurate emulation): four EQs, eleven compressors and limiters, thirteen harmonic character plugins and two channel strips.

To create the content for the perceptual listening tests, I mixed Song 1, titled "Brisbane Girl", to meet the artist's expectations. Mix 5 was deemed the final mix ready for mastering and release, although the track has not been released at the time of publication. Figure 7.3 shows the exported loudness and dynamic range graph for the completed mix.

Once I completed this mix, I set out creating a clean mix, starting by matching the volume and copying automation over to the auxiliary returns with the clean plugin substitutions. Figure 7.4 shows the loudness and dynamic range graph for the clean mix of Song 1, "Brisbane Girl".

While the similarity between the mixes is evident, there are small discrepancies of 0.1 dB in the Peak to Loudness Ratios (PLR), and 0.1 dB in the Momentary Max (M MAX) numbers. I had 59 respondents to the survey, of which 37% (22) claimed music production as their primary source of income. In total, 74.5% of all respondents preferred the artist-approved mix (B), with the remaining 25.5% preferring the clean mix (A). Song 1 is driven by slightly dirty guitars with a "retro" feel and arrangement. Focusing on the respondents who identified as professionals

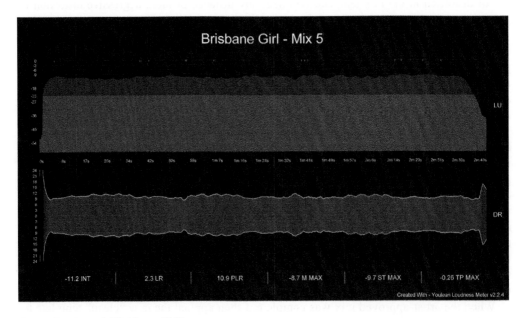

Figure 7.3 Song 1, "Brisbane Girl".

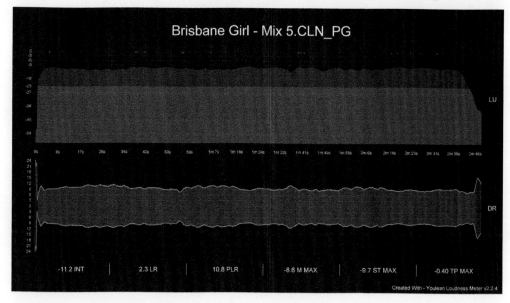

Figure 7.4 "Brisbane Girl"—clean mix.

and used professional monitor speakers who chose Mix B (in the survey), here are a selection of comments:

- The second work feels more dynamic while also having a vocal further forward and more excited.
- [Mix A is] not as glued together, drums and percussive elements sound to (sic) disconnected from the rest if (sic) the music. Bass and overall bottom end energy if (sic) the song is lacking in Mix A [compared to Mix B].
- I liked what the saturation/edge had done for the drums within the mix. It gave the mix energy.

Only three respondents identified as professionals and used professional monitor speakers but didn't choose Mix B:

- Mix B is a bit cloudy in the louder sections, too much distortion perhaps.
- Without knowing exactly what terminology to use, the second mix seems to have a harshness that I can discern its origin, the track is not bad, to me there is almost a low-level white noise or something. More a perception than a distinct sound.
- I Prefer the Cleaner Tones In This Mix [mix A], It Suits The Song Better.

Song 2, titled "Oxford Park", is driven by acoustic guitars. After mix revisions, Mix 4 was deemed the final mix ready for mastering and release by the artist, although the track has not been released at the time of publication. Figure 7.5 shows the exported loudness and dynamic range graph for the completed mix for Song 2.

Again, I created a clean mix using the same method and approach to Song 1. Figure 7.6 shows the loudness and dynamic range graph for the clean mix of Song 2, "Oxford Park".

These mixes are less similar than Song 1. While there is only a 0.1 dB difference in the Peak to Loudness Ratios (PLR), there is a 0.7 dB difference in the Momentary Max

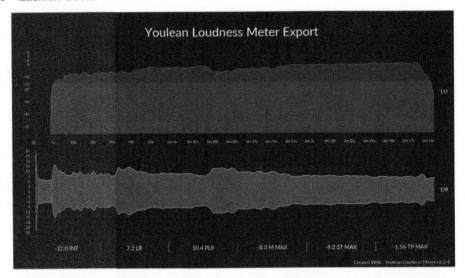

Figure 7.5 Song 2, "Oxford Park".

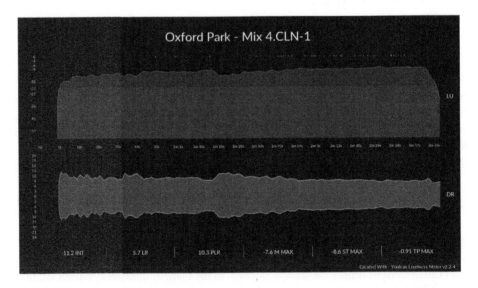

Figure 7.6 Song 2, "Oxford Park"—clean mix.

(M MAX) numbers, 0.8 dB in the Integrated LUFS level and 1.5 dB less loudness range. In future iterations of this research, I would look to match the LUFS level as much as possible, which would've led to turning the clean mix down 0.8 dB. For Song 2, 56% of all respondents preferred Mix A, with the remaining 44% preferring the clean mix (B). While the artist was again chasing a "retro" feel, the mix stands up well without distortion and sounds louder in places, particularly in the first verse. Those respondents who didn't listen to the entire track may have quickly decided on the clean mix. I labelled the artist-approved mix with distortion as Mix A in the survey. Focusing on the professionals using studio monitors, some respondents said of the artist-approved mix:

- The first mix seems to have a more appropriate sonic aesthetic. The transients in this work seem more pleasing and the vocal sits better in the mix.
- In the second mix [mix B] there is a boost somewhere in the low-mids I feel. There's something creating a reverb effect that does not sound natural, the overall tone sounds like a single speaker playing in a small round fishbowl.
- I liked Mix A (sic) openness. I felt Mix B had a density (particularly with the ac guitar) that although interesting, was at the cost of the vocals cutting through within the mix

Only one respondent who identified as a professional and used professional monitor speakers but didn't choose Mix A had this to say:

- Mix A Sounds Messy, the vocals are lost.

I'll now focus my results on the comparative listening on both tracks, delineating between professional and non-professional respondents. The tables compare the respondent's playback system to the pair of mixes they chose, remembering that the artist-selected mixes with distortion were Mix B + A. Some respondents used two playback systems, but I only chose to display the most professional of those, not discounting a professional approach of listening on a lower-grade system. The results are displayed in the Tables 7.1 and 7.2.

Table 7.1 Respondents Who Identify as Professional: 22

Playback medium	Mix A + A	Mix B + A	Mix B + B	Mix A + B	Totals
Pro speakers	2	5	1	1	9
Pro headphones	1	5	2	1	9
Headphones		1			1
Smartphone		1	1		2
Earbuds	1				1
Totals	4	12	4	2	22

Table 7.2 Respondents Who Identify as Non-professional: 37

Playback medium	Mix A + A	Mix B + A	Mix B + B	Mix A + B	Totals
Pro speakers		4	2	1	7
Hi-Fi speaker			2		2
Pro headphones		2	5	1	8
Headphones	1	1	2	1	5
Smartphone			1		1
Earbuds	1	3		1	5
Bluetooth speaker		1			1
Computer speaker	1	2	3	2	8
Totals	3	13	15	6	37

122 *Lachlan Goold*

What is evident in these tables is the greater accuracy that the professionals chose with the artist-selected mix. The non-professionals, while many have professional playback systems, mostly chose Mix B with Song 1 and Mix B with Song 2. The reasons for this are most likely due to the louder volume, as discussed earlier. Interestingly, the most popular playback system was headphones (23), with professional monitors second (16).

Using Matlab, I was able to lay the waveform of each mix together, to reveal the direct differences. Figure 7.7 shows the artist mix in orange and red and the clean mix in blue and green. It seems as though the artist mix has higher true peaks and more dynamics.

Figure 7.8 shows the same comparison for Song 2, with the artist mix in red and blue, and the clean mix in pink and green. The clean mix average RMS volume has dominated the artist mix

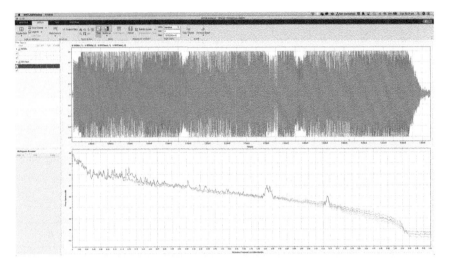

Figure 7.7 Song 1, "Brisbane Girl"—compared mixes.

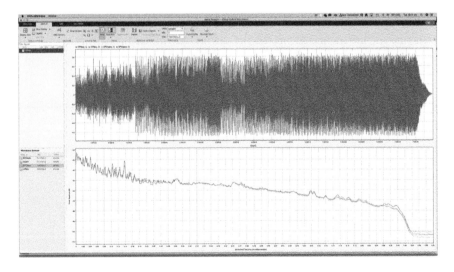

Figure 7.8 Song 2, "Oxford Park"—compared mixes.

An Ecological Approach to Distortion in Mixing Audio   123

in this instance, making the artist mix difficult to see, although the true peaks seem to be similar. This is further evidence displaying the volume discrepancies.

Some respondents, all non-professional, commented on the subtleness between the mixes. In particular, Song 2 generated many comments; these nine respondents chose Mix A:

- It was hard to choose but I think A sounds more exciting. And the kick in B felt a bit obtrusive.......
- No, they were very close.
- Not really . . . except it seems more 'mixed'.
- It's subtly warmer. These two were very similar to me—harder to distinguish any difference.
- These two [are] closer. I think the first [Mix A] seems more dramatic.
- I actually liked both mixes equally.
- These two sounded very similar and it was difficult to discern a specific difference, Mix A sounded slightly less muddy, but the difference is so subtle that I might have imagined it.
- I actually can't hear a difference between the two tracks, but maybe mix A is warmer?
- [I] can't hear any difference.

These four respondents chose mix B:

- Hard to tell the difference. I prefer the opening acoustic gtr (sic) on A, as it is less thick/close. After that I struggle to tell them apart.
- Same as above [mix A, it] just sounds more live. But I don't notice the difference between these two mixes [Song 2] as much as the other two.
- Not sure.
- I can't hear any difference.

The extent to which respondents declared to hear only a subtle difference in Song 2 demonstrates a difference between the performance styles of the two mixes. Song 2 has much more space, being acoustic guitar-driven, and features lighter performances. As described earlier, the amount of distortion apparent in a mix is dependent on the relative levels of instruments fed through each summing stage. The lighter nature of this song and any changes in the mix ecology is less evident than in the denser Song 1 mix.

7.6   Discussion and Conclusion

Initially my question in this study was to determine if distortion is an expected, rather than unwanted artefact? While this research did not answer that question, it has revealed some valid methodological and theoretical approaches to this question. With only two test songs, this study can be no more than a pilot. With this in mind, I'll discuss some of my findings from this project.

My literature review revealed that low-order harmonic distortion is difficult to audibly perceive. This is evident in my method, substituting 30 plugins from the artist-approved mix to create a clean mix. In the case of Song 2, 14 respondents (all non-professional) claimed to h only subtle differences in the mixes. I proposed using a reductive ecological approach to allow a listener to determine if distortion was an essential part of an ecology of a recorded artefact. While there are some limitations in this study, the approach seems valid and suited to further refinement and a more extensive study.

Song 1 had a clear preference for the mix including distortion by most respondents. In Song 1, I was able to create good comparative mixes. Song 2 also had a preference for the mix including distortion, although with a much smaller margin. The mixes for Song 2 had some overall loudness differences, despite many respondents claiming to notice little difference between the two mixes. Additionally, visible in Figures 7.3, 7.4 and 7.7 it is clear that the use of distortion enabled a more dynamic mix with the same average loudness of the clean mix.

The comparisons between the participants who chose the two artist mixes and those who did not was the most interesting: 55% of the professional participants were able to choose both mixes with distortion, with only two respondents choosing neither mix with distortion. This is most likely due to the respondents reading the survey preamble and then choosing which song had distortion present over their actual preferred mix. Additionally, they would be more trained at hearing the nuances of distortion and have the appropriate playback equipment to detect it. Thoresen's [62] analysis of Schaeffer's typomorphology revealed that perception would correspond to listening intentions and seek pertinent features, just as the trained listeners have done here, whereby the desire for distortion is a personal preference. Alternatively, the non-professional respondents marginally favoured the artist mix for Song 1 and the clean mix for Song 2. This may be due to the loudness discrepancies and the less professional playback devices responding more favourably to a marginally louder mix.

What is striking in this research is the current controllability of distortion. In the analogue world, distortion was an unwanted artefact that was a reluctant part of any signal chain. Digital recording allowed recorded signal to be free from distortion and was quickly dismissed as inferior in the domestic recording market. The currently growing DIY recording paradigm partly owes its success to the controllability of distortion. Its presence in a completed artefact is now personal choice rather than unwilling acceptance. That personal choice is most likely part of a semiotic resource that is useful in the meaning-making specific to each artefact.

Finally, I would like to add that although results of this study were not conclusive, I'd like to use the framework for further work exploring more genres of music, with the goal to involve more respondents. While there is room to improve some aspects of volume matching in the method, as a pilot study, this research will help recording practitioners of the professional and academic world to better understand of the role of distortion in mixing music. Overall, a reductive ecological approach to this study was successful. Additionally, I'd like to add emphasis on the value in conducting these types of experiments in non-laboratory-type conditions. The recording industry has always been at the mercy of the consumer, and it is the consumer ultimately determines the ongoing acceptance of any audio practice.

References

[1] Fenton S, Wakefield J P and Fazenda B. Objective measurement of music quality using inter-band relationship analysis. In *Proceedings of the 130th Audio Engineering Society Convention*, Los Angeles, 2011.

[2] Cole S J. The prosumer and the project studio: The battle for distinction in the field of music recording. *Sociology*, Vol. 45(3), 2011, pp. 447–463.

[3] White P. *Alesis ADAT XT20 digital 8-track recorder*. SOS Publications Group, Cambridge, 1998, http://www.soundonsound.com/sos/may98/articles/adatxt20.html.

[4] Doherty D. *Is a lack of warmth an inevitable trait of digital recording?* SOS Publications Group, Cambridge, 2012.

[5] Milner G. *Perfecting sound forever: An aural history of recorded music*. Farrar, Straus and Giroux, New York, 2009, p. 416.

[6] Bourbon A. Hit hard or go home: An exploration of distortion on the perceived impact of sound on a mix. In *Proceedings of the 12th Art of Record Production Conference Mono: Stereo: Multi*, Stockholm, 2019, pp. 19–36.

[7] Moore A. Dynamic range compression and the semantic descriptor aggressive. *Applied Sciences*, Vol. 10(7), 2020, p. 2350.
[8] Hartmann W M. *Signals, sound, and sensation*. Springer, New York, 2004.
[9] Moore A. An investigation into non-linear sonic signatures with a focus on dynamic range compression and the 1176 fet compressor, Doctoral dissertation, University of Huddersfield, Huddersfield, 2017.
[10] Tingen P. *Spike stent: The work of a top-flight mixer*. SOS Publications Group, Cambridge, 1999.
[11] Parallel distortion—Tchad Blake, Accessed April 2021 from www.youtube.com/watch?v=yi3Cm2DtMmg.
[12] Murray S. Interviews: Tchad Blake. Tape Op, 2000, 16.
[13] Crane L. Marcella Araica: The incredible journey of Ms. Lago. Tape Op, 2019, 133, pp. 14–20.
[14] McIntyre P. Tradition and innovation in creative studio practice: The use of older gear, processes and ideas in conjunction with digital technologies. *Journal on the Art of Record Production*, Vol. 9, 2015.
[15] White P. *Waves scheps omni channel*. SOS Publications Group, Cambridge, 2018.
[16] Chris lord-alge gets put to the test: The quest to create CLA. *MixHub*, Accessed April 2021 from https://sonicscoop.com/2019/03/04/chris-lord-alge-gets-put-to-the-test-the-quest-to-create-cla-mixhub/.
[17] De Man B and Reiss J D. Adaptive control of amplitude distortion effects. In *Proceedings of the 53rd Audio Engineering Society Conference,* Semantic audio, Los Angeles, 2014.
[18] Zagorski-Thomas S. *The musicology of record production*. Cambridge University Press, Cambridge, 2014.
[19] 3 subtle ways to use distortion in your mixes. Accessed April 2021 from *Waves.com*, www.waves.com/subtle-ways-to-use-distortion-in-your-mixes.
[20] Czerwinski E, Voishvillo A, Alexandrov S and Terekhov A. Multitone testing of sound system components' some results and conclusions, part 1: History and theory. *Journal of the Audio Engineering Society*, Vol. 49(11), 2001, pp. 1011–1048.
[21] Toulson R, Paterson J and Campbell W. Evaluating harmonic and intermodulation distortion of mixed signals processed with dynamic range compression. In *Innovation in music*. Future Technology Press, Shoreham-by-Sea, 2014, pp. 224–246.
[22] Moore B C, Tan C-T, Zacharov N and Mattila V-V. Measuring and predicting the perceived quality of music and speech subjected to combined linear and nonlinear distortion. *Journal of the Audio Engineering Society*, Vol. 52(12), 2004, pp. 1228–1244.
[23] On the bench: Transformers, Accessed April 2021 from www.proharmonic.com/articles/AT55%20ON%20THE%20BENCH.pdf.
[24] Whitlock B. Audio transformers. In *Handbook for sound engineers*. Routledge, Oxfordshire, 2001, pp. 285–319.
[25] Campbell W, Paterson J and Van der Linde I. Listener preferences for alternative dynamic-range-compressed audio configurations. *Journal of the Audio Engineering Society*, Vol. 65(7/8), 2017, pp. 540–551.
[26] De Man B. Towards a better understanding of mix engineering, Doctoral Dissertation, Queen Mary University of London, London, 2017.
[27] Goold L and Graham P. The uncertain future of the large-format recording studio, In *Proceedings of the 12th Art of Record Production Conference Mono: Stereo: Multi*, Stockholm, 2019, pp. 119–136.
[28] Leyshon A. The software slump?: Digital music, the democratisation of technology, and the decline of the recording studio sector within the musical economy. *Environment and Planning A*, Vol. 41(6), 2009, pp. 1309–1331.
[29] Théberge P. The end of the world as we know it: The changing role of the studio in the age of the internet. *The Art of Record Production: An Introductory Reader for a New Academic Field*, 2012, pp. 77–90.
[30] Homer M. Beyond the studio: The impact of home recording technologies on music creation and consumption. *Nebula*, Vol. 6(3), 2009, pp. 85–99.
[31] Enderby S and Baracskai Z. Harmonic instability of digital soft clipping algorithms. In *Proceedings of the 15th International Conference On Digital Audio Effects (DAFx-12)*, York, 2012.

[32] Saturation: The secret weapon of mixing (+ 5 best free saturation plugins), Accessed April 2021 from www.audio-issues.com/music-mixing/saturation-the-secret-weapon-of-mixing-5-best-free-saturation-plugins/#disqus_thread.
[33] Chalupper J. Aural exciter and loudness maximizer: What's psychoacoustic about" psychoacoustic processors"? In *Proceedings of the 109th AES Convention*, Los Angeles.
[34] Loudness war, Accessed April 2021 from https://en.wikipedia.org/wiki/Loudness_war.
[35] IFPI. Music listening 2019, Accessed April 2021 from https://www.ifpi.org/wp-content/uploads/2020/07/Music-Listening-2019-1.pdf, 2019.
[36] Using templates for tracking and mixing will speed up your workflow—Learn from Steve Genewick, Mick Guzauski and Sylvia Massy, Accessed April 2021 from www.pro-tools-expert.com/production-expert-1/2019/9/13/using-templates-for-tracking-and-mixing-will-speed-up-your-workflow-dont-just-take-our-word-for-it.
[37] Keogh B. On the limitations of music ecology. *Journal of Music Research Online*, Vol. 4, 2013.
[38] Archer W K. On the ecology of music. *Ethnomusicology*, Vol. 8(1), 1964, pp. 28–33.
[39] Gibson J J. *The ecological approach to visual perception*. Psychology Press, New York, 1986.
[40] Clarke E F. *Ways of listening: An ecological approach to the perception of musical meaning*. Oxford University Press, Oxford, 2005.
[41] Schafer R M. *The soundscape: Our sonic environment and the tuning of the world*. Destiny Books, Rochester, VT, 1993.
[42] Lefford M N. Information, (inter) action and collaboration in record production environments, In *The Bloomsbury Handbook of Music Production*. Bloomsbury Publishing, London, 2020, pp. 145–160.
[43] Ecology, Accessed April 2021 from www.dictionary.com/browse/ecology?s=t.
[44] Bourbon A and Zagorski-Thomas S. The ecological approach to mixing audio: Agency, activity and environment in the process of audio staging. *Journal on the Art of Record Production*, Vol. 11, 2017.
[45] Bourbon A and Zagorski-Thomas S. Sonic cartoons and semantic audio processing: Using invariant properties to create schematic representations of acoustic phenomena, In: *2nd Audio Engineering Society Workshop on Intelligent Music Production*, London, 2016.
[46] Stake R E. *Qualitative research: Studying how things work*. Guilford Press, New York, 2010.
[47] Moore A. All buttons in: An investigation into the use of the 1176 FET compressor in popular music production. *Journal on the Art of Record Production*, Vol. 6, 2012.
[48] Lipshitz S P and Vanderkooy J. The great debate: Subjective evaluation. *Journal of the Audio Engineering Society*, Vol. 29(7/8), 1981, pp. 482–491.
[49] Fetters M D, Curry L A and Creswell J W. Achieving integration in mixed methods designs—principles and practices. *Health Services Research*, Vol. 48(6pt2), 2013, pp. 2134–2156.
[50] Lincoln Y S and Guba E G. *Naturalistic inquiry*. Sage, Los Angeles, Vol. 75, 1985.
[51] Denzin N. *The research act: A theoretical introduction to sociological methods*. Transaction Publishers, New York, 1973.
[52] Youlean, Youlean Loudness Meter 2—MANUAL, Accessed April 2021 from https://youlean.co/wp-content/uploads/2019/04/Youlean-Loudness-Meter-2-V2.2.1-MANUAL.pdf.
[53] Graham P. Paradigmatic considerations for creative practice in Creative Industries research: The case of Australia's Indie 100. *Creative Industries Journal*, Vol. 9(1), 2016, pp. 47–65.
[54] Fab Filter Home page, Accessed on April 2021 from www.fabfilter.com/.
[55] DC8C advanced compressor, Accessed on April 2021 from https://klanghelm.com/contents/products/DC8C.html.
[56] L1 ultramaximizer, Accessed on April 2021 from www.waves.com/plugins/l1-ultramaximizer.
[57] Oxford Inflator V3, Accessed on April 2021 from https://www.sonnox.com/plugin/oxford-inflator.
[58] Andrew scheps mixing template tutorial & session files [Trailer], Accessed on April 2021 from www.youtube.com/watch?v=WyJe7VB1k3Q.
[59] Applewood magoo, Accessed April 2021 from https://soundcloud.com/applewoodmagoo.
[60] An ecological approach to distortion in mixing audio: Survey, Accessed April 2021 from https://forms.office.com/Pages/ResponsePage.aspx?id=UOywkz_87UeTRZ_8YBJUHOZbPU7rTp1Dh6VNQAl-M4lUQkVSVjJDS1A1VEpIQktDOVg5MTAxNUQ4TS4u.

[61] De Man B and Reiss J D. APE: Audio perceptual evaluation toolbox for MATLAB. In *Proceedings of the 136th Audio Engineering Society Convention*, Los Angeles, 2014.
[62] Thoresen L and Hedman A. Spectromorphological analysis of sound objects: An adaptation of Pierre Schaeffer's typomorphology. *Organised Sound*, Vol. 12(2), 2007, pp. 129–141.
[63] Taylor R W. Hyper-compression in music production; agency, structure and the myth that louder is better. *Journal on the Art of Record Production*, Vol. 11, 2017.

8 Towards a Lexicon for Distortion Pedals

Tom Rice and Austin Moore

8.1 Introduction

When Grady Martin's broken amplifier produced the now ubiquitous, distorted guitar tone at a Marty Robbins session in 1961, he accidentally stumbled on an effect that many would attempt to emulate [1]. Merely a year later, the session engineer, Glenn Snoddy, developed the effect into a pedal format and sold the design to the Gibson Guitar Corporation, who produced the first commercial distortion pedal, the Gibson Maestro FZ-1 Fuzz-Tone. Snoddy and Hobbs' germanium transistor circuit, powered by two 1.5v batteries, was pioneering, and an entire generation of guitarists rushed to obtain this radical new effect. When Keith Richards used the FZ-1 in the Rolling Stones' 1965 hit 'I Can't Get No (Satisfaction)', its influence grew exponentially.

Now, 60 years after Martin's accidental foray into distortion, it is an effect that is omnipresent and widely available to musicians of any level. In the timescale since its inception, four specific pedals have grown in stature, to the point of iconic status, and established themselves as benchmarks of the effect: Ibanez Tube Screamer, Boss DS-1, ProCo RAT, and Electro Harmonix Big Muff. These pedals have been used by guitar players including Stevie Ray Vaughan, Kurt Cobain, Dave Grohl, and Dave Gilmour and range the full gamut of distortion from subtle overdrive to intense fuzz. Thus, the authors of this chapter propose that their sonic signature is worthy of academic study. Additionally, these pedals utilise different forms of clipping to achieve distortion, making a comparison of the sound quality intriguing from objective and subjective perspectives. The Tube Screamer employs symmetrical soft clipping, the DS-1 uses asymmetrical hard clipping, the RAT implements symmetrical hard clipping, and the Big Muff features two cascading soft clipping stages. Comprehensive circuit analysis of the four pedals is beyond the scope of this chapter but can be found on the Electro Smash website [2]. The current study investigates the sound quality of the four pedals by text mining online sources with a focus on semantics. Subsequently audio examples of the pedals are used in a qualitative experiment to create a lexicon of descriptors, related to the four distortion pedals and distortion in general. This study aims to gather descriptors relating to the distortion characteristics of each pedal and ascribes an objective meaning to these terms. The research will benefit musicians, academics, and pedal developers in the hardware and software domains by providing them with a better understanding of the lexicon.

8.2 Literature Review

A sizable portion of the current literature concerning guitars and distortion is in the domain of metal studies or focuses on metal genres. Often this work investigates 'heaviness' and how it is inherently tied to the timbre of the distorted rhythm guitar. The work of Berger [3] demonstrates the extent of distortion's influence on genre definition. Lilja [4] further validates these notions, as well as expanding upon them, and of special interest is the assertion that distortion intrinsically defines power chords. The research demonstrates how distortion has permeated deeply into the fabric of popular music and that basic musical components are viewed as incomplete without it. Overall, the study provides insight into the practical usage of distortion and its contemporary place within popular music theory, specifically metal and its subgenres. Herbst's work [5, 6] falls into a similar thematic domain and directly relates to Lilja's research into power chords and the effect of distortion on their perceptual quality. Herbst establishes that harmonically simple chords produced less unpleasant acoustic features than harmonically complex chords when played with distortion. He posits that this may be one of the reasons harmonically simple chords (particularly power chords) are commonplace within rock music styles.

Other related research explores electronic circuit design. Sunnerberg [7] provides a comprehensive overview of analogue distortion circuits. The study offers a thorough analysis of various distortion circuits, including differentiating and comparing germanium diode limiters and class B push/pull non-linearities. Building upon these ideas is the work by Murthy et al. [8], which is more focused in its approach but details an analysis of general distortion circuits. This includes an articulate breakdown of the components comprising a three-stage distortion circuit. Schneiderman and Sarisky [9] concentrate specifically on the Boss DS-1, a pedal also investigated in this chapter. They explore the objective changes afforded by modifying the circuit of the pedal and the effect these changes have on guitar players' performance. Their conclusions are pertinent to this study as they argue for the development of a realistic language to describe distortion.

Moving from circuit design into perception, two studies by Tsumoto [10, 11] are arguably the most closely related to this chapter. The identification of terms such as thin-thick, sharp-dull, and dark-bright as perceptual descriptors is rudimentary but a good indicator for the current study's validity. Like Schneiderman and Sarisky, Tsumoto contends that further work should be done to define the sonic qualities of distortion.

Thus, this chapter fills a gap in the literature by investigating the semantics of distortion pedals and developing a lexicon to describe their sonic signature.

8.3 Overarching Methodology

The use of grounded theory [12] provided a fundamental underpinning for this research. The intent was to devise a set of descriptive terms for distortion pedals, redefined to be contextually accurate and forming a lexicon for distortion primarily pertaining to guitar pedals. However, this can easily be expanded to all forms of audio distortion. The value of grounded theory became apparent early in the research. It is an inductive method, leading to conclusions not being drawn until the end of the study. They are based on retrospective observations made throughout the process and are linked with the overarching intent of the research. The adaptability of grounded theory also made it an attractive proposition; the methodology had room to adapt to fit where the investigation was heading. Therefore, a grounded theory approach was deemed the most suitable theoretical methodology for this research. This method was embellished by Empirical Discourse Analysis (EDA), the purpose of which is to critically analyse the function of language in social or genre-specific settings [13].

8.4 Study 1: Content Analysis and Text Mining Method

The first study established the key communicative language used by musicians when describing the sonic properties of the four distortion pedals. Content analysis was deemed as the most effective way of collecting this data due to its ability to systematically summarise written communication in a quantitative manner [14] and was achieved through text mining. To begin, a data retrieval method called 'web scraping' [15] was used to obtain large amounts of semantic content, which would enable subsequent analysis. The chosen material was gathered from several sources including academic texts, equipment reviews, and social media posts. The choice to target a range of sources was deliberate, as the authors' aim was to gather an expansive and diverse vocabulary, including populist colloquialisms, as well as the academic equivalents.

To analyse the textual data, RStudio and various relevant packages [16] were used. A bespoke script scraped the online sources, and the authors manually sifted through the texts to extract descriptors. In addition, the texts were read fully by the authors to get a better understanding of the contextual meaning of the descriptors. Subsequently, RStudio was used to compute all the amalgamated text data and perform frequency analysis, which helped quantify many of the terms, in respect to applicability, based on their usage frequency. The aim of the word frequency analysis was to discover common parlance; therefore, only words counted more than once were included in the amalgamated lexicon. However, single-use descriptors were also recorded and stored separately for further inspection as they may reveal interesting anomalous data or relevant synonyms for use later. Once the general lexicon had been established, the semantic body was subdivided by contextual source, in this case by pedal. These new subsets were then re-analysed to discover pedal-specific descriptors. Once the general and pedal-specific word frequency analysis was completed, manual adjustment was required to remove descriptors which were not describing the sonic character of the sound. Examples of such words are 'expensive' and 'versatile', which are not indicative of any specific sonic trait.

8.4.1 Initial List of Descriptors

Figure 8.1 reveals 'crunchy' is the most popular descriptor, and it became the focus of some speculation, specifically, why it is so common. One reason may be due to the crunch channel found on many amplifiers. Typically, this channel yields a lower gain style of distortion (as opposed to the high-gain lead channel), so it is plausible this is at least one of the reasons for the descriptor's ubiquity. The popularity of 'noisy' also proved to be of some interest, although it was suspected it was linked to specific pedals or specific traits only present in select devices, such as a poor signal-to-noise ratio. This was confirmed by analysis of the textual sources, which revealed that 'noisy' was indeed used to describe a poor signal-to-noise ratio, particularly under high amounts of gain with the Boss DS-1. However, there were also instances where this term was used to describe the character of distortion, which is unsurprising given its noise-like characteristics. Warm is the third most common term, which again is foreseeable as this term is standard vernacular when describing all forms of analogue audio equipment. However, despite some informal agreement on its meaning (an emphasis on the low end with a balanced or attenuated top end), no work has been done to objectively define this term accurately. Similarly, 'dirty', while also somewhat of an expected term, is vague and lacks objective meaning. One can posit its origins come from the fact that distortion has coloured and tainted an audio signal, but is 'dirty' distortion the same as 'crunchy' or 'warm' distortion? Analysis of the textual sources showed that 'dirty' is associated with a high-gain style of distortion that is close to 'grunge' music rather than high-gain heavy metal. 'Creamy' is also highly subjective, and from a sensory perspective one could argue this term is describing a smooth, full, rich distortion much like one might describe food or drink with a creamy consistency. Analysis of online discussions [17, 18] and the textual sources used for text mining suggest this is the case. A 'creamy' distortion is smooth and full with no harshness or graininess. Moreover, this style of distortion is

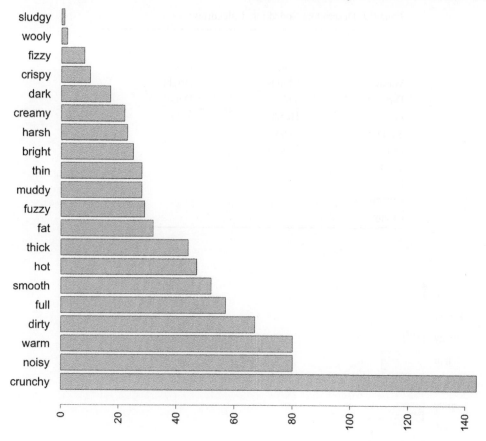

Figure 8.1 List of descriptors from text mining.

associated with low to moderate gain and, from a performer's perspective, manifests in single-note lines and solos.

Table 8.1 codes the descriptors into three categories, relating to positive descriptors, negative descriptors, and descriptors which could be positive or negative depending upon the context. Coding the descriptors in this way potentially highlights the nature of distortion which guitar players prefer. From looking at the table, it appears that undesirable distortion is excessive in the top end ('thin', 'harsh', 'fizzy', possibly 'noisy') and results in a lack of definition in the mid-range ('muddy' and 'wooly'). On the other hand, it could be posited that desirable distortion is focused on the mid-range ('crunchy' and 'crispy'), has bottom end weight ('warm', 'full', 'thick', 'fat'), and is of low to moderate gain and lacking in a coarse timbre ('dirty', 'smooth', 'creamy', and 'hot'). However, more work needs to be done to quantify these terms, and the authors have explored this research in an impending study.

8.4.2 The Tube Screamer Descriptors

Figure 8.2 shows the word frequency chart for the Tube Screamer with 'crunchy' as the most regularly used. This term is subjective in definition and tricky to quantify. Therefore, there is a paradoxical element: why is a descriptor that lacks contextual meaning and is individually

Table 8.1 Descriptors Coded Into Categories

Positive	Negative	Positive or Negative
Crunchy	Noisy	Fuzzy
Warm	Muddy	Bright
Dirty	Thin	Dark
Full	Harsh	Sludgy
Smooth	Fizzy	
Hot	Wooly	
Thick		
Fat		
Creamy		
Crispy		

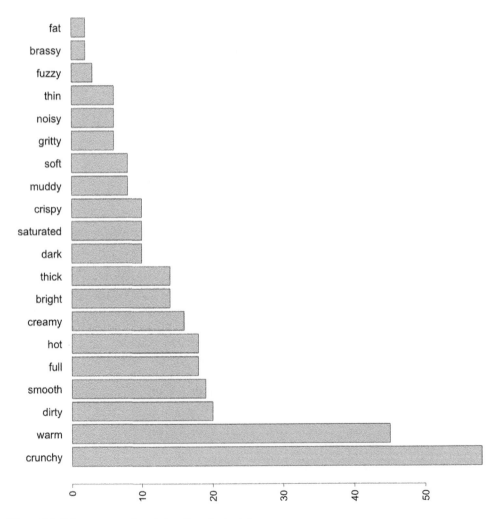

Figure 8.2 Tube screamer descriptors from text mining.

Towards a Lexicon for Distortion Pedals 133

subjective used so frequently, as opposed to more specific terms? As noted previously, its popularity is in part due to amp manufacturers such as Marshall and Peavey labelling low-drive channels on their amps as crunch channels. But it is also highly likely this term was employed by amp manufacturers because of its popularity in the first instance. The term is a good example of onomatopoeia, which potentially goes some way to explaining the populist preference over more technical terminology. The 'ch' digraph in 'crunch' has obvious phonetic ties to the sound of a distorted guitar played with palm muting; the phonetic pronunciation of the voiceless postalveolar affricate mimics its sonic signature. Other examples of onomatopoeic terms that emerged from text mining include 'raspy', 'fizzy', 'boomy', and 'growly'. Aside from 'crunchy', the most frequent descriptor for the Tube Screamer is 'warm', and from the authors' perspective, this is a reasonably accurate description of the pedal's sound. Generally, the Tube Screamer, which uses soft clipping, is seen as the least aggressive of the four pedals in this study, and so the choice of 'warm' from both a phonetic and descriptive perspective seems apt. Equally popular descriptive terms for this pedal included 'smooth', 'full', and 'hot', which are again descriptors indicating a (paradoxically) refined distortion.

8.4.3 The Boss DS-1 Descriptors

Figure 8.3 shows the ten most popular descriptors for the Boss DS-1. As evidenced, 'noisy' is the prevalent descriptor, possibly due to the amount of noise this pedal generates, particularly in higher gain settings. Thus, this descriptor is likely to describe the pedal's signal-to-noise ratio, a recognised

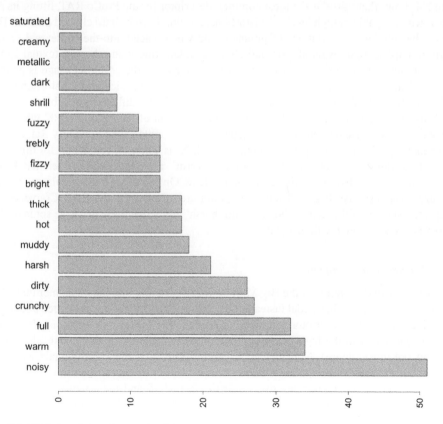

Figure 8.3 DS-1 descriptors from text mining.

facet of its sonic signature, albeit a negative feature. The popularity of 'harsh', 'hot', and 'muddy' further reinforce the impression that this pedal holds certain negative sonic traits and may relate to its hard clipping approach. 'Warm' was the second most popular descriptor for the DS-1, but it emerged much more regularly than any other similar terms such as 'dark', for example, which was used less than five times. This irregularity, coupled with the emergence of the term 'harsh', a stark contrast to notions of warmth, lends credence to the idea that 'warm' may have been used out of familiarity, simply because this word is common parlance when describing virtually any distortion effect. One could argue the word 'warm' is the default descriptor for distortion of all kinds and is put forward with little consideration to its actual meaning. The authors investigated the textual data for the DS-1 and discovered that 14% of the time 'warm' was used to highlight the pedal did not exhibit this quality. Therefore, when taking this into account, 'warm' drops behind 'full' to become the third most used descriptor. Without further testing the reason for this term's popularity is speculative, but it does appear out of place when compared with 'harsh', 'bright', 'fizzy', 'trebly', and 'shrill'. One possible reason is due to the DS-1's tone control. Looking at a circuit analysis [19] illustrates how this control creates a scooped mid-range around the 500Hz point. Perceptually this results in the low and high end of the frequency spectrum sounding more present. Therefore, depending on the position of the tone control, the pedal may exhibit more warmth or more harshness (and fizziness and brightness). The Tube Screamer and ProCo RAT both use a tone circuit which accentuates the mid-range. Therefore, this appears to be one of the reasons why terms pertaining to brightness and harshness were not as frequently used for these pedals.

8.4.4 The ProCo RAT Descriptors

Figure 8.4 shows that 'dirty' is the most common descriptor for the ProCo RAT, fitting its reputation as a unit capable of high levels of distortion, resulting from its hard clipped diodes. 'Dirty' naturally has obvious connotations of impurity, and when placed into the contextual realm of distortion, implies high levels of gain and clipping. Interestingly, in other contextual spheres, 'dirty' is an intrinsically negative word, but when referring to distortion its usage is generally used to indicate a positive aspect of the timbre. This fits with the basic idea of distortion that one is deliberately impurifying an audio signal to create a radically different yet subjectively positive change in sound. Intriguingly, ProCo released a version of the pedal in 2002 which had a clean and dirty switch and released a subsequent version in 2004 called You Dirty Rat. So, along with its name (RAT), this may partly explain why 'dirty' is common vernacular for this pedal.

Closely trailing 'dirty' are 'thick', 'smooth', 'warm', and 'fat', which infer that ProCo's pedal creates an objectively positive distortion palette. On top of this, the regularity of 'thick' and 'fuzzy' suggests that higher levels of clipping are afforded by the RAT. Conversely, the infrequent instances of the terms 'abrasive' and 'harsh' imply that these negative sonic traits are not commonly associated with this pedal.

8.4.5 The Big Muff Descriptors

The most frequent descriptor for the Big Muff is 'fuzzy', leaning into the misnomer that Electro Harmonix's creation is a fuzz pedal instead of the distortion/sustainer it was created and labelled as. The inclusion of the descriptor 'grinding' (recorded only twice) was of passing interest and may be connected with the high-gain, fuzz-like distortion offered by this pedal (and possibly the Grindcore music genre). Nonetheless, it demonstrates that many individuals will extend their vocabulary outside of popular terms to attempt to describe subjective sonic traits. 'Bassy' and 'thick' emerged as popular descriptors for the Big Muff, and these words likely describe the pedal's distortion character, which is achieved by limiting the higher-order harmonics and

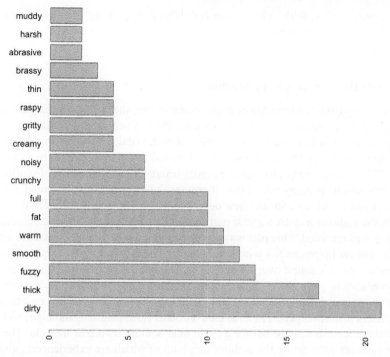

Figure 8.4 DS-1 descriptors from text mining.

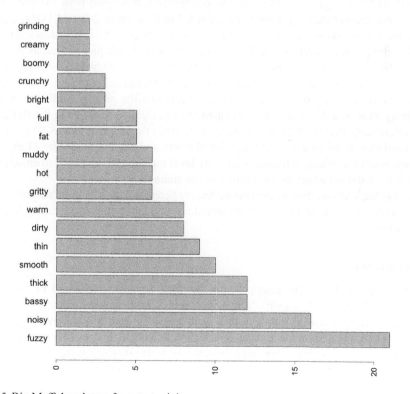

Figure 8.5 Big Muff descriptors from text mining.

boosting the lows and low mids. Also of note is the fact that 'crunchy' is much less common for the Big Muff. Again, this is due mainly to its clipping circuit, where the filtering is not as mid-range focused as the others.

8.5 Study 2: Descriptive Survey Method

The previous study gathered descriptors from sources where the online commenters were probably describing each pedal's sound from memory. Thus, a second study was devised, which presented participants with blind audio examples of each pedal. The audio was presented in the form of an online survey, with embedded audio clips and comment boxes for participants to add descriptors. This second study also afforded cross-reference with the results from the previous study, and this was its primary role. Thus, if any results were repeated, it would validate their existence as a part of the lexicon, and new descriptors could be added accordingly.

To create the audio examples, a guitar part consisting of power chords, open chords, and single-note lines was recorded. This part was performed twice on a Fender Telecaster with single-coil pickups and an Epiphone SG with humbuckers. The musical part's varying nature aimed to showcase the pedal's sound over a range of typical rock performance styles, utilizing open chords, power chords, palm mutes, and single-note lines. The two pickup styles were recorded to offer participants a choice of tonal presentation. The recordings were then re-amped through the four different distortion pedals at three gain amounts (low, medium, and high) and input into a Fender Deluxe Reverb Reissue set for a clean tone with no reverb or tremolo. The three different gain amounts were set by the authors' ear, both of which are experienced guitar players. The motivation behind the different gain settings was to observe how descriptors change as a function of gain. The signals were captured with the industry standard Shure SM57 microphone and an API 3124+ mic preamp. The mic was positioned half an inch from the grill cloth and slightly off the speaker dust cap centre and recorded to Protools at 24bit 44.1kHz. During the survey, each subject was presented with 12 audio pieces (each of the four pedals recorded with three gain settings) and asked to provide three descriptors for each piece of audio. As listening fatigue can be an issue, the participants were not presented with both pickup recordings. Instead, the subjects were offered the choice between single-coil or humbucker pickups at the start of the survey and were encouraged to select the more aurally familiar pickup. Additionally, the start of the survey showed a disclaimer that specified the descriptors should be in the sonic domain and not referencing irrelevant factors such as cost, aesthetic, or genre usage. The survey was designed and shared online using the Qualtrics platform and was completed by 11 experienced guitar players internationally. All audio files were level matched to -15 LUFS to ensure that differences in level did not affect the perception of the audio files.

The following sections discuss the results first by pedal and then by gain amount. Finally, a general discussion of the descriptors is presented with an aim to ascribe meaning to the more nebulous terms.

8.5.1 Results by Pedal

'Crunchy' is the most popular descriptor for the Tube Screamer, followed by 'thin', 'compressed', 'warm', 'mids', 'low-gain', 'dirty', and 'fizzy'. Despite descriptors such as 'muffled' and 'muddy' being used a small number of times, it can be argued that words evoking a low-gain style of distortion, focused on the mid-range, are associated with this pedal's sound quality. The Tube Screamer's soft clipping distortion relates to words that denote a softer type of distortion. Except for 'fizzy' and 'fuzzy', few words suggest an abrasive sonic signature. Moreover, some of the single-count terms include 'mild', 'easy', 'natural', 'soft', and 'laidback', which again represents a subtler form of distortion. It is interesting to consider these results, as it should be kept in mind that participants in the experiment were presented with the audio in a blind test scenario. Thus, they did not have preconceived expectations about sound quality based on the name of the pedals. Comparison of the descriptors between this study and the text mining reveal some consistency.

Conversely, the DS-1 pedal is associated with descriptors that evoke a harder, more grainy timbre. 'Fizzy' is the most used descriptor, followed by 'saturated', 'distorted', 'metallic', 'warm', 'scooped', 'dark', 'driven', 'rich', 'high-gain', and 'fat'. These descriptors point toward a bright, high-gain style of distortion that is harmonically dense due to hard clipping. 'Warm' may seem like the antithesis of this sound, but it was only used to describe this pedal with low-gain settings, whereas 'fizzy' was used across all settings. As with the Tube Screamer, the descriptors gathered in this study and the text mining are similar. It is also worth noting that under this test, 'crunchy' was not used to describe the sound quality of the DS-1 audio examples. In the previous study, it was the fourth most popular descriptor, and one reason for this may be because of these terms' ubiquity. Consequently, the terms used for each pedal in this test may be more accurate descriptions of their sonic signature. Recalling the sonics of a piece of equipment

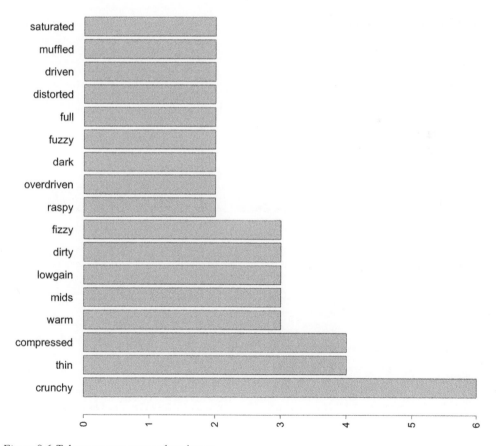

Figure 8.6 Tube screamer survey descriptors.

138 *Tom Rice and Austin Moore*

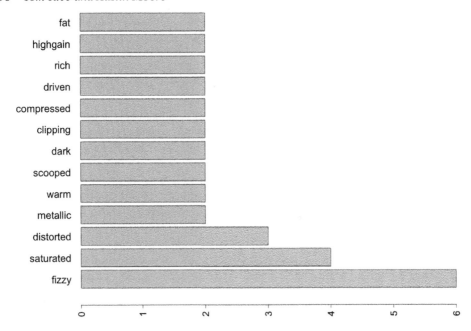

Figure 8.7 DS-1 survey descriptors.

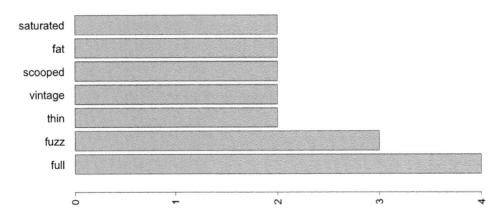

Figure 8.8 Rat survey descriptors.

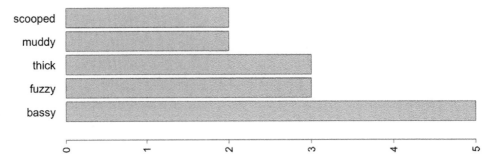

Figure 8.9 Muff survey descriptors.

from memory has many biases that will influence the result. These biases are not present, however, under double-blind testing.

'Full' is the prevalent descriptor for the Rat pedal, followed by 'fuzz', 'thin', 'vintage', 'fat', and 'saturated'. 'Thin' was only used to describe the low-gain setting of this pedal, and 'fuzz' described the high-gain settings. Therefore, it can be proposed that this pedal's sonic signature under moderate gain settings is perceptually full, which thins out or saturates into fuzz under light and high-gain settings. Notably, 'crunchy' and 'fizzy', the most used descriptors for the Tube Screamer and DS-1 respectively, were used only once to describe the Rat and only under low-gain settings. Thus, it appears that the Rat does not exhibit as much mid-range emphasis (crunch) or the high-order distortion artefacts (fizz) manifested in the Tube Screamer and DS-1. It is worth considering that the descriptor 'dirty' was not used at all in this study, whereas in the text mining study, it was the most common term. Therefore, it can be posited that this term was used more out of familiarity than as an actual true description of this pedal's sound quality when given an audio reference.

The Muff's sonic signature was described mostly with the word 'bassy', followed by 'fuzzy', 'thick', 'muddy', and 'scooped'. No descriptors suggested a bright timbre (such as 'fizzy') or mid-range focus (such as 'crunchy'); therefore, the Muff's distortion character is summarised as a thick, fuzzy distortion focused on the low end of the frequency spectrum. The descriptors from both studies are broadly in line with one another.

8.5.2 Results by Gain Setting

Initially, it was hoped that analysis could be broken down into the three different gain amounts per pedal. However, a comprehensive analysis was not possible as some pedal/gain combinations had low response rates. Therefore, it was decided to combine the results for all four pedals to create a new dataset of descriptors for low-, medium-, and high-gain settings. While this approach has its limitations (largely due to the variation in maximum gain offered by the pedals), it reveals some insight into how the respondents' vernacular changed as a function of gain.

The low-gain settings show that the words 'warm', 'overdrive', and 'crunchy' are the most popular, illustrating that low-gain distortion is associated with a 'crunchy' sound quality that is not overly bright. Listeners could likely discern the low-gain audio examples and were using 'overdrive', a word commonly associated with low-gain distortion, out of familiarity. The word 'driven' is the prevalent descriptor for the medium-gain settings, and again this may be used out of familiarity (some guitar amps offer a medium-gain 'drive' channel). 'Full', 'compressed', and 'fizzy' were also commonly used and reveal some interesting insights. It appears the listeners could discern that aggressive clipping can compress the signal and increase harmonics, especially higher-order harmonics. However, one can argue that 'full' is the only accurate subjective descriptor of all the terms collected for medium-gain settings. Thus, a moderately clipped guitar signal with compression and a broad range of harmonics is perceived as a full guitar tone. It is curious to consider that high-gain guitar sounds in this study were not perceived as full. One would think that perceived fullness rises as a function of gain, but here this is not the case. The most common descriptors for high-gain settings are 'saturated' and 'fuzzy'; 'full' was not used, but 'thick' did appear twice. It should be stated that 'fuzzy' was not only used for the Big Muff pedal as one might have expected. It was used several times to describe all pedals under high gain. Therefore, the results suggest that certain high-gain guitar tones are perceived as saturated and fuzzy.

140 Tom Rice and Austin Moore

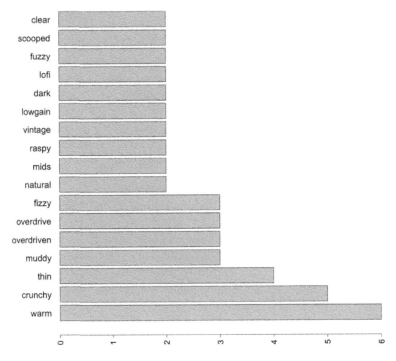

Figure 8.10 Low-gain survey descriptors.

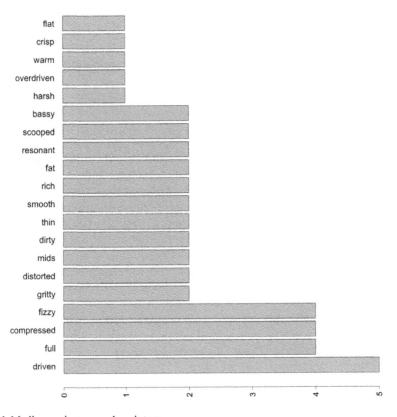

Figure 8.11 Medium-gain survey descriptors.

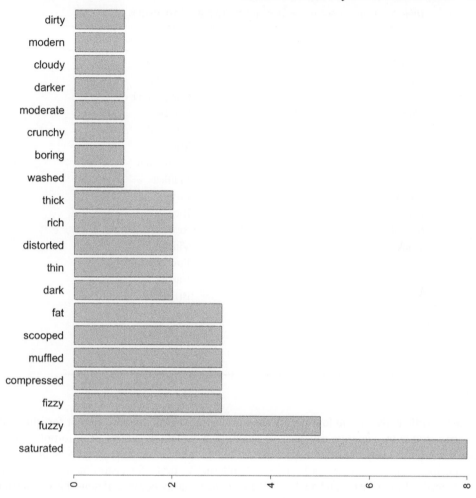

Figure 8.12 High-gain survey descriptors.

8.5.3 Comparison of Descriptors

When comparing the two studies, several terms like 'crunchy', 'warm, and 'fuzzy' emerged in both. Moreover, the high frequency of these descriptors in both studies corroborated the notion that these terms were ubiquitous when describing the sonic signature of the four pedals. The popularity of some descriptors changed between the two studies, and the reasons for this are likely to be multi-variate: presentation of audio in one study and not in the other, the potential for a more critical perspective when listening to audio examples, the prospect that listeners were comparing audio files against one another. These final two examples could explain why the words 'fizzy' and 'thin' are the second and third most prominent descriptors in study two, while they are much less favoured in study one. However, the primary role of the study was to cross-validate descriptors to affirm their importance, so in this sense, their popularity ranking is of little meaning. Table 8.2 presents the descriptors in two different groups, common to both studies and unique to one study. One can posit the terms common to both studies have more

Table 8.2 Comparison of the Descriptors from the Two Studies

Common to Both Studies	Unique to One Study
Crunchy	Saturated
Fizzy	Scooped
Fuzzy	Compressed
Thin	Driven
Warm	Distorted
Full	Mid
Dark	Overdriven
Fat	Muffled
Muddy	Bassy
Dirty	Rich
Smooth	Raspy
Thick	Vintage
	Low-gain
	Harsh
	Bright
	Hot
	Crispy
	Creamy
	Noisy

credence, particularly for the four pedals used in this study. Additionally, except for 'thin', all the terms used in the common category are attempts to explain the timbre of distortion. On the other hand, only 'raspy', 'hot', 'crispy', and 'creamy' from the unique category appear to be attempts to do this. The other words are merely alternative terms for distortion or highlighting other objective traits of its sound quality that are not necessarily because of non-linear distortion (compressed and scooped, for example).

8.5.4   Nebulous Descriptors

A small number of terms emerged that were used infrequently but were interesting enough to be worth exploring. Firstly, 'brassy' and 'tubular'; both terms have apparent ties to brass musical instruments, which, when played loudly, are known for their projection capabilities and potentially harsh, piercing timbres. Moreover, brass instruments produce odd-order harmonics that are associated with square waves, which is in turn associated with clipping. Thus, these terms are describing the brassy nature of the pedals that produce odd-order harmonics and an inspection of the descriptors revealed this to be true. The terms were used to describe the Tube Screamer and Muff, which both produce odd-order harmonics. Similarly, the term 'raspy' was used several times in the survey, and this term can again be associated with wind instruments, this time the clarinet. 'Raspiness' is related to many aspects of the clarinet's timbre, and often it can be a positive description of the instrument's timbre, but it is also used to describe a poorly played note or issues with the reed, which can result in a distorted and noise-like tone. Therefore, it is understandable why 'raspy' has been used in the context of distortion pedals. Again, it is worth considering that the clarinet produces primarily odd-order harmonics, particularly in low registers.

8.6 Conclusions and Further Work

This research set out to develop a working lexicon for distortion pedals, and the results presented in this chapter are a solid starting point. The two studies used different methodologies, but several common descriptors suggest that many of the terms are common parlance among guitar players and sound engineers. However, these terms are often highly subjective, and this chapter has proposed objective meanings for several of the descriptors, particularly some of the vaguer examples. Of particular interest are the different descriptors used to elucidate (often via onomatopoeia) the various styles of clipping and filtering and how some of the terms appear to be describing positive and negative sonic traits. Finally, the descriptors gathered will provide scholars and equipment developers with a rich lexicon of distortion, which can be used for future studies and industry-focused research and development. As noted previously, the authors of this chapter have completed further work, which used a consensus vocabulary development process [20] to develop more robust definitions of the descriptors and a unique 'distortion wheel' similar to the SCAA and WCR Coffee Taster's Flavor Wheel. The work will be published in due course.

References

[1] Kosser, M.: *How Nashville became Music City, U.S.A: 50 Years of Music Row*. Hal Leonard, Milwaukee, WI (2006)
[2] ElectroSmash—Electronics for audio circuits, www.electrosmash.com/
[3] Berger, H.M.: *Metal, Rock, and Jazz: Perception and the Phenomenology of Musical Experience*. Wesleyan University Press, Middletown, CT (1999)
[4] Lilja, E.: Theory and analysis of classic heavy metal harmony. In: *Advanced Musicology*, Vol. 1, IAML Finland, Helskini (2009)
[5] Herbst, J.-P.: Distortion and rock guitar harmony: The influence of distortion level and structural complexity on acoustic features and perceived pleasantness of guitar chords. *Music Perception: An Interdisciplinary Journal*. 36, 335–352 (2019)
[6] Herbst, J.-P.: Heaviness and the electric guitar: Considering the interaction between distortion and harmonic structures. *Metal Music Studies*. 4, 95–113 (2018)
[7] Sunnerberg, T.D.: Analog musical distortion circuits for electric guitars, Thesis, Rochester Institute of Technology (2019). Accessed from https://scholarworks.rit.edu/theses/10066
[8] Murthy, A.A., Rao, N., Beemaiah, Y.R., Shandilya, S.D., Siddegowda, R.B.: Design and construction of arduino-hacked variable gating distortion pedal. *IEEE Access*. 2, 1409–1417 (2014)
[9] Schneiderman, M., Sarisky, M.: Analysis of a modified boss DS-1 distortion pedal. In: *Audio Engineering Society Convention 126*. Audio Engineering Society, New York (2009)
[10] Tsumoto, K., Marui, A., Kamekawa, T.: The effect of harmonic overtones in relation to "sharpness" for perception of brightness of distorted guitar timbre. In: *Proceedings of Meetings on Acoustics 172ASA*. p. 035002. Acoustical Society of America (2016)
[11] Tsumoto, K., Marui, A., Kamekawa, T.: Predictors for the perception of "wildness" and "heaviness" in distorted guitar timbre. In: *Audio Engineering Society Convention 142*. Audio Engineering Society, New York (2017)
[12] Strauss, A., Corbin, J.M.: *Grounded Theory in Practice*. Sage Publications, Thousand Oaks, CA (1997)
[13] Phillips, N., Hardy, C.: *Discourse Analysis: Investigating Processes of Social Construction*. Sage Publications, Thousand Oaks, CA (2002)
[14] Hsieh, H.-F., Shannon, S.E.: Three approaches to qualitative content analysis. *Qualitative Health Research*. 15, 1277–1288 (2005)

[15] Zhao, B.: Web scraping. *Encyclopedia of Big Data*. 1–3 (2017)
[16] Feinerer, I.: An introduction to text mining in R. *R News*. 8, 19–22 (2008)
[17] How to make your guitar sound creamy: 5 more confusing audio terms, explained, https://blog.sonicbids.com/how-to-make-your-guitar-sound-creamy-5-more-confusing-audio-terms-explained
[18] Overdrive and distortion, www.gilmourish.com/?page_id=262
[19] ElectroSmash—Boss DS-1 distortion analysis, www.electrosmash.com/boss-ds1-analysis#tone-level-frequency
[20] Zacharov, N. ed.: *Sensory Evaluation of Sound*. CRC Press, Taylor & Francis Group, Boca Raton, FL (2018)

Part III

Retrospective Perspectives of Distortion

9 Hit Hardware

Vintage Processing Technologies and the Modern Recordist

Niall Coghlan

9.1 Introduction

The history of recorded music and the evolution of technology are inextricably intertwined, from the earliest recordings direct to wax cylinder and phonograph, to electronic capture and processing via microphones, amplifiers and mixing desks to the advent of magnetic tape recording, electronic synthesisers and the modern digital era of sampling and emulation of audio tools and technologies (for comprehensive accounts of these histories, see Katz [1] and Milner [2]). As the commercial appetite for recorded music has grown, so too have the technological requirements of the typical recording studio, along with the market supplying these technologies with access to high-quality recording technologies available even to amateur or so-called prosumer recordists [3]. Varying perspectives on these evolutions are found in the users of the day, from enthusiastically embracing these new technologies, sticking with 'tried and trusted' approaches, gradual integration of workflow efficiencies or renouncing them altogether and seeking to only use vintage or historic technologies [4, 5]. However, despite a seemingly ever-increasing stream of 'improved' technologies and the efficiencies offered by 'in-the-box' computer processing, certain pieces of vintage hardware (across the spectrum of recording and musicking technologies) remain highly desirable, are extensively used in the modern music production landscape and have achieved significant degree of 'brand' recognition. Given the ready availability of software emulations, supposedly improved technology and changing models of content production, this chapter seeks to understand the factors that lead to professional adoption of vintage hardware tools, often leading to 'classic' status for those tools.

9.2 The Rise of the Machines

The field of music production has often been presented in popular culture in quasi-mystical terms, with words such as magic, x-factor etc. obfuscating the processes and procedures that contribute to the development of a recorded artefact [6–8]. Many of these cultish vintage devices have been caught up in this mystic fog, with discussion of their benefits (both actual and supposed) proliferating in magazines, blogs and online fora. As a result, the market for them is such that prices have increased dramatically over recent decades, and there is a thriving market in software and, more recently, hardware emulations. Many of these 'classic' units predate the digital device boom of the 80s and 90s, though even at that point their appeal was building [9–11]. Recording industry focused discussions, interviews and academic articles frequently name-check a variety of devices as being 'classics', such as the Urei 1176, Fairchild 670, Pultec EQP-1, Roland Space Echo, SSL and Neve mixing desks, though a comprehensive account of all devices considered to be classics is outside the scope of this chapter [12–15].

DOI: 10.4324/9780429356841-12

Patterson [16] identifies the overlapping roles of sound engineer and producer and for the purposes of this chapter, and its focus on vintage hardware, I have addressed these roles under the title of 'recordist', an approach suggested by Bennett [17]. It is not the aim to try to quantify the changes or distortions introduced by specific units (for an example of work of this nature, see Moore et al.'s [18] analysis of dynamic range compressors) but more to explore the reasons as to why recordists choose to use these vintage tools, despite the availability of more current and supposedly improved technologies and how the distortions introduced by these devices contribute to their appeal. It should also be noted that in the context of this chapter distortion is considered as a change to the original input signal, rather than solely concerning the 'special effect' distortion, also commonly known as clipping, fuzz, overdrive etc. [19]. While these processes may also form part of the signal changes induced by these units, other distortions could include frequency, temporal or transient distortion, altering the spectral, harmonic, time-course or amplitude content of the input signal [20] [18]. These low-order distortions of the original input signal are frequently subtle rather than transformative, however, when combined together in a signal 'chain' can offer noticeable cumulative enhancements to the source audio. In general, introducing any piece of equipment into the recording chain creates an opportunity to introduce distortions, both desired and undesired. The nature of these distortions can vary from device to device, encompassing the intended functionality and 'maverick' applications [17]. Technologies associated with these distortions include magnetic recording tape, transformers and inductors, active circuitry such as valves and transistors and non-linearities introduced through circuit design [11] [20]. While there are many ways of capturing and manipulating an audio signal, desirable vintage units typically belong to one (or more) of the categories, pre-amplifier, compressor/limiter, equaliser [9], along with 'special effect' units designed to dramatically process the sound, such as echo, reverberation, fuzz, chorus etc. [21]. However, not all vintage processors were created equal and only a select few have proven their longevity in the recording industry.

9.2.1 Recordists and Technology

Although the original goal of recordists was to capture audio faithful to the original performances, and to some degree the accompanying acoustic environment [22], the second half of the 20th century saw the emergence of approaches that manipulated recorded elements to present an exaggerated or 'hyperreal' dimension in the presentation/staging of these elements [23], the 'sonic cartoons' described by Bourbon and Zagorski-Thomas [24]. This allowed the recordist a creative input to the recorded artefact, via the application of technological processes across the various phases of the recording and mixing process. Frith and Zagorski-Thomas illustrate the relationship between technology and artistic outcomes when they state [25]: 'in the studio technical decisions are aesthetic, aesthetic decisions are technical, and all such decisions are musical' (p. 3), and Moore [26] illustrates how the technical application of processors is interwoven with aesthetic outcomes. This reshaping of the source audio furthered the concept of producer as creative artist [27] and shaper of sound recordings, the technological wizard who applied the 'secret weapon' to the source material [28].

In recent years the core of the typical recording studio has become the computer [29]. This transition to 'in-the-box' recording and mixing, along with advancements in computer power, has removed many of the constraints imposed by a hardware-based approach, such as track counts, limited instances of signal processors, destructive editing, maintenance etc. Reviews and user discussion of software offerings are largely positive regarding sound quality, and the transfer of digital material between stakeholders is trivial (with internet access and compatible systems) compared to analogue systems such as multi-track tape. Yet there remains a large market for both new hardware signal processors and specific vintage units [30].

9.3 Methods

In order to explore the motivations behind using vintage processing technologies, a grounded theory methodology was used [31]. This approach allowed for initial data to be collected from an analysis of recordist-centric documentary sources, comprising equipment reviews and overviews, 'technique' articles and interviews with recordists in industry publications. This resulted in identifying a number of significant factors relating to vintage processing technologies. At this point these factors were revised and grouped into categories, with reference to existing literature in the field, much of which is noted in this chapter. These identified areas were then used as the basis for discussion in six semi-structured interviews carried out by the author with six professional music recordists in order to further explore the identified factors. On completion of the interviews, they were transcribed and annotated using data coding techniques, then analysed for themes relating to the interviewee's experiences with using vintage hardware, with these factors then cross-referenced among the interviews, the previously identified categories and with the existing literature. The resulting factors are addressed in subsequent sections of this chapter and discussed with reference to the literature.

The interview participants were recruited through personal connection with the author, interviews lasted around an hour, were conducted via audio call and recorded. All participants had 18+ years of experience in sound engineering and music production roles, across a variety of genres, were currently working in that role and were based on the island of Ireland at the time of interview (2020), though all worked regularly with international clients. There were five male participants and one female participant. All participants agreed to be identified, and quotes from the interviews have been included in the following sections, to illustrate the identified factors.

9.3.1 Factors Identified

From the combined documentary, academic literature and interview analyses, four main categories for adoption of these technologies were identified, those relating to technocultural choice, sonics, workflow and iconicity. These categories also contain sub-factors and are discussed in detail below.

9.4 Technocultural Choice

Katz [32] describes 'technocultural choice', a process of decisions regarding what technologies to use and how to use them, including using analogue technologies, because they allow them to engage with music in a more immediate and immersive way, with additional trappings of authenticity. Technologies build on what has come before and bring with them particular logics, procedures or codes based on their predecessors [33], with the decision to apply a particular technology to a recording process incorporating 'historical, cultural and technological meanings' [17] (p. 77). This can manifest as a production 'ethos' in which the recordist makes the decision to mandate or reject specific technologies or approaches, which may in turn create technical or performance constraints. 'I made my records sound like that because of the tools that I've chosen over the years' (McLarnon, J., personal communication, 24/07/20).

Buy-in to the vintage hardware ethos can be beneficial to recordists seeking to situate their practice in the continuum of recorded music, providing a form of validation to the recordist and collaborators [34]. However, there is a 'professionalism paradox' [30] associated with the use of hardware; users are seen to be more 'pro' or belong to a niche elite, whereas those whose day-to-day living depends on the creation of music recordings must work within deadlines and frameworks that allow for quick turnarounds and recall of mixes, both processes advantaged by software, whereas those in the hobbyist or 'prosumer' sphere are working in an environment that may have less of those overheads. 'It's not about the gear, it's about your ears and the room and the player, you don't need that much gear to make some great records' (McLarnon, J., ibid.).

It should be noted that the professional recordists' technocultural choice to use vintage equipment is not equivalent to technostalgia or retromania, but instead driven by one or more practical motivational factors, as detailed herein and noted in Bennett [35]. While there is undoubtedly at least some element of exclusivity or elite status conferred by using vintage, 'classic' equipment, in that the typical financial overhead required is significant, this does not seem to manifest as a motivating factor in the professional users of these tools interviewed for this research. 'I don't believe it's an emotional thing for me . . . it's just like "does that sound better than this?" and if it does then I should get that' (O'Reilly, R., personal communication, 15/07/20).

However, Thorén and Kitzmann [36] found that owners of analogue equipment frequently used it as a component of identity construction in their own mythos. Using vintage tools in their professional practice can be a form of self-expression or statement of artistic identity, in the same way that a painter might choose to specialise in a particular medium. 'I've found that whatever is in the rack, that's one of the most important aspects of my sonic palette, the accumulation of the differences introduced by the equipment' (O'Reilly, R., ibid.).

All the recordists interviewed for this piece stressed that vintage tools were not necessary to their practice, because their roles could be filled by other similar sound processors. 'At the end of the day if it's doing what I need it to do I don't care what it says on the badge' (Ó'Sealbhaigh, F., personal communication, 08/07/20).

9.5 Sonics

9.5.1 Colour

The primary context for using these processors is the effect they have on the sound to which they are applied, be that for subtle manipulation of dynamic range, radical transformation of the source signal or, more typically, somewhere in between [11] [20]. 'We want what that piece of equipment is adding to the initial and "pure" sound as a bolt-on' (McCaul, P., personal communication, 24/07/20). These tools are typically applied for corrective or creative reasons; corrective relating to 'fixing' perceived issues with the source performance or recording such as excess dynamic range variability, with creative relating to desired sonic aesthetic and colouration of the source signal. However, these processes are intertwined, the technical and creative axes noted by Frith and Zagorski-Thomas [25].

> You're really looking for anything that gives you an exciting sound or a cool sound in a musical way, that's not harsh, not distorting in an unpleasant way, giving you something unique that would take you a long time to get using other combinations of equipment.
> (Lynch, K., personal communication, 23/07/20)

The recordists I interviewed all viewed their application of hardware processing tools as aesthetically enhancing the material on which they were working, using metaphors such as a chef's use of seasoning, a painter's use of a particular colour (colour being a frequently used term to describe post-processing distortions) or photographic processes such as lens filters. 'They're just colourations, I think of them as like lens filters like if you were taking an analogue photograph, they're just different types of colourations that bring a flavour to the audio' (McLarnon, J., ibid.). This viewpoint aligns with Bourbon and Zagorski-Thomas' aforementioned 'sonic cartoons' [24], transforming the source material into a larger-than-life, hyperreal version of the original. 'Some of this vintage gear ... they do to audio what Technicolour did to film, they just make it more beautiful, more unreal and otherworldly (McLarnon, J., ibid.). The vintage approach is not without risk, however, because along with the desired effects can come undesirable distortions such as added noise and a 'loss of definition and focus' [37] (pp. 330), though it should be noted that these definitions are context dependent.

All of the recordists I interviewed made explicit that any given recording is typically the product of a complex signal chain, beginning with the musician in the acoustic space of the recording, followed by a series of preamps, processors, signal distribution devices and the recording and playback media themselves. At any given point in the chain, a processor is interacting with the product of all the previous interactions of devices and components in that chain, producing a complex and somewhat unpredictable outcome, though within expected boundaries in most cases. 'I believe that every item you buy has 2–3 amazing uses ... that's why I have a stack of outboard, but then when you take all of those numbers and multiply them together your options become much greater' (O'Reilly, R., ibid.).

Specific devices are used for the unique sound qualities they impart, and while some devices can be seen as a one-trick pony, the devices that become classics often do that trick across multiple use-case scenarios. 'The classics for me, the sound that they impart on to the signal would be part of their character, working within their scope' (Slevin, D., personal communication, 09/07/20).

A frequently expressed motivation among recordists for their application of processing technologies is to bring something special to or instil unique sonic characteristics in the recorded material. 'I'm searching for a thing that has never existed, whether that's a message that hasn't been said or a sound that's never been made, in order to captivate people' (O'Reilly, R., ibid.). Lynch summarises this well: 'It's just that extra 5% that people long for' (Lynch, K., ibid.). As highlighted in the preceding paragraphs, application of these processes has a cumulative and interactive effect, enriching the source audio with 'beneficial' distortions.

9.5.2 Unpredictability

There is a narrative within music production of digital recordings being sterile or 'cold' sounding, that representing sound waves with bits and bytes, ones and zeros leads to a form of inauthenticity in the recorded material, and that using analogue equipment can counteract these drawbacks [11]. 'When digital came along there was the whole thing with "everything needs to be warm again" because digital is too cold' (Ó'Sealbhaigh, F., ibid.). Thorén and Kitzmann [36] situate hardware in the analogue versus digital debate as 'desirable vs. undesirable imperfections', the unstable and unpredictable by-products of analogue technology such as noise and distortion seen as desirable imperfections and the lack of these by-products in digitally manipulated signals as undesirable imperfections. 'These were perceived as faults or problems or bad side effects of the equipment back in the day but now it's those unique artefacts that they produce that have become desirable' (Ó'Sealbhaigh, F., ibid.).

Electronic components are often specified within a given tolerance range, which, when considered as part of a complex interactive system, can lead to hardware devices having subtly different characters or behaviours. It seems some recordists use these tools precisely because of these instabilities, and variance among specific units can lead to some having a character that sets them apart from other units of the same model [38].

> We all like to think we're in control of these things but we're not really ... they do their own thing. You can kind of learn to 'provoke' them but there is an element of unpredictability about it. Which is great!
>
> (McCaul, P., ibid.)

Some so-called golden units have become even further fetishized or lauded as being particularly special. Specific model numbers or production runs can also gain a further cachet with users; there were at least 13 revisions of the 1176 compressor/limiter, and users have expressed preferences for some revisions over others [39]. 'I think there's a quirk in nearly everything I own, everything has something that's a bit mad in it but I absolutely love it' (O'Reilly, R., ibid.).

9.5.3 Replication of Iconic Sounds

For many users the motivator to use a specific piece of equipment is linked to the desire to recreate a particular sound from a specific song or record. When there are so many variables in the recording chain, some element of cross-production consistency may be helpful in moving towards recreating a 'classic' record's sound. Factors such as the recording space, mic position, performer input etc. may be outside the ability of the recordist to recreate (assuming they are known at all), but the technological factors provide a degree of consistency in the recording chain. 'I'm almost trusting the choices of the people who've made the albums that have had a massive impact on me' (O'Reilly, R., ibid.).

There has been a canonisation of specific aspects of specific recordings or producer's practices, and the self-referential tendencies of the recorded musical continuum lead contemporary producers to acknowledge and/or mimic the sounds (and by extension the techniques and processors) of previous eras. To authentically replicate or reference a given sonic aspect of a previous recording, it may also require the producer to use the same technical setup or equipment. Kaiser [34] highlights the contemporary move to not only emulate specific pieces of classic equipment but now also the processing chains and, by extension, 'sound' of famous producers. 'If you want that sound, get that item' (O'Reilly, R., ibid.).

9.6 Workflow

A hybrid approach, incorporating contemporary and vintage equipment, is typical for most of the recordists' work examined for this research, with exceptions being one who stated a preference for working entirely with analogue equipment (eschewing computer-based recording as a preference, although this was still typically offered as a service to clients). Digitally based workflows with multiple levels of undo and version control can lead to delayed decision making and a 'fix it in the mix' attitude, the 'option dilemma' noted in Duignan [40]. 'The problem with that is that if everyone understands that everything is replaceable, everything is redo-able you can end up in a situation where there's no finality to it' (Ó'Sealbhaigh, F., ibid.).

Several of the recordists interviewed as part of this research noted that modern multi-track DAW-based workflows allowed them to get more use out of vintage processors, by re-processing and recording audio files in multiple passes, allowing for a specific unit to be reused across multiple instruments or recordings and increasing its value to the recordist.

> Due to the modern workflow a producer may get much more use from a single piece of classic equipment than previously, they no longer need a rack full of a particular item but may use that same item multiple times across the production.
> (Slevin, D., ibid.

Some participants also noted the risk of sonic homogeneity that can arise from use of 'industry standard' tools or multiple instances of similar processors, both virtual and physical.

People say 'oh look at this you can have a GML8200 on everything' when the reality was that studios would have it on one thing and then you'd move on to a different box or a different thing and you'd end up with a whole patchwork of great colour and variety of colour.

(McCaul, P., ibid.)

9.6.1 Interface and Tangibility

Most of the vintage hardware tools discussed for this chapter share a limited set of controls or functions, particularly when compared to the powerful and complex software processors available today. However, these very constraints are among the factors prized by users of these tools, simplifying the complex decision-making processes inherent in the recording process. Recordists who use vintage hardware tools often express a preference for being able to reach out and twist a knob in order to adjust device parameters, rather than using a mouse or other data entry device.

> With hardware, you can just grab a knob and tweak it, where it might not be so intuitive when you're grabbing a software rendition of a knob on a computer screen, or if you're working with a control surface and you have to scroll to the right channel and the right knob to find what you're looking for.
>
> (Ó'Sealbhaigh, F., ibid.)

The performative aspect of working with hardware matters to some recordists [41], treating the signal processor as one might an instrument and 'playing' it. This can be particularly relevant when working as part of a collaborative process, creating opportunities for those other than the recordist to participate in a hands-on way, unlike typical software-based approaches [30] (p. 99).

> Engineers and people who work in studios have learned that what they do by pushing levels into a piece of equipment, there's an interaction with their actions and the colour and the sound that's produced by that piece of equipment. It's not linear, it's very non-linear.
>
> (McCaul, P., ibid.)

Kaiser [34] discusses the 'credibility gap' of software emulations of hardware, because they lack the tactile and haptic feedback provided by hardware, along with olfactory sensations such as the smell of heated tubes and process-oriented workflow considerations such as changing tape reels or waiting for tubes to stabilise. This tangibility and physical nature of interaction with a tool's controls, and the feedback received, appears to be of importance particularly when considered in light of 'performing' a mix, the ability to rapidly access multiple control parameters on the fly, along with the direct access provided by the typical one knob per function layout of many of these classic devices.

> It's a different way of making a record, when you had a tape machine and you had a console and outboard equipment, you really felt that you were making something, as opposed to editing something. You really felt that you were constructing music, that you were putting music together. It was just there and your brain works in a different way, it's so much more satisfying.
>
> (Lynch, K., ibid.)

9.6.2 Familiarity

With software and digital control providing complete replicability of sound and settings (each instance having identical sonic and operational possibilities), using hardware, with its inherent instabilities and individual units' character due to small differences in electronic tolerances, introduces an element of chaos or unpredictability into the sonic outcome. While these subtle non-linearities (such as oscillator tuning drift or wow and flutter in the output of a tape-based delay unit) can be programmed into digital units (either at the manufacturer or user level), this can present an additional layer of complexity and cognitive load into the production process. A software plugin may offer control over parameters such as distortion or wow and flutter, but now the user must decide how much wow and flutter is appropriate, potentially creating distraction from the creative process. Many users seem to adopt hardware tools because they allow them to get the results they want more quickly than using other tools. 'A lot of the time you're working under pressure so you have to be able to jump straight to what you know is going to work' (Ó'Sealbhaigh, F., ibid.).

User familiarity is important, both in terms of the anticipated sonic outcome and the process/operation used to obtain it; recordists know what to expect from the sonics and how to get that sound quickly. The typical recording studio session is usually constrained by budgets and by extension time [29], so that the ability to rapidly shape aspects of a sound source so that a specific technical or creative goal must be achieved rapidly. Familiarity with the functionality of a tool facilitates this process, with the relationship between time available and the end result noted by all interview participants. 'You kind of have to know how these things are going to affect the sound because the budgets aren't there, you don't have time to be experimenting' (Ó'Sealbhaigh, F., ibid.).

Another facet to this is that many of these devices have particular interface or design quirks that the user must be aware of or have prior knowledge of to access; for example, the Urei 1176 'Attack' control operates in the opposite direction to most similar devices, something that is not clear from the front panel labelling, or the hidden 'all buttons in' mode. 'It's tricks you've learned from other producers and engineers about how they got these sounds' (McLarnon, J., ibid.).

9.6.3 Misuse

A regularly identified theme in the documentary analysis was that of 'misusing' vintage hardware, via unorthodox signal chain placement, operation outside normal parameters or the original design intention of the device, the 'creative abuse' of Keep [42]. This is perhaps most readily apparent in the case of the Pultec EQP-1, the manual for which states that its frequency boosting and attenuating controls should not be used simultaneously, a technique for which the unit has become well known and prized [43]. Another frequently raised example of this 'misuse' of equipment is the Urei 1176 compressor's 'All Buttons In' mode, in which the four buttons used to apply specific compression ratios to the incoming signal are simultaneously selected, resulting in a fifth 'hidden' setting that was not part of the original specification [26]. Many of the vintage items aspiring to the 'classic' title share this sonically pleasing response to pushing the equipment beyond its original design parameters, overdriving transformers, simultaneous cut and boost, 'improper' signal chain positioning or 'hidden' tricks. This potential for creative misuse appears to be important to users, though playing more of a bonus feature role than as a prime motivator to use a particular tool. 'If I was looking for something, if I hadn't got a direction, I might start putting stuff through different boxes, just to see what will happen. You're looking for a happy mistake basically' (Slevin, D., ibid.).

Interview participants were asked about their engagement with this kind of creative abuse as to whether it typically was part of their practice. All referred to the previously acknowledged commercial overheads involved in this kind of experimentation, the need to achieve specific goals with limited time versus the more open-ended processes involved in experimentation. However, they also expressed a desire to explore the parameters of their equipment, even to the extent of 'misuse'. 'If you're faced with a new piece of equipment then you always want to push all its buttons and see if it can do anything interesting' (McLarnon, J., ibid.). This 'creative abuse' is strongly correlated to the application of processors for sonic colouration:

> Sometimes it's a case where you need to experiment, you're looking to get new sounds, you're changing what you have to create new tones, to create new sounds. That's where the fun is, when you start deliberately trying to misuse equipment in order to get completely different sounds, unexpected sounds.
>
> (Ó'Sealbhaigh, F., ibid.)

As an extension of this idea, interview participants were asked about their attitudes toward equipment that had developed a fault that led to a pleasing sonic outcome, and whether they would choose to repair the fault or prioritise the 'new' sound. In all cases the commercial imperative of having working and reliable equipment was identified as the first priority, though most had also utilised 'broken' equipment in specific scenarios. 'Sometimes things will throw up a fault that you get the flavour, but even then, you should fix your faults and remember the flavour and how to get it if you need to' (McLarnon, J., ibid.). Most respondents also noted the specific sonic character inherent in most faulty processors, 'I've had gear that messes up and gives you a sound but I can only use it in that one unique situation but if I have a working piece of equipment, I can use it all the time' (Lynch, K., ibid.). However, this character can also be prized as a unique component of their sonic palette: 'I would probably fix the tool, and then make another that did the same thing to have as a standalone. And if that didn't work, I might go back to breaking the other thing!' (McCaul, P., ibid.).

Only one of the interview participants gave an example of a faulty piece of equipment in which they had preserved the fault. 'I gave a "wonky" Echoplex unit to our technician and he asked me "do you want me to fix this up?" and I was "no way" I've used that item on loads of tracks!' (O'Reilly, R., ibid.). In this case the participant stressed that the fault was preserved because of a specific sonic character imparted by the fault but that this was possible because he had duplicate devices that fulfilled the original functionality.

9.7 Iconicity

As noted, a large number of vintage or precursor devices have become 'legendary', 'classic' or 'iconic' [35], with Meynell even going so far as to suggest deification [30] (p. 102). Association with significant artists, recordings, events or places can add to the reputation of vintage devices, a phenomenon linked to the replication of sounds on specific recordings noted previously. The famous Abbey Road Studios 'Sale of the Century' saw surplus recording equipment promoted as Beatles memorabilia over its technological functionality, and likewise, when Kraftwerk's prototype vocoder came on the market its purchaser compared it to owning a Jimi Hendrix guitar [44]. While there is recognition of these devices from studio clients [35], they are rarely found to be a motivator in and of themselves for a client to work with the owner, a finding supported by the interviews carried out for this research. However, ownership of these items may suggest a degree of credibility that can pay dividends for the client experience, such as a source of inspiration or to coax an improved performance from a musician/client. 'I think partly it's driven by expectation, if you have these iconic pieces of equipment, and you have them in the right hands, then your record is going to sound good . . . that kind of expectation' (Lynch, K., ibid.).

O'Reilly uses the example of working with a young musician inspired by a veteran act, in which O'Reilly recreated the same guitar pedal setup as that used by the veteran act in order to instil feelings of importance and authenticity in the musician, with the intent of leading to an improved performance:

> When he's standing making his debut album as a teenager, and he steps on a fuzz pedal, does he want to step on an empty silver box, or does he want to step on the pedal that his hero played?
>
> (O'Reilly, R., ibid.)

O'Grady [8] notes the 'mystery' associated with parts of the production process due to the complex and expensive equipment and processes that take place behind the scenes, and perhaps it is this lack of clarity about the specifics of the engineering and production processes that has in part contributed to the fame of these 'classic' machines. Specialist media have fed this appetite for narratives regarding the recording and production of music (also see previous discussion regarding iconic sounds), and information about the techniques and technologies used, and at least for some the relationship between the technologies used and the perceived quality of the resulting product have become conflated. 'You're looking for a marquee or brand. . . . I think the pedigree is as much of a selling point as the sound of the thing' (Slevin, D., ibid.).

9.7.1 Investment

The digital environment is one of fast-paced technological change, rapidly rendering programmes, plugins and operating systems obsolete. The vintage hardware discussed here was often over-engineered by today's standards, with many of these multi-decade-old pieces of equipment still in daily use and with proven build quality. This vintage equipment has demonstrably stood the test of time and short of failure could still be in use for decades to come, unlike even the most current computer operating system. 'If it's well maintained its way more reliable than people ever realised or think, it's very reliable' (McLarnon, J., ibid.).

As already noted, and documented in Meynell [30], prices for these iconic vintage technologies have risen steadily over recent decades. For some recordists, vintage equipment is seen as an investment, units with already proven worth over decades of use and unlikely to suffer the obsolescence and depreciation associated with much modern (frequently digital) equipment. 'It's like buying classic guitars or vintage wine, it only goes up in value. You have nothing to lose if you have the money to invest' (Lynch, K., ibid.). This helps to offset some of the depreciation associated with other studio technologies, such as computers and software. 'The resale value is as much as why they are popular as the sound of it, it's a low-risk investment' (Slevin, D., ibid.). However, original components may no longer be available or are in limited supply, making maintenance or replacement difficult when required [35] [45]:

> I guess the reason they are still there is because of the original design, so many people have tried to make clones of them, but if you don't have the original components it's very hard to get the original sound from those things. Even down to the very basic components, you have to keep them as original as possible.
>
> (Lynch, K., ibid.)

9.8 Conclusion

The primary motivation for using vintage technologies among recordists interviewed for this research was undoubtedly the sonic character imparted (in the context of this chapter 'distortions'), often referred to as 'colour' but also prizing the chaotic and interactive results of combining these processors. The 'historic' and self-referencing aspects of popular recording also have a part to play, with recordists sometimes seeking to replicate classic *sounds* by using the same tools used to create them initially.

The sonic dimension is closely followed by advantages imparted to a recordist's workflow by using these tools, leveraging the advantages of modern digital technologies (such as low noise, non-linear editing/recording) in conjunction with classic technologies for a hybrid approach. The tangible nature of the control surfaces of much of these classic technologies affords a hands-on approach to signal processing that many recordists find to be a refreshing alternative to the 'in-the-box', mouse-driven approach ubiquitous in the contemporary field. This tangibility also affords opportunities to learn and 'perform' on the equipment, comparable to that of a performer on a musical instrument. The modular nature of combined stand-alone processors also affords 'misuse' of the equipment as a tool for creative inspiration and sonic outcomes that are not always possible with contemporary equipment (often designed with protective measures to prevent 'accidental' misuse).

The use of vintage and precursor technologies represents a 'technocultural choice' for recordists, with motivations ranging from a desire to use the 'best' tool for the job to an ethos of *only* using vintage technologies or as an expression of creative identity. The iconic nature of many of these classic vintage tools also plays a part here, sometimes providing an element of 'legitimisation' from the perspective of clients (though professional recordists usually stress that vintage tools are not necessary to carry out high-quality work). This iconicity also has a peripheral benefit, in that the desirability of these tools has led to an active secondary market, with many of them appreciating in value over recent decades.

The professional longevity of vintage devices appears to be a product of multiple factors, as considered in the preceding text and exemplified in Meynell's [30] consideration of the changing role of the Pultec EQP-1, from technical tool to sonic flavour to emulated icon. This versatility, the ability to play different roles in the recording and production process and beyond, would appear to be a core determinant in a vintage unit's 'classic' status, along with the longevity imparted by high build quality not often found in contemporary equipment, allowing these devices to survive long enough to become classics. Even the 'one-trick ponies' that perform a specific sonic function are simultaneously playing other roles, as workflow enablers, creative inspiration or investments. In most cases it is not that they possess qualities that make them radically different from comparable vintage or modern units, but that they impart that 'extra 5%' to the sound, workflow and experience that is sought out at high levels of recordist practice.

References

[1] Katz, M. *Capturing Sound: How Technology Has Changed Music*. Berkeley, University of California Press (2004).
[2] Milner, G. *Perfecting Sound Forever: The Story of Recorded Music*. New York, Farrar Straus & Giroux (2009).
[3] Théberge, P. *Any Sound You Can Imagine*. Middletown, CT, Wesleyan University Press (1997).
[4] Tingen, P. *Steve Albini: Sound Engineer Extraordinaire*. Accessed December 2020 from www.soundonsound.com/people/steve-albini
[5] Bennet, S. *Revolution Sacrilege! Examining the Technological Divide Among Record Producers in the Late 1980's*. Accessed November 2020 from www.arpjournal.com/asarpwp/revolution-sacrilege-examining-the-technological-divide-among-record-producers-in-the-late-1980s/

[6] Reilly, B. *The History of Pultec and the Storied EQP-1*. Accessed November 2020 from https://vintageking.com/blog/2017/11/pultec-eqp-1-equalizer-history/
[7] Reumers, R. *Demystifying the 'Magic' of the Pultec EQ*. Accessed November 2020 from https://abbeyroadinstitute.nl/blog/demystifying-the-pultec/
[8] O'Grady, P. The master of mystery: Technology, legitimacy and status in audio mastering. *Journal of Popular Music Studies*, Vol. 31, No. 2. University of California Press, CA (2019) pp. 147–164.
[9] Fletcher, A. Reader's guide to vintage (. . . er, old) gear. *Mix Magazine*, Vol. 20, No. 11 (1996) pp. 84–100.
[10] Poss, R. M. Distortion is truth. *Leonardo Music Journal*. Vol. 8. MIT Press (1998) pp. 45–48.
[11] Simons, D. *Analog Recording: Using Analog Gear in Today's Home Studio*. San Francisco, Backbeat Books (2006).
[12] Senior, M. *Classic Compressors: Choosing the Right Compressor for the Job*. Accessed December 2020 from www.soundonsound.com/techniques/classic-compressors
[13] *Computer Music Magazine*. Accessed August 2020 from www.musicradar.com/tuition/tech/7-ways-to-get-more-out-of-your-classic-compressor-605700
[14] Felton, D. *Top 20 Best Compressors of All Time*. Accessed August 2020 from www.attackmagazine.com/reviews/the-best/top-20-best-hardware-compressors-ever-made/
[15] Bieger, H. *Studio File: Sunset Sound*. Los Angeles. Accessed August 2020 from www.soundonsound.com/music-business/sunset-sound-los-angeles
[16] Patterson, J. Mixing in the box. In: R. Hepworth-Sawyer & J. Hodgson (eds.) *Mixing Music, Perspectives on Music Production*. London, Routledge (2017) pp. 77–93.
[17] Bennett, S. *Modern Records, Maverick Methods: Technology and Process in Popular Music Record Production 1978–2000*. London, Bloomsbury (2019).
[18] Moore, A., Till, R. & Wakefield, J. P. An investigation into the sonic signature of three classic dynamic range compressors. *140th AES Convention Paper*. Paris, France (2016).
[19] Augoyard, J.-F. & Torgue, H. *Sonic Experience: A Guide to Everyday Sounds*. Montreal, Kingston & London, McGill-Queen's University Press (2005).
[20] Robjohns, H. *Analogue Warmth: The Sound of Tubes, Tape and Transformers*. Accessed December 2020 from www.soundonsound.com/techniques/analogue-warmth
[21] White, P. *Basic Effects and Processors*. Maryland, New Amsterdam Books (2003).
[22] Frith, S. *Sound Effects. Youth, Leisure and the Politics of Rock 'N' Roll*. Suffolk, Bury St. Edmunds (1983).
[23] Roquer, J. Sound Hyperreality in Popular Music: On the Influence of Audio Production in Our Sound Expectations. *Sound in Motion*. Newcastle upon Tyne, Cambridge Scholars Publishing (2018).
[24] Bourbon, A. & Zagorski-Thomas, S. Sonic cartoons and semantic audio processing: Using invariant properties to create schematic representations of acoustic phenomena. In *Proceedings of the 2nd AES Workshop on Intelligent Music Production*. The 2nd Audio Engineering Society Workshop on Intelligent Music Production, London (13 September 2016).
[25] Frith, S. & Zagorski-Thomas, S. *The Art of Record Production: An Introductory Reader for a New Academic Field*. Oxfordshire, Routledge (2012).
[26] Moore, A. All buttons in: An investigation into the use of the 1176 FET compressor in popular music production. *Journal on the Art of Record Production*, Vol. 6 [Online] June 2012. Accessed August 2020 from www.arpjournal.com/asarpwp/all-buttons-in-an-investigation-into-the-use-of-the-1176-fet-compressor-in-popular-music-production/
[27] Gillett, C. The producer as artist. In: H. W. Hitchcock (ed.) *The Phonograph and Our Musical Life*, ISAM Monograph, No. 14. New York, City University, NYC (1977) pp. 51–56.
[28] Hitchins, R. *Vibe Merchants: The Sound Creators of Jamaican Popular Music*. Oxfordshire, Routledge (2014).
[29] Pras, A., Guastavino, C. & Lavoie, M. The impact of technological advances on recording studio practices. *Journal of the American Society for Information Science and Technology*, Vol. 64, No. 3 (2013) pp. 612–626.
[30] Meyell, A. How Does Vintage Equipment Fit into a Modern Working Process? In: S. Zagorski-Thomas & A. Bourbon (eds.) *Bloomsbury Handbook of Music Production*. London, Bloomsbury (2020).

[31] Strauss, A. & Corbin, J. Grounded theory methodology: An overview. In: N. Denzin & Y. Lincoln (eds.) *Handbook of Qualitative Research*, 1st edition. Sage (1994) pp. 273–284.
[32] Katz, M. *Musical Listening in the Age of Technological Reproduction*. Oxfordshire, Routledge (2015).
[33] Greene, P. D. & Porcello, T. (eds.). *Wired for Sound: Engineering and Technologies in Sonic Cultures*. Middletown, CT, Wesleyan University Press (2005).
[34] Kaiser, C. Analog distinction—Music production processes and social inequality. *Journal on the Art of Record Production*, No. 11. Accessed August 2020 from www.arpjournal.com/asarpwp/analog-distinction-music-production-processes-and-social-inequality/
[35] Bennett, S. Endless analogue: Situating vintage technologies in the contemporary recording & production workplace. *Journal on the Art of Record Production*, Vol. 7. Accessed November 2020 from www.arpjournal.com/asarpwp/endless-analogue-situating-vintage-technologies-in-the-contemporary-recording-production-workplace/
[36] Thorén, C. & Kitzmann, A. Replicants, imposters and the real deal: Issues of non-use and technology resistance in vintage and software instruments. *First Monday*, Vol. 20 (2015) pp. 11–12. Accessed December 2020 from https://firstmonday.org/ojs/index.php/fm/article/view/6302/5134
[37] Mynett, M. *The Metal Music Manual*. Oxfordshire, Routledge (2017).
[38] Frahm, N. Nils Frahm on his favourite vintage music gear. *Roland Blog*, October 2015. Accessed December 2020 from www.roland.co.uk/blog/nils-frahm-on-his-favourite-vintage-music-gear/
[39] White, P. Industry talk: Modelling classic hardware in software. *Sound on Sound Magazine*. Accessed August 2020 from www.soundonsound.com/people/uad-modelling-classic-hardware-software
[40] Duignan, M. *Computer Mediated Music Production: A Study of Abstraction and Activity*. Thesis. Victoria University of Wellington, NZ. Accessed December 2020 from https://core.ac.uk/download/pdf/41335961.pdf
[41] Anthony, B. Mixing as a performance: Educating tertiary students in the art of playing audio equipment whilst mixing popular music. *Journal of Music Technology and Education*, Vol. 11, No. 1 (2018) pp. 103–122.
[42] Keep, A. Does creative abuse drive developments in record production? *Paper presented at the First Art of Record Production Conference*, London, University of Westminster. Accessed December 2020 from www.artofrecordproduction.com/aorpjoom/arp-conferences/arp-2005/17-arp-2005/72-keep-2005
[43] Pulse Techniques Pultec Manual EQP-1. West Englewood, NJ, Preservation Sound. Accessed December 2020 from www.preservationsound.com/wp-content/uploads/2012/02/Pultec_EQP-1.pdf
[44] Synth Britannia (2009) *Directed by Ben Whalley*. England, British Broadcasting Corporation. Viewed September 2019
[45] Robjohns, H. *Pulse Techniques EQP-1A: Analogue Passive Equaliser*. Accessed December 2020 from www.soundonsound.com/reviews/pulse-techniques-eqp-1a

10 Even Better than the Real Thing

A Comparison of Traditional and Software-Emulated Distortion in the Contemporary Audio Production Workflow

Doug Bielmeier

10.1 Introduction

Audio production industry professionals are responsible for recording, mixing, and mastering of the music recordings and audio productions sold within the music industry [1]. Contemporary audio professionals, when working toward an aesthetic, aim to mimic the warmth and character of traditional analog recording techniques using ever-evolving hardware and software. Specifically, audio professionals seek to incorporate the harmonic distortion (HD) of traditional non-linear recording hardware [2, 3, 4]. Until recently, to obtain this 'analog sound' recording engineers used hybrid digital/analog techniques that required expensive hardware units including compressors, parametric equalizers, and large-format analog consoles [5, 6]. Over the last decade, low-cost software emulations of these classic hardware devices have become available, thus improving access to and incorporation of HD and non-linearity into contemporary digital recordings [7]. However, the authenticity and usefulness of these hardware emulators (i.e. software) is extensively debated throughout the industry. Understanding how and when to use hardware and software devices to create HD is needed to assist audio production industry professionals in establishing contemporary workflows.

This chapter will (a) define HD and non-linearity in traditional hardware devices, (b) discuss the authenticity of HD emulators, (c) compare the advantages and disadvantages of using software and hardware to create HD, and (d) analyze the results of the Harmonic Distortion Survey. The survey will provide insight into how, when, and where audio production professionals create HD using hardware and software emulators in their audio production workflows.

10.2 Hardware Created THD

In audio hardware circuit design, groups of non-linear components are combined to create complex audio processing units [8]. Incoming signals travel through these non-linear components which alter the signal's fundamental frequency, and thus distort the output signal [9]. The wave shaping, which occurs in the time domain, results in additional harmonic components, which were not present into the input signal [10].

Total harmonic distortion (THD) refers to the deviations in voltage created by these non-linear, complex circuits. Mathematically, THD is defined as the ratio of the root mean square (RMS) amplitude of a set of higher harmonic frequencies to the RMS amplitude of the fundamental frequency [11]. When calculating the THD of an electronic device, a function generator is used to input a sinusoidal signal into the device. An oscilloscope with a Fast Fourier Transform (FFT) function or a spectrum analyzer can be used to monitor the output waveform. The

DOI: 10.4324/9780429356841-13

voltage amplitude, *A*, of the signal can be measured in decibels (dB). The RMS amplitude for the one-time period of the sine wave, *Vrms*, can be calculated using equation (1).

$$V_{rms} = 10^{\frac{A}{20}} \qquad (1)$$

By monitoring the signal, the amplitude of the first five or six harmonics can be found and the V_{rms} can be calculated for each harmonic, where *V1* is the first harmonic and *V2* is the second order harmonic and so on. The THD can be calculated using equation (2) which expresses THD as a percentage [9].

$$THD = \frac{\sqrt{V_2 + V_3 + V_4 \ldots + V_n}}{V_1} * 100\% \qquad (2)$$

Understanding THD and how it is calculated is critical for understanding the benefits and character of classic audio hardware devices and how these devices can be emulated via software. Specifically, the classic UREI 1176 compressor is a highly coveted hardware compressor because of its function and unique characteristic sound [12]. In 1967, the UREI 1176 Revision A was the very first 1176 compressor that was produced by Bill Putman [13]. Revision A became the basis for all the other revisions, but it was also the most unique. Revision A is often referred to as 'vintage' due to its higher distortion from a lack of filters to reduce low noise. The sound of Revision A has been described as the "juiciest, noisiest, vintagey-ist" [12]. No low noise filters were incorporated into the device until 1970 with the release of Revision D. While the Revision D circuitry is very similar to Revision A, the Revision D is known for retaining the classic character while having lower noise/distortion [13]. Currently, these 1176 Revisions are also available via software plug-ins.

In the last decade, it is increasingly possible either to emulate legacy audio devices and effects or to create new ones using digital signal processing [7]. What is often debated is how accurate these software emulations can simulate the HD and non-linearities of the original component [14].

10.3 Authenticity of THD Emulators

Software that emulates classic hardware has become commonplace in the audio production workflow [15, 16]. Some software developers maintain that writing software is a much more fluid engineering method and, by its very nature, more flexible than a fixed hardware design. Furthermore, the notion that 'classic' analog gear can do something that the average computer cannot do is outdated [16]. Universal Digital Signal Processing (DSP) objects (e.g. plug-ins) can be designed to recreate vintage processors for use in commercially available Digital Audio Workstations (DAWs). These designs are created by software designers, new to the industry or employed by the original hardware manufactures, to recreate this processing in the digital domain.

Most developers use a combination of physical signal testing of the original device while incorporating circuit diagram generated DSP routines, prior modelling experience, and intensive listening sessions [15]. Software developers assert that most classic equipment can be physically modeled in the digital world [15, 17, 18]; however, there exist three main challenges to creating authentic THD emulators. First, the biggest challenge to THD emulator development is computing power. Analog equipment that exhibits high-bandwidth, non-linear behavior requires large amounts of computing power because as the non-linearity increases, the computing power required for simulation increases. As a result, some companies have developed turnkey or outboard processing units that connect to computers via USB or FireWire [19]. The outboard units offload processing power from the computer and increase the available processing workload of DSP models. The second challenge for THD emulator development is characterization of all non-linear frequencies. Typically, software developers have exposed these non-linearities by developing a set of test signals for specific frequencies [17]. However, for equipment with unknown non-linearities, like a vintage analog compressor, a full characterization cannot be made with a finite number of test signals. The third challenge for THD emulator development is not being able to capture device interaction. In an analog audio workflow, devices can interact with one another, which results in unexpected changes in frequency response or distortion characteristics, which are difficult to model [20].

Alternatively, other software developers focus only on the mathematics of sound and opine that all models are simply complex differential equations. Thus, the goal for these developers is to match output signals of a software emulation with the output signals of the hardware device [21]. This is achieved by using a circuit diagram of the original hardware to develop a multi-dimensional differential equation that relates the output levels to variations in input levels and control parameters. Using software such as SPICE (Simulation Program with Integrated Circuit Emphasis), the integrity of circuit designs can be evaluated. Furthermore, SPICE can predict and alter component behavior by entering different parameters, and circuit performance can be evaluated by various input/output plots and derivation of mathematical transfer functions [22]. The development of digital models of hardware components is possible because all analog devices were made from components that have mostly defined behaviors.

Furthermore, with a detailed digital circuit model design, software developers can "fix" typical hardware errors, such as inaccurate distortion compressor attack times. For instance, there existed a small difference in the way Softube's 1176 hardware distorted for fast release times. By looking at the digital circuit model, the developers were able to track down a 2 mV bias difference in the detector circuit of the hardware. As a result, software developers inserted 2 mV into the virtual model and improved performance in the THD emulation [15].

A limitation of the mathematical approach to the development of HD emulators is that it cannot reproduce the drifting, shifting, and unpredictable behavior of components (e.g. capacitors) as they age [23]. In this approach, developers tend to evaluate only the basic qualities of frequency response, linearity, and THD, and ignore "side-effect sounds", such as noise and hum; however, these side-effect sounds have a critical effect on how a signal sounds to listeners [15]. Thus, the most authentic software emulators undergo many iterations of measuring and listening to identify the strengths and weaknesses of different models as compared to the original hardware [15, 16]. Recently, Volterra series (VS), a mathematical model for non-linear circuits, can include time-domain simulation with X-parameters. This model can be a better supplement for blackbox macromodeling methods for non-linear circuits [24].

As another alternative, some software developers seek to bring the best analog sound features to the digital domain without modelling existing analog equipment or limiting specific hardware [23]. Some designers and producer question the efficacy of classic hardware, stating that it is next to impossible to successfully and persuasively model an analog compressor-limiter [15].

Debate over the precise degree of accuracy of these emulations remains, even while many software developers offer clear and transparent methods and reasoning for their emulation's efficacy. Desire for the coveted hardware originals continues, but they are often rare and expensive. Modeled plug-ins provide a means for every DAW user to access their distinctive character and offer additional advantages over their hardware progenitors.

10.4 Comparison of the Advantages and Disadvantages of Software and Hardware THD

For some THD emulators and hardware, there exists little or negligible sonic difference between each approach [15–22]. Thus, sound quality cannot be the sole factor when deciding to use a software or hardware approach to creating THD. Table 10.1 shows the advantages of incorporating hardware or software into the audio production workflow as defined in the literature [15, 17, 18, 20–22]. Software plug-ins offer several unique features that are not available in hardware.

A main advantage of software as compared to hardware is cost. An individual plug-in is significantly less expensive than its hardware component. More importantly, it costs nothing for additional instances or insertion of the plug-in into the audio production workflow. For example, the Warm Audio WA76 Discrete FET Compressor, a hardware clone of the classic UREI 1176 compressor, costs about $600 USD per mono unit and would cost the same to add more units to work in real-time [25]. Alternatively, the cost to purchase the Waves CLA-76 Compressor/Limiter Plug-in, which includes Revisions A, B, and D, is $30 USD for unlimited instances of the software [26].

Another advantage of software as compared to hardware is the ability to continuously edit and automate the workflow. When working with hardware, audio professionals will "bounce" each analog instance into their DAW. These bounces are permanent/destructive and time consuming. Alternatively, in software audio professionals can create another instance of the plug-in and use preset parameters to speed setup of the plug-in. When loading a session in a DAW from a previous session, plug-ins will load to their previous saved settings allowing audio professionals to pick-up where they left off, which saves time in the studio for clients and producers. Also, when changing parameters over the course of a production, plug-ins within a DAW allow for precision of editable automation levels. Congruently, these automation levels can be assessed and changed quickly by the click of a mouse.

When evaluating the advantages of hardware, there are three main advantages to consider. First, the circuit component-based non-linearities are unique. Not only can these devices create these characteristic tones, but often specific devices have very specific uses in the production process. Second, some specialized hardware has unparalleled advantages that cannot

Table 10.1 Advantages of Hardware and Software Created Harmonic Distortion

Advantages of Hardware	*Advantages of Software*
Vintage or characteristic tone/sound	Multiple instances
Harmonic distortion/non-linearity	Recall/presets
Real-time processing	Cost
Tactile experience/knobs	Ease of use
Designed for specific use	Digital or DAW based routing
Absence of aliasing	Automation
Ease of use	Precision
Minimal obsolescence	
Physical routing	

be captured by software. For example, the classic LA-2A compressor has been a "go-to" for audio professionals compressing lead vocals for decades [13]. The device has set attack and release time parameters that are optimized for compressing vocals [20]. Third, some hardware has an enduring quality that reduces or eliminates obsolescence. A great functioning hardware compressor made today will be useful regardless of advancements in recording and computing software. However, software plug-ins will need to be updated or become unusable when DAWs and operating systems are upgraded.

There are many advantages and disadvantages to creating THD via hardware and software. Within the audio production industry, both new hardware units and plug-in emulations are continually made and revised. Therefore, understanding how audio professionals incorporate harmonic distortion into their workflow becomes equally, if not more, important as their authenticity and advantages.

10.5 The Harmonic Distortion Survey

The Harmonic Distortion Survey was conducted to better understand the creation of THD by hardware and software in the workflows of audio professionals. The goal of the survey was to understand: (a) how, why, and where audio professionals are using THD in their production workflows and (b) why audio professionals choose a hardware or software approach to THD creation.

The online survey was conducted during the second half of 2020 via a convenience sampling of the author's industry connections and online discussion groups and forums. The 100 survey respondents had between 1 to 20 years of experience working professionally in the audio production industry. Professional work was defined as work for financial compensation for clients on commercial recordings, productions, video, or other media. The respondents identified as having roles as Audio Engineers, Mixing Engineers, Mastering Engineers, ADR/Foley/Film Engineers, Live Sound Engineers, Broadcast Engineers, Editors, and Producers. Figure 10.1 shows the experience level and type of work for respondents. More than half of respondents were audio engineers and a majority of respondents had more than 10 years of experience.

The audio professionals were asked about whether they incorporate harmonic distortion into their audio production workflow. Figure 10.2 shows the number of audio professionals in each audio production role and their use of harmonic distortion via hardware, software, both, or none. More than two-thirds of audio professionals (N=59, P 40, 67.8%) reported the use of a combination of hardware and software emulated harmonic distortion. Surprisingly, only one audio professional (N=59, P 1, 1.7%) used no harmonic distortion in their audio workflow. The audio engineers, mixing engineers, and producers of this survey were most likely to create THD using both hardware and software.

The audio professionals were asked to identify the greatest advantage of using hardware devices and software in an audio production workflow. Table 10.2 shows the reported percentage of indicated advantages for both hardware and software. The results strongly suggest that the main reason why audio professionals use hardware to create THD is because of a vintage classic tone (N=59, P25, 42.4%). In addition, the results suggest that audio professionals use software to create THD because of the ability to use multiple instances (N=59, P16, 27.6%) and recall/presets (N=59, P15, 27.6%). Another interesting result of the survey was found in the write-in categories for both hardware and software THD creation. For creation of THD via hardware, audio professionals identified tactile control and the absence of aliasing as additional advantages. Surprisingly, three respondents argued that there is no advantage to the use of hardware. For creation of THD via software, audio professionals identified precision, customization, and repeatability as additional advantages. Similarly, the three respondents who indicated that

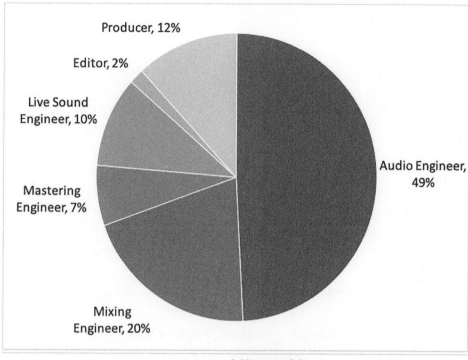
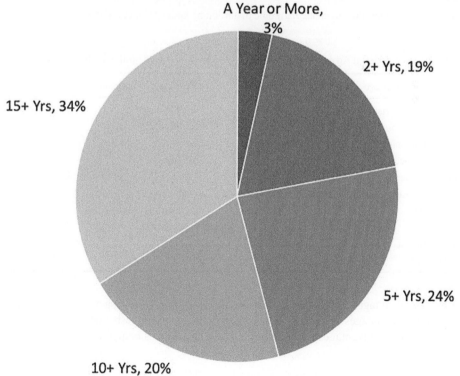

Figure 10.1 Survey respondents' a) years of experience and b) audio production role.

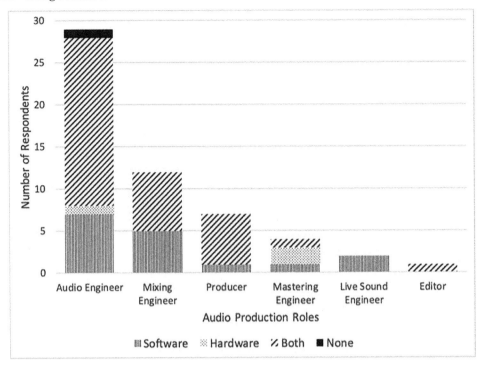

Figure 10.2 Audio workflow to create harmonic distortion by audio production role.

Table 10.2 Most Important Advantages of Hardware and Software as Identified by Respondents

Advantages of Hardware	Responses	Advantages of Software	Responses
Vintage or characteristic tone/sound	25	Multiple instances	16
Harmonic distortion/non-linearity	9	Recall/presets	15
Real-time processing	5	Cost	9
Tactile experience/knobs	4	Ease of use	6
Designed for specific use	4	Digital or DAW based routing	3
Absence of aliasing	3	Automation	2
No advantage	3	Precision	2
Ease of use	2	No advantage	1
Minimal obsolescence	2		
Physical routing	1		

there was no advantage to hardware lauded the use of software and disagreed with the premise that either had advantages or disadvantages. Overall, this data suggests that audio professionals feel there are many advantages for creating TDH via software as compared to hardware.

10.5.1 Audio Professionals' Perspectives about Using THD in Their Workflows

In an open-ended survey question, audio professionals were asked to explain their reasons for using harmonic distortion in their workflow via hardware or software. Using a purposeful coding method, the most common reasons for using harmonic distortion were identified as tone and character, tradition, and inexplicable.

Audio professionals cited the enhancement of harmonics in audio production affected the tone and character of sound by allowing the music to resonate, fill up space, create excitement, create energy, and translate. One audio professional stated that the "inclusion of harmonic distortion is a way of adding warmth, in some case controlling transients, and add a certain 'glue' when applied to multiple sources (like your stereo buss)." Another audio professional added that harmonic distortion gives a "fuller thicker sound, a vibey tone that will sound less sterile. Sometimes, I like to push things harder on certain instruments (drums, drum machines, synths, samples) to get a nasty badass aggressive sound."

Some audio professionals cited tradition as their reason for using harmonic distortion in their workflow. One audio professional opined that harmonic distortion is responsible for the sound of a lot of what people love in recorded music and "Early recording gear had terrible THD specs compared to today, but music listeners' collective ear seems to just be tuned for that and it's hard to get away from that entirely and get a pleasing result." Another audio professional agreed that humans have adapted to hearing the rich, expanded tones of imperfect equipment from the past. Yet another audio professional maintains harmonic distortion makes sounds more relatable and claimed,

> You wouldn't get the same sense of nostalgia listening to an old Billie Holiday record if her voice and all the instruments suddenly sounded as pristine as Steely Dan. I find the sound of Frank Sinatra's or Aretha Franklin's crackly vocal recordings more relatable than the more pristine recordings of their colleagues.

Meanwhile, other audio professionals had little to no reason for using harmonic distortion in their workflow. Responses from audio professionals included the reason is mostly creative-driven and less technical and "makes things sound good," and "sometimes you just need to." One audio professional stated that they just like the way it sounds better when they use it.

Collectively, the responses suggest that audio professionals choose to create TDH via hardware because of the resulting sound. Furthermore, the responses support the survey finding that audio professionals overwhelmingly create THD via hardware because of the vintage sounds.

10.5.2 Audio Professionals' Perspectives about Hardware vs. Software

In another open-ended question, the audio professionals were asked to provide their general thoughts and opinions about the use of hardware versus software to create harmonic distortion. Using a purposeful coding method, preferences for hardware, software, both, and neither were identified.

Some of the audio professionals preferred the harmonic distortion created by hardware. One professional stated, "Hardware wins almost every time. I've had the good fortune of using a lot of vintage equipment and comparing it to digital emulations and it seems to consistently win." Another professional concurred: "software currently does not compete for me, tubes and transformers and out-of-the-box clipping software recreations I've tried fall flat every time and distort either too much or too little; no sweet spot or vibe gained." Other professionals cited the absence of aliasing artefacts and the enjoyment of tactile control as something unique to working with hardware. Another professional added:

> From my own experience, when I have A/B'd hardware to software harmonic distortion tools, both are relatively similar. The point of divergence is when you really drive the units. Software is going to reach a linear upper limit, but hardware, when driven, is going to exhibit a unique/nuanced character specific to the unit.

One professional defended the limitation of not being able to undo an edit was extremely helpful in that it limited choice and prevented endless revisions on a project. Another professional did concede that, "Hardware is always better, but being better doesn't equate to being more useful in most professional audio situations that don't involve the hardware being installed already."

Some of the audio professionals preferred software emulated harmonic distortion. One audio professional opined, "Software is significantly higher quality and more flexible for the costs. To get the same quality from hardware you are spending 10x as much or more." Other professionals were confident that the accuracy of software emulations is indistinguishable from their hardware progenitors. One professional explained,

> Digital modeling of analog circuits has become nearly indistinguishable from the real thing. When you factor in that some hardware can just be finicky, the reliability of DSP is often overlooked. The bottom line these days is you get 90% of the sound for 10% of the price.

In contrast, another professional argued that what defines hardware versus software is often a blurred line. The professional explained,

> A lot of hardware is completely digital, sometimes even running on Windows or Linux. Hardware vs software is often just a difference in form factor, where analog vs digital is the difference in design. Sometimes the harmonics occur in the analog stage of a digital hardware processors.

Meanwhile, other audio professionals had no preference for either software emulated or hardware created harmonic distortion. One professional asserted, "Whatever tool works. Different tools. Both can suck, both can be great." Another professional opined:

> I love both! I love driving hardware hard, cranking inputs, and getting the benefits of the slight compression that comes with different levels of distortion. It makes things feel more even and yet more energetic. So many great companies have made excellent-sounding software with multiple instances and instant precise recall.

A specific professional who identified their role as a Live Sound Engineer maintained, "It's all about convenience in live sound. If I want to bring an analog device, I either must own it, or someone must rent it for me. Software is easier to manage in live sound."

Collectively, the responses suggest that the use of hardware and software to create THD is still debated by the audio professionals of this survey. Preference, availability, cost, and tool specialization are all key reasons for and against using either system. Surprisingly, audio professionals discuss using hardware in non-optimal configurations to create unique sounds.

10.5.3 The Impacts of Harmonic Distortion on the Audio Production Workflow, Final Product, and Client Experience

In a final open-ended question, audio professionals were asked to explain how the use of harmonic distortion impacts their workflow, deliverable product, and client's experience. When asked about their workflow, audio professionals identified its use as an important step in the larger recording and mixing process. During the recording process, a few professionals mentioned using harmonic distortion prior to reaching the DAW. One professional described, "I love to process individual sounds on the way in, while I track, to commit to a certain sound, and then mix from those intuitions and choices, which somewhat restrict my options, but in a nice way." Another professional agreed, "I especially like distorting things before they hit digital converters, because analog distortion is less reliably predictable. I like the chaos of it, and find it often creates happy accidents that I get less often in the digital world." In contrast, some professionals reported a hybrid use of harmonic distortion in their workflow. One professional explained, "I like to use hardware distortion while tracking (overloading preamps, tube hardware etc.) and use software for parallel processing: distortion as an audible effect rather than as a subtle coloration." Another professional agreed, "I personally use a hybrid set up: most of my analog stuff is for recording purposes and the mixing/mastering portion is mostly all in-the-box." The professional also opined, "You can get your Grammy in your mom's basement about as well as you could at Abbey Road, but I would rather be in Abbey Road than in the basement."

When asked, many audio professionals reported harmonic distortion having specific effects on their final deliverable product. Some audio professionals reported that harmonic distortion can achieve certain objectives quickly, which done differently can take more time and effort. Other professionals asserted that harmonic distortion can add 5–10% more color, making their final product sound better and more competitive. Other professionals agreed but maintained that harmonic distortion is an integral tool alongside spectral and dynamics processing.

When describing their client's experience and expectations, most audio professionals agreed that the client often has little knowledge or understanding of what harmonic distortion is and its impact. One professional stated, "Rarely do clients know when or if I've used any harmonic distortion, but I would like to think they appreciate the results." Congruently, professionals asserted that the auditory result or sound is important to their clients. One audio professional opined, "My clients don't care about harmonic distortion, they care about the mix." However, other professionals claimed that the use of harmonic distortion is essential to the client experience. One professional asserted,

> Harmonic distortion impacts my workflow as well as clients' experience. The earlier in the process I start using it, the more the client can tell what the finished recording will sound like. Having a vision of the end result influences and inspires the performer; taking a bland performance and making it something special.

Another audio professional warned, "Sometimes distorting something too much while recording, and the client decides later they don't like it, you have to jump through hoops to fix it or explain that it was recorded that way and take responsibility for the mistake." While another professional confessed, "In my job I don't use it much and when I do it's subtle. In most cases I keep it as a secret ingredient. Most of my clients would possibly frown upon the idea that I am distorting something." From a mastering engineer perspective, one professional asserted, "Clients are impressed with apparent loudness, when using harmonic distortion, as compared to actual integrated LUFS measurements."

Collectively, the responses suggest that some audio professionals of this survey choose to create HD via software and hardware based on what stage they are at in their audio production workflow to benefit from the advantages of each method. Furthermore, audio professionals of this survey agree that HD is an essential tool in their audio workflow. Alternatively, these audio professionals have mixed feelings about how HD is perceived by clients. Specifically, there is no consensus on how much and when to add HD to mixes heard by clients. The variety of methods of implementing HD into audio production workflow may contribute to the ongoing debate between the creation of HD via hardware and/or software.

10.6 Discussion

The Harmonic Distortion Survey underscored that these audio professionals agreed harmonic distortion is an essential tool in their audio workflow. However, how and at what stage THD should be implemented is still strongly debated. Audio engineers, mixing engineers, and producers of this study are most likely to create THD using both hardware and software. Some of these audio professionals feel that there are many more advantages for creating TDH via software as compared to hardware, which is generally focused on recreating vintage sound. While the use of hardware and software to create THD is still debated throughout the industry, the key reasons for and against using either system are preference, availability, cost, and tool specialization. Some audio professionals of this study choose to create THD via software and hardware based on what stage they are at in their audio production workflow to benefit from the advantages of each method. Furthermore, there was no agreement in this study as to how clients perceive THD and what stage THD should be heard by clients. Overall, this survey highlights the variety of opinions in the ongoing debate over the use and implementation of THD. However, further insight may be gained by comparison with literature research.

The results of the Harmonic Distortion Survey and literature were compared and two major themes emerged: a generational affinity for software and a continued interest in hardware by audio professionals in their audio production workflows. While these themes appear to be disparate, they are both related by generational differences and prior experience. Some mix engineers view plug-ins as new tools using their knowledge of the hardware progenitors as only a reference or starting point.

10.6.1 A Generational Affinity for THD Software

A majority of survey respondents identified as audio engineers, and more than half claimed to use a hybrid workflow of both hardware and software emulated harmonic distortion. These results were then compared to the experience level of the audio professionals. The data suggests that three-quarters of audio engineers with less than five years of experience working in the industry preferred software over hardware.

The explanation of why this study's less-experienced audio engineers prefer software is threefold. First, more-experienced audio engineers may have had greater access to and experience with hardware. Furthermore, more-experienced audio engineers may have undergone training in audio production facilities that incorporated mostly audio hardware, which was more common a decade ago [4, 6, 23]. This explanation is supported by the responses of audio professionals in open-ended questions, who described their use of hardware in their audio production workflows. Specifically, more-experienced audio professionals described specific sounds and qualities created by overdriving specific hardware models they had spent a large amount of time operating. Conversely, less-experienced audio professionals have not had this amount of time and experience with hardware.

Second, this trend may reflect the type of audio training the respondents had based on the period in which they gained formal audio production training [27]. In a recent study, new hires in the audio production industry reported attending specific recording and audio programs at four-year institutions that the previous generations of audio production professionals had not attended [28, 29]. Furthermore, they were less likely to work as an apprentice or intern at a recording studio. Thus, more-experienced audio professionals may have trained in larger multi-million-dollar recording studios, and less-experienced professionals may have trained in smaller audio production facilities and at universities. These disparate learning environments may be influencing their willingness to work with alternative workflows.

Third, this trend may be influenced by changes in modern audio production work. Over the last two decades, the audio industry has seen a rise in the digitization and democratization of the audio production workflow [30]. Thus, less-experienced audio professionals may be responding to the current needs of the industry without trying to harken back the sound and process of multi-million-dollar recording spaces of past generations.

10.6.2 A Continued Interest in Hardware

Many of the audio professionals surveyed displayed a continued interest in hardware created harmonic distortion in their audio production workflows. While many acknowledged the disadvantages of cost, maintenance, operation, and incorporation in the audio workflow, almost half cited its importance and advocated its use. Many maintained that the difference in sound quality and performance is significant even though this is in direct contradiction to software emulation documentation and research [3, 5, 8, 16, 19]. A major factor in this belief could relate to how and why these audio professionals are using these hardware units.

Audio professionals use hardware created THD because of the character or tone created when overdriving or purposely misoperating the hardware devices. Specifically, audio professionals described "driving hardware hard and cranking inputs" and that "the real point of divergence in software and hardware is when you really drive the units." These sound effects may not be possible to be recreated by software emulators that strive to recreate high-fidelity outputs. Furthermore, the literature suggests that software emulation development includes a combination of physical signal testing, circuit diagram generated DSP routines, prior modelling experience, and intensive listening sessions [15, 17, 18]. However, the literature does not explicitly state which operational methods are being recreated. Also, it is unknown how software developers approach emulating overdriven or misused hardware to achieve specific sonic outcomes. Overall, this could suggest that there is a disconnect between what sounds and performance characteristics software is designed to emulate and the types of sounds audio engineers want to create by using and misusing the hardware.

This reverence for audio hardware may be further fostered by nostalgia, myth, and past experience. From the perspective of those audio professionals who have had the opportunity to work with hardware, larger format consoles, and in multi-million-dollar recording studios, the concept that all that can be replicated in a computer may seem absurd and like science fiction. In addition, for a generation whose most beloved recordings and media had been created, mixed, and mastered in similar situations, a strong sense of tradition and nostalgia can dictate workflow and practice.

10.7   Conclusion

This chapter defined basic THD and non-linearity, including the formula for calculating the percentage of error this non-linearity can create. Non-linearity has become such a ubiquitous part of the modern recording sound, and therefore THD hardware devices are relatively expensive and highly coveted. In comparison, software emulations, or plug-ins, can be obtained inexpensively and can provide audio professionals with the tones and character of the original hardware. Furthermore, THD software emulators have many advantages over hardware and offer flexibility within the digital realm. The Harmonic Distortion Survey detailed the preferences of audio professionals related to harmonic distortion, who feel that harmonic distortion is an essential tool in their audio workflow. However, how and at what stage THD should be implemented is still strongly debated. Audio engineers, mixing engineers, and producers are most likely to create THD using both hardware and software. Audio professionals feel that there are many more reasons for creating TDH via software as compared to hardware, which is generally focused on recreating vintage sound. While the use of hardware and software to create THD is still debated throughout the industry, the key reasons for and against using either system are preference, availability, cost, and tool specialization. Some audio professionals choose to create THD via software and hardware based on what stage they are at in their audio production workflow to benefit from the advantages of each method. Furthermore, there is not agreement as to how clients perceive THD and when mixes with THD should be heard by clients.

A comparison of the Harmonic Distortion Survey results and literature was conducted and two major themes emerged: a generational affinity for software and a continued interest in hardware by audio professionals in their audio production workflows. Generational affinity for software is most likely influenced by less-experienced engineers' having less experience with hardware in training/education and the workplace. A continued interest in hardware by more-experienced audio engineers may be a result of tradition and knowledge of how to misuse devices to create distinctive sounds not intended by the original hardware design and not emulated in the software.

There are many myths about what both hardware and software can and cannot do. Some of the audio professionals in the Harmonic Distortion Survey confided that this is reinforced by fellow professionals, less-informed clients, and enthusiasts. Nevertheless, one commonality is that almost all the audio professionals of this survey agreed that harmonic distortion is an essential tool in their workflow. Moreover, harmonic distortion is a large part of recording culture, and its use is paramount to the larger production process.

References

[1] "Industry, international federation of the phonographic: Facts & stats," 2016 [Online]. Available: www.ifpi.org/facts-and-stats.php.
[2] L. Tronchin, "The emulation of nonlinear time-invariant audio systems with memory by means of Volterra series," *Journal of the Audio Engineering Society*, vol. 60, no. 12, pp. 984–996, 2013.
[3] J. &. T. S. Vanderkooy, "Harmonic distortion measurement for nonlinear system identification," in *140th Audio Engineering Society Convention*, Paris, 2016.
[4] J. Strawn, "Technological change: The challenge to the audio and music industries," *Journal of the Audio Engineering Society*, vol. 25, no. 12, pp. 170–184, 1997.
[5] V. Limsukhawat, "Analog mixer digitally controlled via plugin," in *137th Audio Engineering Society Convention*, Los Angeles, 2014.
[6] C. Holbrow, "Turning the DAW inside out," in *146th Audio Engineering Society Convention*, Dublin, 2019.
[7] F. Rumsey, "Digital audio effects and simulations," *Journal of the Audio Engineering Society*, vol. 58, no. 5, pp. 420–427, 2010.
[8] P. Audio, "Introduction to the basic six audio tests," 2016 [Online]. Available: www.ap.com/solutions/introtoaudiotest.
[9] "Total harmonic distortion and effects in electrical power systems," *Associated Power Technologies*, 2011 [Online]. Available: www.aptsources.com/wp-content/uploads/pdfs/Total-Harmonic-Distortion-and-Effects-in-Electrical-Power-Systems.pdf.
[10] A. Moore, "An investigation into non-linear sonic signatures with a focus on dynamic range compression and the 1176 fet compressor," Doctoral thesis, University of Huddersfield, 2017.
[11] S. L. A. M. R. P. Burrascano, "Accuracy analysis of harmonic distortion estimation through exponential chirp pulse compression," in *2019 International Conference on Control, Automation and Diagnosis*, Grenoble, 2019.

[12] DIY Recording Equipment, "The complete guide to UREI/Universal audio 1176 revisions," 2011 [Online]. Available: www.diyrecordingequipment.com/blogs/news/15851664-the-complete-guide-to-urei-universal-audio-1176-revisions.
[13] R. B. D. Jeffs, "Dynamics processors," 2005 [Online]. Available: www.rane.com/note155.html.
[14] A. Moore, "Dynamic range compression and the semantic descriptor aggressive," *Applied Sciences*, vol. 10, no. 7, p. 2350, 2020.
[15] M. Lambert, "Have you ever wondered exactly what goes on when classic hardware is recreated in plug-in form?," 2010 [Online]. Available: www.soundonsound.com/sos/aug10/articles/modelling-plugins.htm.
[16] C. McDowell, "Apparatus, a system and a method of creating modifiable analog processing," *USA Patent 10,241,745,* 26 March 2019 [Online]. Available: https://www.uspto.gov/patent/10241745.
[17] J. Sterne, *Testing Hearing: The Making of Modern Aurality*, Oxford: Oxford University Press, 2020, p. 159.
[18] G. Wright, *Peabody Computer Music: 46 Years of Looking to the Future*, ICMC, University of North Texas, CEMI, 2015 [Online]. Available: http://hdl.handle.net/2027/spo.bbp2372.2015.031.
[19] S. H. Hawley, B. Colburn and S. Mimilakis, *SignalTrain: Profiling Audio Compressors with Deep Neural Networks*, Ithaca, NY: Cornell Press, 2019.
[20] J. Loar, *The Sound System Design Primer*, London: Routledge, 2019.
[21] R. Toulson, "Evaluating analog reconstruction performance of transient digital audio workstation signals at high-and standard-resolution sample frequencies," *Innovation in Music: Performance, Production, Technology, and Business*, p. 385, 2019.
[22] M.-W. E. A. Kwon, "Simulation program with integrated circuit emphasis compact modeling of a dual-gate positive-feedback field-effect transistor for circuit simulations," *Journal of Nanoscience and Nanotechnology*, vol. 19, no. 10, pp. 6417–6421, 2019.
[23] R. Izhaki, *Mixing Audio: Concepts, Practices and Tools*, London: Taylor & Francis, 2013.
[24] Z. Xiong, "Volterra series-based time-domain macromodeling of nonlinear circuits," *Transactions on Components, Packaging and Manufacturing Technology*, vol. 7, no. 1, pp. 39–49, 2017. http://doi.org/10.1109/TCPMT.2016.2627601
[25] Warm Audio, "WA76 discrete compressor," *Warm Audio*, 2020 [Online]. Available: https://warmaudio.com/wa76/.
[26] Waves Audio, "CLA-76 compressor/limiter," *Waves Audio*, 2020 [Online]. Available: www.waves.com/plugins/cla-76-compressor-limiter.
[27] D. Bielmeier, "Apprenticeship skills in audio education: A comparison of classroom and institutional focus as reported by educators," in *137th Audio Engineering Society Convention*, New York, 2014.
[28] D. Bielmeier, "Audio recording and production education: Skills new hires have and where they reported," *Journal of the Audio Engineering Society*, vol. 64, no. 4, pp. 218–228, 2016.
[29] D. Bielmeier, "Future educational goals for audio recording and production (ARP) programs: A decade of supporting research," in *145th Audio Engineering Society Convention*, New York, 2018.
[30] D. Tough, "Shaping audio engineering curriculum: An expert panel's view of the future," in *129th Audio Engineering Society Convention*, San Francisco, 2010.

11 'It Just Is My Inner Refusal'
Innovation and Conservatism in Guitar Amplification Technology

Jan-Peter Herbst

11.1 Introduction

In the Western world, technological development is generally seen as positive for creative practices (Niu and Sternberg 2006). In music, however, new technologies must be considered in a more multi-faceted way. Instruments such as keyboards have benefitted from technological advances through increased processing power and additional functionality like polyphony on emulations of monophonic analogue synthesisers. Similarly, electronic drum kits have become more realistic because they contain sounds from multiple samples recorded at various velocities. In music production, there has been a dual development: While modern DAWs have extended functions and improved workflow (Watson 2019), vintage analogue gear is as popular as ever, and the market for emulations is growing rapidly (Bourbon 2019). Regardless of the innovative technologies used to introduce more advanced instruments or emulate analogue hardware more authentically, equipment manufacturers present innovation as something positive. Technological developments are likely an improvement when viewed free of emotions and without the various implications and meanings associated with them, be it social status (Cole 2011), nostalgic longings (Williams 2015) or iconicity (Bennett 2012). This is not to say that older analogue gear, often associated with 'warmth' (Bennett 2012), tactility and intuitive handling, may not be more pleasing, but newer digital technology usually excels in functionality and workflow (Pinch and Reinecke 2009). Evaluating the analogue and digital divide in gear discourse, Crowdy (2013, p. 152) reminds us that 'sound aesthetics [are] effectively acting as a proxy for other areas of opinion and value. These may include appearance, usability, brand loyalty, peer pressure and justification of expense'. Musical equipment is linked to social distinction (Kaiser 2017).

Electric guitarists have always been traditionalists tending to reject technological development, which distinguishes them from other instrumentalists, engineers and producers (Herbst et al. 2018, Herbst 2019a, 2019b). Classic guitar shapes such as Les Paul, Stratocaster, Telecaster and Flying V are still the most popular today, especially from their original manufacturers Gibson and Fender, who serve the market demand with rebuilds and relic series. Alternatives to traditional valve amplification, fundamental to the guitar's electrification in the early 1930s, were explored more widely in the 1980s with hybrid designs of valve preamps and transistor power amplifiers in larger rack units, but this did not permanently change guitarists' preferences. With the nu metal genre in the 1990s, hybrid systems became popular again for a short time, before the general trend went back to the classic valve amplifier sounds of the 1960s and 1970s. The popularity of low-power amplifiers, capable of producing valve distortion at low volumes, demonstrates this. Rejecting technologies other than valve is mainly owed to the importance of distorted sound. Guitarists commonly believe that simulation and modelling amplifiers are sufficient for practising and live performances, yet they are hardly suitable for serious recording. Musicians interviewed by Pinch and Reinecke (2009) were flexible about digital recording equipment and instruments but refused to use guitar amplification simulations in the DAW. Similarly, while Mynett (2017, p. 57) in his *Metal Music Manual* acknowledges the high quality of modern simulations, he summarises common criticisms, such as perceived 'lack of natural "air"' and the technology's inability to 'deliver a musician's unique tonal identity to a production'.

In 2011, the German company Kemper GmbH launched a guitar amplifier based on a new 'profiling' technology as an alternative to modelling approaches that apply digital signal processing to simulate the behaviour of components of a physical system such as amplifiers, cabinets, microphones and microphone positions. Inventor Christoph Kemper explains the difference between modelling and profiling:

> Modeling . . . is bringing the physics of the real world into a virtual world by defining formulas for the real world and letting them calculate on a real-time computer. . . . Profiling is an automated approach for reaching a result that is probably too complex and multidimensional to achieve by ear, or by capturing the behavior of individual components in isolation. . . . 'modeling' was used as a marketing term by some companies. It says: 'Here is a valid virtual copy of a valuable original'. What I have rarely seen is an A/B comparison between the original and the virtual version. Why is that? Profiling . . . is a promise to create a virtual version of your original, but with the ability to qualify the results by a fair A/B comparison.
> (Collins 2011)

Profiling creates a 'sonic fingerprint' of a valve amplifier that captures and distinguishes between the various sound-forming stages, the amplifier, cabinet and microphones used. Consequently, it does not represent the sound of a random Marshall JCM800 but of a particular Marshall JCM800 device recorded in a specific room with a certain microphone and preamplifier. A properly created profile should have no perceptible differences between the original and its copy.

Despite the enormous impact profiling has had on the guitar and recording scene, there is surprisingly little research on it. In their pioneering study, Herbst et al. (2018) quantitatively tested the authenticity of profiles based on an analysis of low-level timbre features. The results suggest negligible deviations between the profile and the original amplifier. Recently, Düvel et al. (2020) conducted a listening test confirming that listeners, even music experts, cannot reliably distinguish the original amplifier from its profile in direct A/B comparison. This result is consistent with countless listening tests online. Despite its apparent quality, profiling technology was confronted with the guitarists' scepticisms, as a review in *Guitarist* shows: 'Our culture is bound up in ritual, superstition and myth—and we like it that way. We know great tone and it sure as hell doesn't come from ones and zeroes' (Vinnicombe 2012, p. 119). Such adverse reactions were not limited to guitarists; even producers like Lasse Lammert, creator of an official Kemper rig pack, rejected the technology, not because of its inferior quality but out of tradition-conscious conviction, stating 'it just is my inner refusal' (Herbst 2019a, p. 57). Profiled sounds are often seen as 'very close to "real" amps' (Anderton 2013), rarely as amplifiers in their own right. But the perception gradually improved, and many now consider the profiler a 'truly revolutionary piece of kit for serious recording guitarists and producers' (Vinnicombe 2012, p. 122). For studying the views of guitarists and producers of guitar amplification, profiling technology is ideally suited because it creates authentically analogue, even if digital, clean and distorted sounds. By studying the company's website and marketing strategy, this chapter analyses from the perspective of innovation research why guitarists and record producers, who for decades rejected amplification not based on traditional valves, gave in and gradually adopted profiling technology.

11.2 Diffusion of Innovation

In his influential work *Diffusion of Innovation*, Rogers (2003, p. 12) defines innovation as an 'idea, practice, or object that is perceived as new by an individual', the desirability of which depends on the individual and their situation. A characteristic feature of innovation is uncertainty individuals wish to reduce through information-seeking activities, which eventually lead to a decision whether or not to adopt it (Rogers 2003, p. 14). This process takes five steps (Rogers 2003, pp. 169–189): (1) An individual learns about an innovation and seeks to understand it, which may result in deeper interest ('knowledge'). (2) The individual forms an attitude toward the innovation as an emotional reaction ('persuasion'). (3) A decision is made, often after a trial period or through observing another person testing the innovation. This can lead to adoption ('decision'). (4) The individual uses the innovation to further reduce uncertainty ('implementation'). (5) The individual seeks reinforcement of the decision by evaluation ('confirmation').

The adoption of an innovation depends on five attributes (Rogers 2003, pp. 15–16): (1) The innovation needs to be perceived as better than the previous technology ('relative advantage'). (2) The innovation must be consistent with existing values, experiences and needs ('compatibility'). (3) It must not be perceived as overly difficult to use or understand ('complexity'). (4) Adoption is more likely if the innovation can be tested ('trialability'). (5) Chances of adoption are higher if the innovation is visible to others ('observability'). Also important are recommendations from people who share characteristics and beliefs (Rogers 2003, p. 19), for example, word-of-mouth communication among individuals without commercial interests (Arndt 1967). Similarly influential are 'opinion leaders', people with 'technical competence, social accessibility, and conformity to the system's norms' (Rogers 2003, p. 27). Endorsements are one way to strategically deploy opinion leaders. The main function of endorsed branding is to add credibility through the celebrity endorser's expertise (Masterman 2007, p. 109). That can be an 'implicit' endorsement by association, 'explicit' endorsement or 'imperative' endorsement, whereby artists actively recommend using the product (Masterman 2007, p. 107), which is most effective in video format (Mowat 2017). Online brand communities can further increase the acceptance of an innovation, especially if they involve consumers as 'active partners to the product development process, as a kind of co-producers' (Arvidsson 2006, p. 70).

11.3 The Kemper Profiling Universe

11.3.1 Method

The company's website is its official communication channel. Besides product information, written artist statements and video interviews are at the centre of the online appearance. These were analysed with Mayring's (2014) *summarising content analysis* method, in which the essence of qualitative data is condensed into hierarchical categories that are formed inductively. With this method, 152 quotations were extracted from the 57 written statements and 39 videos of artists and producers (approximately 6 hours' duration). Statements about touring and live performances were ignored to focus the analysis on record production. The quotes resulted in 219 codes, divided into seven categories: 'attitudes and beliefs' (31), 'adoption' (15), 'sound quality' (46), 'attributes' (66), 'production aspects' (35), 'community' (6) and 'significance of innovation' (20). Timecodes show that an artist quote is from a video interview (Kemper 2020g), while the written statements come from the artist gallery (Kemper 2020f).

11.3.2 Overview of the Website

On the start page, the technology is boldly described as '*the* leading-edge digital guitar amplifier and all-in-one effects processor. Hailed as a game-changer by guitarists the world over, the PROFILER™ is the first digital guitar amp to *really nail* the full and dynamic sound of a guitar or bass amp' (Kemper 2020a). With profiling technology, for the 'first time in history, guitar players are free to create their most individual tones . . . and then capture these tones exactly into the digital domain with the unique technology' (Kemper 2020a). This text attracts attention and motivates visitors to continue reading. The following section promises access to the 'rarest and finest amps on the planet' and a constantly growing library of amplifiers. Finally, the profiler's use is described in detail for three scenarios: the studio, the stage and practice at home. Every visitor, be it a guitarist, an engineer or a producer, will see an application for this innovation. The product page (Kemper 2020b) gives an overview of the profiler's different versions—head, rack, power rack and power-head—tailored precisely to the three scenarios of playing, recording and producing. This introduction fulfils the first step of diffusion of innovation: awareness, interest and 'knowledge' (Rogers 2003, pp. 169–174).

The community area consists of three parts. The first is the 'Rig Exchange' (Kemper 2020c), where more than 15,000 user profiles can be downloaded, most of which are user-rated. This platform offers an unlimited supply of amplifier profiles from mainstream to boutique devices. In the 'User Performances' area (Kemper 2020e), users can present their sound and playing. In theory, it acts as a proxy for 'trialability' (Rogers 2003, p. 16) from 'regular' peers, but the small number of 12 videos makes it less effective. The forum (Kemper 2020d), however, is a thriving brand community with about 500,000 posts. However, only 500 threads are dedicated to recording, and those discussions are rather basic. Professional recording artists and audio experts do not seem to participate in this community. The subforum 'Feature Requests' contains 34,000 posts, which confirms the importance of consumers as co-producers for the acceptance and development of an innovation (Arvidsson 2006, p. 70).

The most effective areas for diffusion are 'Videos' (Kemper 2020g) and 'Artists' (Kemper 2020f). The 'Artists Gallery' showcases the most renowned endorsers with written statements about the profiler, the 'Hall of Fame' lists hundreds of endorsers categorised by continent and the discography links directly to albums supposedly recorded with the Kemper. The video area comprises four parts. The 'Tutorials and Demos' fulfil the function of 'knowledge' (Rogers 2003, pp. 171–174) and demonstrate low 'complexity' (Rogers 2003, p. 16). The other three areas, 'Artist Talk', 'Rig Check' and 'Artist Performance', present, to some extent, endorsement videos, in which potential adopters can see celebrities expressing their views on the innovation or demonstrating it in action.

The website fulfils all requirements for successful diffusion of innovation and provides a brand community for consumer support and participation in further product development. The 'regular' adopters are useful agents for word-of-mouth recommendations, and the celebrities act as respected opinion leaders.

11.3.3 Attitudes and Beliefs

The interviewed artists' attitudes towards profiling technology reflect the general scepticism of guitarists to digital technologies: 'First of all, I thought it's just another digital amp' (Lionel Loueke 0:47–1:00), or 'We obviously had experience with digital stuff, but we never would have considered that for recording' (Trivium 1:12–1:28). This scepticism is relatable to guitarists and adds credibility to the re-evaluation of digital technologies. It is an important part of the narrative because it shows that the professionals share doubts about digital technology, as is the case with producer Andy Sneap's interview (2:57–4:30),

> I've always been a bit dubious about these profiling amps because you don't want it to be true, you don't want it to work, you don't want to put your sound into a box. . . . I don't want it to work, but it does work.

In many interviews, the professionals express an unexpected positive surprise that appeals to the emotional attitude of an individual interested in the innovation in the 'persuasion' phase (Rogers 2003, pp. 174–177). The change from rejection to naming profiling technology as one of the greatest innovations for music production sends a strong message to sceptics, as producer Michael Wagener (4:55–7:50) shows:

> I have to say I had a lot of modellers in the studio . . . but I haven't found one that I liked. I did not want to record them. . . . When the KPA first entered my studio, I was very sceptical, and I heard it, and I was like, 'oh no, this is amazing!' Hands down, this is the best machine entering my studio in the last fifteen years. This is the biggest innovation for recording in at least the last fifteen years.

Many professionals were principally receptive to digital technology because of its better functionality, but the perceived inferior sound quality, especially of distorted tones, has prevented this so far (Herbst 2019a). With the profiler, the situation changed for many, including guitarist Marco Heubaum (Kemper 2020f):

> The Profiler is the guitar device I was always hoping for. . . . I was becoming so sick of a valve amp stack being difficult to align with effect equipment in a reliable, comfortable, and reproducible way on the one hand, and on the other hand of the imperfection of digital amp modeler's sounds and usability. But this is history now, as the Profiler wipes away the drawbacks of both worlds, creating a wonderful combination of natural sound and all the comfort that is needed.

The profiler is perceived to combine the strengths of analogue and digital without the drawbacks of either. Pinch and Reinecke (2009, p. 164) describe in their study of 'technostalgia' how musicians value a connection to the past by paying tribute to their instrument's history. That is certainly one of the reasons why many guitar players hesitated to adopt newer technologies. Yet, the profiler offers the potential to bridge the worlds, as producer Alex Silva (10:16–11:40) reflects:

> I think companies like Kemper are on the cutting edge of being able to provide musicians with the quality that they need without feeling that they're kind of losing connection to the past of guitar playing. . . . I don't think one thing suddenly replaces the other. But I think they are certainly two things that co-exist and benefit each other in a way.

In this respect, two elements are crucial for the profiler's perception and narrative. The professionals confirm the analogue valve's unbroken value in a digital world, and in conjunction with this, they affirm the 'realness' of the digital amplifier. In many video interviews, the professionals describe themselves as a 'valve amp snob' (John Huldt, 6:55–7:20), 'very analogue guy' (Lionel Loueke, 4:10–4:38) or 'tone purist' (Alex Skolnick, Kemper 2020f). A list of owned models, usually vintage amplifiers, hand-wired series or boutique models, support these affirmations. Most professionals also explain their recording practices, which is either very puristic (Wolf Hoffmann, Kemper 2020h) or highly sophisticated (Bob Kulick and Brett Chassen, Kemper 2020h). To conceal the (inaudible) digital nature of the amplifier, several professionals reveal to feel better if they record the profiler with their trusted preamps and do not use the amplifier's equaliser section but process the sound on their analogue console (Alex Silva, Kemper 2020h). This step of 'analogization' makes the Kemper more compatible with their traditional analogue convictions.

Closely related to the analogue qualities, the profiler's 'realness' is often emphasised, 'I love how close it comes to an actual amp, and the reaction and all that. It doesn't try to be a modelling thing; it tries to be a real amp . . ., and it does that perfectly' (Timo Somers 1:18–1:33). Not only is it described as *sounding like* a 'real amp', but it also *looks like* an amplifier (Wagener, 10:20–10:25). In most statements, the profiler is considered a *representation* of a 'real amp' but rarely an amplifier. The sound quality is distinctly analogue, and the functionality is better, but few professionals see the profiler as the next step in the development of amplification technology superior to valve amps, especially when it comes to distortion.

11.3.4   Adoption

In Rogers' (2003, p. 19) theory, the personal recommendation of a close person sharing similar beliefs is important for sceptics to consider an innovation. Many interviews involve word-of-mouth recommendations, often in the exact phrase 'a friend of mine told me about it' (Kemper 2020h: Wolf Hoffmann, Dino Cazares, Rem Massingill, Johnny Radtke). This friend is usually a fellow guitar player. Generally, producers are open to new technologies that improve their workflow, and through them, many professional guitarists learned about the profiler (Oliver Drake, Kemper 2020f). This peer recommendation helps overcome the common scepticism. According to Rogers (2003, p. 16), another key factor for adopting an innovation is 'trialability'. Several guitarists and producers state that they learned about profiling technology at music conventions such as NAMM and Musikmesse and developed an interest in it (Kemper 2020h: John Payne, Michael Wagener, John Huldt, Rod Gonzales). The videos suggest that Kemper GmbH sent devices to many professionals to try them out in return for written artist statements and video interviews, most of which were recorded within a short period of time to capture the emotional reaction to the innovation. In rare cases, as with producers Andy Sneap and Sean Beavan (Kemper 2020h), a follow-up evaluation was made several years after their first encounter. The initial excitement captured in various videos is transformed into 'implicit endorsements' (Masterman 2007, p. 107) without the company's intervention:

> It's easy to endorse products that actually do what they claim to do, and in the case of the Profiler, I'm happy to add my name to the list of those who are now believers!! This is a fabulous production tool and now a staple in my Atlanta studio.
> (Jan Smith, Kemper 2020f)

As another strategy to win over interested visitors, the website contains demonstration videos of the profiling process, including A/B comparisons and celebrities playing the profiler (HammerFall, Kemper 2020h). This demonstration usually takes place in a professional studio with vintage consoles and outboard gear to suggest that a digital device like the Kemper does not conflict with an analogue studio environment (Tim Pierce, Kemper 2020f).

11.3.5 Sound Quality

Sound quality has always been the main reason why guitarists and producers rejected transistor and digital amplifiers for distorted sounds. Guitarist John Huldt (3:09–3:25) explains it 'has to sound like a guitar, it can't sound like something *trying* to sound like a guitar amp'. Many statements therefore commend the profiler's tonal qualities and playing feel. John Payne (6:40–7:23) brings it to the point:

> The most important thing at the end of the day is, does it sound good? You can have all those accessibility options as we've had before on pedalboards, and in my mind, many of those fall short other than being great for demoing or whatever, but in a track, I've used this with real amps and the profiling amplifier, and it really holds its sound.

Producers especially emphasise the profiler's quality. As Sneap (6:05–6:15) notes, capturing an amplifier's sound through a carefully selected recording chain creates a profile of the 'amp at its prime'. Experts claiming that the profiles sound better than the original amplifiers does therefore not come as a surprise, 'Once you profiled your amp, you can refine it even more. I think you can sometimes actually beat your tone with the Kemper with the saturation and the compression, the pick attack and everything' (Sneap 12:48–13:08). Rem Massingill (6:45–7:05) points out the possibility of adding characteristics not present in the original amplifier, whereas Payne (1:20–3:50, 6:05–6:35) highlights that the envelope centre provides access to sound qualities similar to those of a synthesiser and that distortion level can be increased beyond what is possible with the original. Ace of Skunk Anansie (12:19–13:39) in a similar context remarks that profiling old amplifiers on the verge of breaking up 're-corrected it into what it was . . . [and] immortalised the amp because it's not gonna last, and it rectifies it so that it was usable'. The profiler thus allows improving vintage amplifiers by retaining the good features while removing the bad ones, and it enables customising the sound of an old amplifier. All these qualities meet Rogers' (2003, p. 15) criterion of the innovation's 'relative advantage'.

Its promise to deliver on an A/B comparison sets the profiler apart from conventional modelling solutions (Collins 2011). Several statements hence highlight the indistinguishable sound. Producers are less affected by the playing feel of the amplifier and judge primarily by the sound quality. Sneap (10:00–10:20) and Wagener (8:30–9:20) by no means consider the profiler inferior under studio conditions:

SNEAP: It does sound that accurate, and if I'm sat here in the studio, and I can't tell the difference between that and the actual amp that I've profiled then . . . If it's good enough for me, in a studio environment, I don't know who is gonna fault it, to be honest.

WAGENER: I think it's right on. I've had a situation where I had a DI track in the computer and send it out to an amp and had a real good tone; then I profiled that amp and punched it into the original track, and I could not hear a difference. And that's good enough for me.

For guitarists, an authentic playing feel, often criticised in transistor amplifiers, simulations and modellers, is crucial. This seems not to be an issue with the profiler (Kemper 2020h: Sean Beavan, Andy Sneap, John Huldt; Kemper 2020f: Tim Stewart).

11.3.6 Attributes

Sound quality is the primary requirement for an amplifier in a professional context, but it needs other attributes as well to make the innovation more attractive than a 'real' analogue amplifier, what Rogers (2003, p. 15) calls a 'relative advantage'. In line with the affordances of digital technology, *convenience* is the most frequently mentioned attribute. A major advantage of the profiler is that the 'mic'ing up' process is required just once in the profiling session. After that, the sound in its best quality can be used in any session (Ace 13:50–13:57, Payne 4:30–6:05). For producers, the profiler 'save[s] hours of time that would otherwise be spent mic'ing up amps, trying different combinations, etc. Now I get to just plug in and go' (Chris Hesse, Kemper 2020f). Especially the degree of distortion can be easily adjusted at any time in a production. As Wagener (20:47–21:14) states, fixing the amplifier's equaliser settings is relatively easy in the mix, whereas a wrong distortion level is more difficult to correct. Profiling technology thus provides distortion control in the mixing phase.

Another convenience-related attribute is *flexibility* in amplifier selection:

> A performance that is all switchable on the fly, You want a 6505 rhythm, but feel the need to have a two rock clean; this is now a possibility, you want a fender twin clean, but with that Friedman organic gain staging? that dream is finally a reality.
>
> (Colin Parks, Kemper 2020f)

Such flexibility is particularly valuable given the increased blurring between recording, mixing and production. Choosing sounds at the flick of a button allows adding or changing guitar parts at any stage of the production process, be it for corrective or creative purposes (Tim Palmer, Kemper 2020f). Auditioning whether a sound works for a form part requires little effort (Sneap 6:13–7:20). Profiling devices provide access to an unlimited number of amplifiers and recording chains, making the recording process more *versatile*. There are producers, Beavan (0:20–0:34) and Wagener (16:25–17:13), even admitting having downloaded sounds from the user community. But unlimited possibilities, as Silva (19:10–19:46) notes, carry the risk of not committing to production decisions, something that proponents of analogue recording have criticised about digital production in general (Watson 2019).

Apart from the technical side of production, these affordances also shape the musical side. Several artists and producers highlight profiling technology to support *inspiration* and *creativity*. Payne (13:50–14:38) explains how the profiler has brought guitarists and keyboard players on par with an immense sound library:

> It's been a wonderful learning curve, and I actually wrote a couple of songs on it. Sounds are inspiring to players. Synthesists and pianists already had that, hit the presets, and go for great sounds. Before, we had pedal boards and stuff. With something like this, you can just go and got a great sound. For that, it has brought us to the same age as keyboard players, the immediate accessibility to sounds that then inspire you to write something different.

Billy Gould (2:56–4:33) confirms the positive effect of immediate access to a variety of sounds on the songwriting process:

> Every amp has its own strengths and weaknesses, and you learn how to play to these amps. . . . And what you can do, is you can start writing and using the strengths of the amplifier, where it sounds better on certain chords than on other amps. When I write now, I write using those

characteristics, the character. . . . That was a huge thing for me because you hear what it's gonna sound, it's not just on a fake distorted little box, and you're hoping that people would understand later what you really wanna do.

From a production perspective, Sneap (6:13–7:20) stresses the creativity that comes from immediately trying out different sounds rather than separating the recording, arranging and mixing phases:

Want to try a different layer, like a cleaner tone within the chorus, you can literally click straight through it and see if it's gonna work. You're not trying to mentally stack ideas up; you can try things out as you go through the day. And also, it breaks things up a little; you can go to a clean tone, do the clean parts for a couple of hours, go back to heavy tones, you're not getting bored of doing the same thing all the time. To me, it makes it more creative.

Also related to creativity is the '*fun* factor' highlighted in various interviews, 'The Profiler is the most fun I've ever had with any guitar related item' (Keith Merrow, Kemper 2020f). Other related intangible attributes commonly expressed are *vibe*, *mojo* and *magic* (Kemper 2020h: Rem Massingill, John Huldt, Chris Henderson, Andy Sneap).

An important criterion in Rogers' (2003, p. 16) theory is an innovation's 'complexity'. In line with the theory, many professionals emphasise the profiler's ease of use. Even technically unexperienced musicians are not overwhelmed,

I'm a dumb guitar player, I'm used to see volume, bass, treble, mids, gain, and that's it. And all of a sudden, you have all these knobs and shining lights and stuff like that, 'what's going on'. But it's all in sections that make a lot of sense; it's really easy to use.

(Huldt 7:46–8:20)

Likewise, producers who work on tight schedules appreciate the ease of use, 'It looks incredibly complex, I thought. Having now used it . . ., it's so simple to use, I never looked in the manual' (Silva 4:30–6:10). The profiler is easy to understand because its interface follows conventional formats, similar to the skeuomorphic approach taken in plugin emulations of analogue studio hardware (Bourbon 2019, p. 215), facilitating acceptance of the digital device.

11.3.7 *Further Production Benefits*

The profiler shares many advantages with other digital production tools, especially *total recall*; writing recall sheets becomes obsolete (Rod Gonzales, 1:40–2:52). Furthermore, Gould (0:10–0:49) explains, 'we could come back and revisit songs later, we could travel, I was out of town a lot. We could work at home on parts, and we had his amp in there'. In this context, the profiler's usefulness for *remote work* is highlighted. Instead of sending re-amped audio files to the collaborator, exchanging the profile and fine-tuning it in the context of a mix is regarded as more effective and creative (Beavan 5:30–6:15, Payne 10:34–12:04). Many producers celebrate the profiler as a *money saver*, given the recording industry's limited budgets (Silva, Kemper 2020h). Beavan (3:00–4:10) recommends renting a high-quality studio for one day to create profiles and record performances in the home studio. Artists like Huldt (1:33–1:57) emphasise that one can record an entire solo guitar album with a profiler at home, which is not possible in a commercial studio due to cost. Payne (9:54–10:34) points out that a 'lot of people have great control rooms, but they don't have great live rooms, so they can't hook up a big setup with multiple speakers and loud volume'. Another benefit seen concerns the *transition from demo to album production*. Ace (2:56–4:33, 8:43–10:44) highlights that recording demos with his personal sound in good quality helps him evaluate how the song would turn out on the record. If an overdub or ensemble performance had the right feel, it could be used on the record instead of having to recreate the feel for the sake of better sound quality.

11.4 Conclusion

In the ten years of its existence, profiling technology has convinced many guitarists and music producers who have traditionally been sceptical about digital amplification of any kind, primarily because of the shortcomings of distorted tone. Kemper succeeded in combining the best of both worlds, creating an analogue sound with digital functionality. For those with a cherished amplifier collection, the profiler provides one-touch access to the sounds of each amplifier, which is useful for performers and producers. For those without extensive resources, the profiler is equipped with a wide range of mainstream and boutique amplifiers that can be expanded through the 'Rig Exchange' on the website. The company also sells signature 'rig packs' from producers such as Michael Wagener that give musicians without expensive amplifiers, recording chains and live rooms access to world-class guitar sounds. The Kemper website analysis suggests that both the company's marketing strategy and technology's convincing quality have led to the successful diffusion of the innovative profiling technology. Guitarists and producers no longer need to decide between analogue and digital; the profiler is equally convenient and convincing. For several of the artists, profiling is 'a dream come true'. Others regard it as nothing short of a 'game changer'. Andy Sneap (8:29–8:33) believes profiling to 'move recording forward the same way as Pro Tools has', and Brett Chassen (6:50–8:56) sees it as the 'future' and an 'indispensable studio tool'. Apart from its potential to produce a better sound than the copied original valve amplifier, the profiler fosters musical innovation in many professionals' eyes. As the artists and producers have expressed, access to a wide variety of high-quality sounds when needed sparks inspiration in songwriting, arrangement, mixing and production. Profiling amplifiers are therefore becoming increasingly common in rehearsal rooms, live stages and recording studios, which shows that guitar player and music producer scepticism toward digital amplification technology is waning.

References

Anderton, C. *Expert Review: Kemper Profiling Amp* (2013). Accessed April 2020 from www.harmonycentral.com/expert-reviews/kemper-profiling-amp

Arvidsson, A. *Brands*. London: Routledge (2006).

Bennett, S. Endless Analogue: Situating Vintage Technologies in the Contemporary Recording & Production Workplace. *Journal on the Art of Record Production*, Vol. 7 (2012). Accessed April 2020 from www.arpjournal.com/asarpwp/endless-analogue-situating-vintage-technologies-in-the-contemporary-recording-production-workplace

Bourbon, A. Plugging in. Exploring Innovation in Plugin Design and Utilization. In: Hepworth-Sawyer, R., Hodgson, J., Paterson, J., Toulson, R. (eds.) *Innovation in Music Production*, pp. 211–225. London: Routledge (2019).

Cole, S. J. The Prosumer and the Project Studio. The Battle for Distinction in the Field of Music Recording. *Sociology*, Vol. 45, No. 3 (2011), pp. 447–463.

Collins, S. *Christoph Kemper of Kemper Amps Talks About Their New Profiling Amp* (2011). Accessed April 2020 from www.guitar-muse.com/kemper-profiling-amp-2949-2949

Crowdy, D. Chasing an Aesthetic Tail. Latent Technological Imperialism in Mainstream Production. In: Baker, S., Bennett, A., Taylor, J. (eds.) *Redefining Mainstream Popular Music*, pp. 150–161. London: Routledge (2013).

Düvel, N., Kopiez, R., Wolf, A. and Weihe, P. Confusingly Similar: Discerning between Hardware Guitar Amplifier Sounds and Simulations with the Kemper Profiling Amp. *Music & Science*, Vol. 3 (2020), pp. 1–16.

Herbst, J.-P. Old Sounds with New Technologies? Examining the Creative Potential of Guitar 'Profiling' Technology and the Future of Metal Music from Producers' Perspectives. *Metal Music Studies*, Vol. 5, No. 1 (2019a), pp. 53–69.

Herbst, J.-P. Empirical Explorations of Guitar Players' Attitudes Towards their Equipment and the Role of Distortion in Rock Music. *Current Musicology*, Vol. 105 (2019b), pp. 75–106.

Herbst, J.-P., Czedik-Eysenberg, I. and Reuter, C. Guitar Profiling Technology in Metal Music Production: Public Reception, Capability, Consequences and Perspectives. *Metal Music Studies*, Vol. 4, No. 3 (2019), pp. 481–506.

Kaiser, C. Analog Distinction—Music Production Processes and Social Inequality. *Journal on the Art of Record Production*, Vol. 11 (2017). Accessed April 2020 from www.arpjournal.com/asarpwp/analog-distinction-music-production-processes-and-social-inequality

Kemper. *Start Page* (2020a). Accessed April 2020 from www.kemper-amps.com/profiler/overview

Kemper. *Products* (2020b). Accessed April 2020 from www.kemper-amps.com/products/profiler/line-up

Kemper. *Rig Exchange* (2020c). Accessed April 2020 from www.kemper-amps.com/rig/exchange

Kemper. *Forum* (2020d). Accessed April 2020 from www.kemper-amps.com/forum/

Kemper. *Users' Corner* (2020e). Accessed April 2020 from www.kemper-amps.com/users-corner

Kemper. *Artist Gallery* (2020f). Accessed April 2020 from www.kemper-amps.com/artist-gallery

Kemper. *Videos* (2020g). Accessed April 2020 from www.kemper-amps.com/video#a-videos-interviews

Masterman, G. *Sponsorship*. Oxford: Butterworth-Heinemann (2007).

Mayring, P. *Qualitative Content Analysis*. Klagenfurt: SSOAR (2014).

Mowat, J. *Video marketing strategy*. London: Kogan Page (2017).

Mynett, M. *Metal Music Manual*. London: Routledge (2017).

Niu, W. and Sternberg, R. The Philosophical Roots of Western and Western Conceptions of Creativity. *Journal of Theoretical and Philosophical Psychology*, Vol. 26 (2006), pp. 18–38.

Pinch, T. and Reinecke, D. Technostalgia: How Old Gear Lives on in New Music. In: Bijsterveld, K., van Dijck, J. (eds.) *Sound Souvenirs*, pp. 152–166. Amsterdam: Amsterdam University Press (2009).

Rogers, E. *Diffusion of Innovations*. New York: Free Press (2003).

Vinnicombe, C. Kemper Profiling Amp. *Guitarist*, Vol. 2012, No. 5 (2012), pp. 118–124.

Watson, J. Committing to Tape. Questioning Progress Narratives in Contemporary Studio Production. In: Hepworth-Sawyer, R., Hodgson, J., Paterson, J., Toulson, R. (eds.) *Innovation in Music Production*, pp. 248–265. London: Routledge (2019).

Williams, A. Technostalgia and the Cry of the Lonely Recordist. *Journal on the Art of Record Production*, Vol. 9 (2015). Accessed April 2020 from www.arpjournal.com/asarpwp/technostalgia-and-the-cry-of-the-lonely-recordist

12 A Saturated Market

Ask Kæreby

12.1 Introduction: It's Not Old, It's 'Vintage'

Classic audio equipment has enjoyed a renaissance similar to that of classic musical instruments. Allow me to offer a few concrete examples: A bass player may lust after a pre-CBS Fender Precision Bass, meaning the instrument was produced when the company was still owned by Leo Fender, before being sold to the Columbia Broadcasting System corporation in January 1965, which brought on a number of (mostly cost-cutting) changes [1]. In the same way, an audio engineer can yearn for a 'brass-ring' AKG C414 EB condenser microphone or an original Teletronix LA-2A compressor. The AKG C414 microphone has been produced in numerous iterations (including different capsules, identified with 'brass' or 'Teflon' [2]), and while later versions improve upon technical performance [3], they do not sound identical to the original [4]. Universal Audio purchased Teletronix in 1967 and retired the product in 1969 [5], and despite reissues being produced from 1999 onwards (p. 12), original LA-2A units command very high prices on the vintage audio market. Many older electronic designs, including those now considered 'classic', typically have higher levels of noise and distortion than those of later decades, so it may seem curious that technically inferior pieces of equipment can be deemed (aesthetically) superior by those with 'golden ears'. A good case in point is the user guide for the reissued LA-2A: the dynamic range is 'better than 70' decibels (p. 1)—in other words, inferior to the quality of a CD and even the best (outer) parts of an LP—yet the compressor receives praise from renowned audio engineers who rely on its characteristic sound for their music productions (pp. 13–16).

While this retro tendency can be partly explained by conservatism, nostalgia and the drive for authenticity (as described by Philip McIntyre in 2015 [6]), there also seem to be audible, aesthetic reasons at play. This chapter investigates both of these lines of reasoning.

12.2 Background

Within music production in the current century, most signal processing, along with mixing, is digital. This processing is increasingly being done with software plug-ins hosted on the tracks or channels of a digital audio workstation, such as Steinberg's Cubase, Avid's Pro Tools and Ableton's Live, which generally performs the duties of audio recording, editing, processing and mixing, along with MIDI sequencing. This tendency is fuelled by the seemingly constant improvement of the capabilities of (home) computers; for a brief discussion of such game-changing equipment, see, for instance, the article '25 Products That Changed Recording' in *Sound on Sound* magazine [7]. Virtually all known—and increasingly, some unknown—functions and processors are available in a software format, yet analogue hardware refuses to go away completely, although reliance on it has been heavily diminished. Even though the costs of maintenance, insurance and power consumption of hardware are often ignored, the initial outlay can be substantial. However, the tactile, hands-on experience of a hardware processor may lead to more immediate results, making for a workflow which is both more enjoyable and more efficient, offsetting at least some of the investment for a working professional. This view is expressed by post-production engineer Murray Allen, who comments on another classic, the UREI 1176 compressor [5]: 'It has a unique sound to it that people like, it's very easy to operate, and it does a great job' (p. 14). However, a discussion of workflow is not the main focus of this chapter. I instead concentrate on exploring audible differences and seek possible explanations for the preference for certain types of distortion.

12.2.1 Technical Aspects

Allow me to briefly describe some interesting features of human hearing: Sounds with a spread-out frequency content activate more of the so-called critical bands of the basilar membrane in our inner ear [8], resulting in a loudness greater than the sum of its parts (p. 100). Consequently, the sound/instrument can be lowered in a mix and remain audible—trading signal level for bandwidth, so to speak. The creation of additional frequencies related to the signal is thus very interesting for mixing engineers and all producers of music, and the generally more useful distortion products come in the form of *harmonic* distortion (i.e. new frequencies at integer multiples of the input frequencies). The harmonic series constitutes octave-transposed versions of the common musical intervals in just intonation, rather than the now ubiquitous equal temperament. With the 1st harmonic being the fundamental, the 2nd is an octave above, the 3rd harmonic an octave plus a fifth above, and the 4th two octaves above, while the 5th harmonic is two octaves plus a major third above the fundamental, etc. As the reader is likely aware, musical instruments and human voices produce not only fundamentals (sine waves, which are generally not musically appreciated on their own, modular synthesiser enthusiasts notwithstanding), but a range of harmonics as well. The ebb and flow of these harmonics largely determine the perceived timbre of sound sources (pp. 240–245), and we can extend the concept of timbre to include the different 'sounds' of processors (for example, compressors [9]). As singing and playing with more intensity produces sounds with higher levels of harmonics, in addition to being louder, a process of increasing the level of harmonics through distortion can add 'intensity' while sounding quite 'natural'—up to a point. The conflict between the harmonic series and equal temperament is also pre-existing, and while piano-tuning seeks to strike a compromise, known as the Railsback curve, between the instrument being in tune with itself (its own harmonics) and adhering to equal temperament, listeners of most musical cultures are rather tolerant of imperfect harmonics—in fact, slightly 'off' harmonics often produce more pleasing results than when they are perfectly aligned, as can be heard in **audio example 1**,[1] containing two times three seconds of a 220 Hz sawtooth wave with harmonics first 1–2 Hz off, then exact. Finally, a quick showcase of subtle harmonic distortion reproduced by Acustica Audio's 'nonlinear convolution' technique can be heard in **audio example 2**,[2] in which a string bass loop alternates between the original recording (two times four measures of 3/4 time) and processed versions by, respectively, an EMI valve preamp (0m14s), a WSW germanium design (0m41s) and a clone of a Neve preamp (1m09s). The corresponding harmonic distortion levels have been measured in Figures 12.1, 12.2 and 12.3.

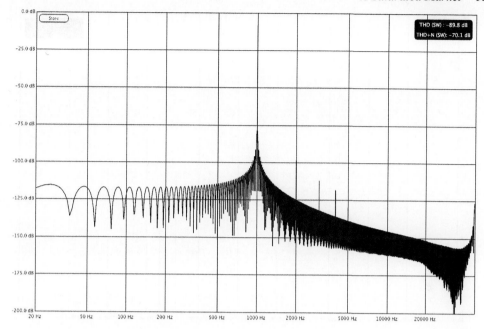

Figure 12.1 Note the predominant 2nd harmonic typical of triode valve designs and the visible noise floor.

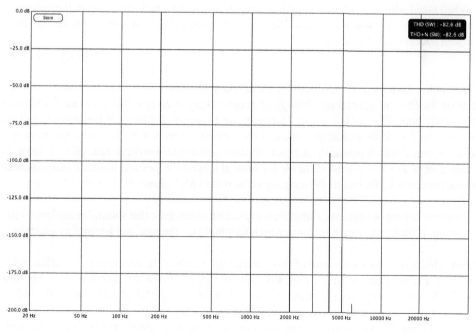

Figure 12.2 Note the predominant 2nd and 4th harmonic of this germanium transistor-based design.

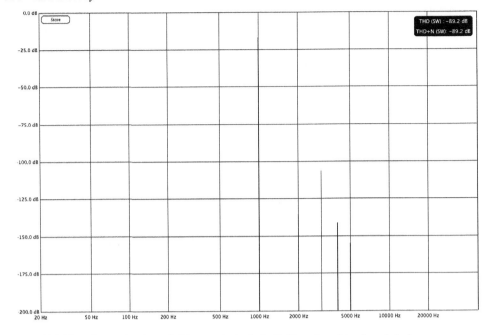

Figure 12.3 Note the predominant 3rd harmonic typical of silicon transistor-based designs.

12.3 Retromania

> When faced with a totally new situation, we tend always to attach ourselves to the objects, to the flavor of the most recent past. We look at the present through a rear-view mirror. We march backwards into the future.
>
> [10] (pp. 74–75)

This rather famous quote by media theorist Marshall McLuhan is one of many bold statements on the life and general psychology of people living in an age of mass media in his bestselling 1967 book *The Medium is the Massage: An Inventory of Effects*. When consulting trade magazines of the 1990s and 2000s, one can see that there seems to be at least some truth to it. *Sound on Sound*, which primarily targets ambitious home studio owners, reacted to a 'craze for all things retro' in 1997, when David Mellor instead longed for forward-thinking manufacturers (and users/readers). His line of thinking seems to reflect McLuhan's [11]:

> Trying to imitate old designs, rather than exploiting techniques that should be ageless, is only holding us back; we need new ideas which will build on the past, not attempt to return to it.

Indeed, Mellor may be well aware of this connection, writing: 'In conclusion, . . . [t]he fallacy of retro is that the past can give us the solutions to today's problems, which it can't any more in audio than it can in politics', seemingly directly referencing McLuhan's own [10]: 'Politics offers yesterday's answers to today's questions.' (p. 22)

Sound on Sound seemed to continue this line when editor-in-chief Paul White, in the leader of the January 2010 issue, declared that 'vintage is a state of mind', describing how highly regarded classic albums of the 1960s and '70s were produced primarily for vinyl discs using traditional engineering values that could be easily applied to the present. He argues that 'records were appreciated mainly for what was going on in the mid-range.... The result is that it is more comfortable to listen to', and counters the many reader questions about which gear will help them get a 'vintage sound' by posing his own question [12]:

> [C]ould it be that the most important piece of equipment needed to recreate vintage-sounding mixes is already installed between your ears, and it's just a matter of getting it recalibrated?

It seems that for the majority of readers, the answer to this question was increasingly in the negative. In fact, White's opinion could be seen as swimming against the tide. We need look no further than the very next issue, in February 2010, to see Hugh Robjohns examining 'analogue warmth' in great detail and explaining to readers how to achieve 'the sound of tubes, tape and transformers' using a plethora of hardware devices and software plugins [13].

This retro-trend can also be found among the audio professionals who comprise *Resolution* magazine's readership. Take, for example, Ashley Styles' 2005 article 'Crafting the Signal', which suggests vintage equalisers and compressors for said crafting, arguing [14]:

> In some circumstances, a pure, unaltered audio signal might be all that is required. However, some 'crafting' of the signal is often required and the signal must travel through another maze of components.

Another example comes from *Attack Magazine*, a publication focusing on electronic music production and culture, which shows that retro is not exclusive to 'traditional' music production based on acoustic and electric instruments. In a 2015 article, Bruce Aisher explores the effects of analogue recording and mixing, in the vein of the curious attraction which 'analogue' seems to have [15]:

> All kinds of technical expertise has been utilised to improve the technical performance of audio recording formats over the last half a century or more, but more than any other point in that period we're now looking back to older hardware to deliver that elusive bit of extra vibe in our music.

The trend can be seen to continue through the March 2019 issue of *Computer Music* (for producers of electronic music—not to be confused with *Computer Music Journal* from MIT Press) [16]. It published a brief online tutorial of '12 sure-fire ways to achieve a vintage sound', following up on a theme of 'vintage effects' featured in the October 2018 issue. From these sources, we can find that the 'retromania' wave became apparent in a wide range of audio communities in the late 1990s.

12.3.1 The Industry Follows Suit

The tendency to cherish older equipment has not been lost on related industries. Detroit-based Vintage King was established in 1993 to refurbish such units and has since grown into a major reseller of not only used but also plenty of new gear, analogue as well as digital [17]. A UK equivalent is found in the perhaps more honestly named Funky Junk Ltd., established in 1991 [18]. The marketing of vintage gear has become a widespread phenomenon among musical instrument retailers. The largest chain in the US, Guitar Center [19], today has a 'vintage' category, separate from 'used' [20]. The trend has also had an impact on rental companies, such as Blackbird Audio Rentals in Nashville, which receives high demand for its vast collection of vintage microphones and other equipment [21], and a clear statement to this effect is offered by Tim Wood of the aptly named Vintage Audio Rentals in Hamburg [22]:

> [B]ecause of the technical perfection & the associated 'sterile' sound of digital, more and more analog devices are finding their way back into the production chain to achieve the classic, warm 'back to the roots' vintage sound. . . . [N]obody cares (at least in popular music) whether your system has a 100 dB signal to noise ratio or 0.001% distortion, it has to rock! And here, a 3% tube- or transformer-based harmonic distortion or a nicely saturated inductor filter can make all the difference!

In the quote, specifications typical of decent hi-fi equipment are equated with 'sterile' perfection, while otherwise exorbitant distortion levels or overloaded components are seen as entry points to making your music 'rock'. It's remarkable to see someone so technically knowledgeable attacking perfection with it's own weapon—specifications and circuit descriptions typical of an engineer. Here, the 'classic, warm . . . vintage sound' is seen not only as separate from technical perfection, but as the inverse, making high, rather than low, distortion levels a hallmark of quality.

Few 'classic audio' companies lived to experience 'retromania', and, in many cases, technology and manufacturing processes have also moved on. When one of the few remaining manufacturers, Georg Neumann GmbH, reintroduced tube electronics to its microphones with the M 149 model in 1995 (a modernised version of the legendary U 47 and M 49 microphones) [23], it was 30 years after the company started to 'transistorise' its products, and the original tubes were long out of production [24]. Interestingly, the M 149 won a TEC award for 'outstanding technical achievement' in 1997, ahead of not only competing retro-models from Røde, Schoeps and Soundelux but also highly accurate microphones from B&K/DPA and Earthworks [25]. The aforementioned Universal Audio was refounded in 1999, specifically to 'faithfully reproduce [the company's] classic analog recording equipment'—and, interestingly [26], 'to design new digital recording tools . . . with the sound and spirit of vintage analog[ue] technology . . . warmth and harmonics in all the right places'. Here we see more hints at the apparent aesthetic value of harmonic distortion—but this time not bound to the components of actual hardware as with Styles and Wood.

The growing interest in vintage equipment prompted two further reactions in manufacturing. Firstly, a number of 'boutique' manufacturers were established during the 1990s to 'clone' or recreate the abandoned classics of other companies (Lydkraft had an early start with its Tube-Tech remake of a Pultec EQ in 1985 [27]). Vintage Rupert Neve–designed products were popular subjects for reissue with these small companies, and Vintech, established 1997 [28], has published distortion measurements that show its product is closer to the vintage original than both competing 'clone' manufacturers' products and those of Siemens subsidiary AMS/Neve [29], which had acquired Rupert Neve's business in 1985 [30].

The other effect was the emergence of a DIY culture of recreating and adapting vintage gear, either to save money, to learn about electronics or to get the otherwise unobtainable. This movement can be considered a form of *maker* culture, fuelled by internet forums such as GroupDIY.com (approaching 30,000 users as of December 2020) [31] and by small-scale manufacturers of printed circuit boards like PCB Grinder [32]. These developments cannot have gone unnoticed by Bryce Young, who in 2011 founded Warm Audio [33]—a company that quickly became known as a cost-efficient producer of 'classic' designs popular with DIYers. As one customer testimonial puts it [34]: Warm Audio's products aim to provide 'that warm and fuzzy feeling of real and authentic analogue sound, without having to deal with old equipment'. But perhaps no one has taken the vintage obsession further than Chris and Yoli Mara, who opened the Nashville recording studio Welcome to 1979 in the year 2008 [35]: 'a time capsule of vintage vibe and gear ... able to do entirely analog tracking and mixing', with 1979 being the average year of manufacture of their studio equipment.

All of this points to 'retromania' as a very real phenomenon, found not only in trade magazines and literature but also in the production, retail, rental and recording sectors of the audio industry.

12.3.2 Some Possible Explanations for the Retro-trend

I have outlined numerous examples of retro-affection, but how did this trend become established among musicians, producers and recording engineers, when superior equipment is available and clearly has better specifications?

In the documentary film *Sound City*, musician Dave Grohl, of Nirvana and Foo Fighters fame, tells the story of Sound City Studios, opened in 1969 in Los Angeles. It was, at least for a time, a budget-friendly choice with a very relaxed atmosphere—apparently you could urinate in a corner and no one would care! Throughout the film, Grohl praises the (now vintage) Neve 8028 mixing console that was installed in Studio A four years after Sound City opened, and which he eventually purchased for his own studio. However, the installation of a state-of-the art Neve console in 1973 did not bring clients in droves; rather, the hit records which happened to be recorded in the studio, by for example Tom Petty and the Heartbreakers and later Nirvana, did [36]. This is perhaps less surprising when considering how working musicians typically communicate: the word-of-mouth that kept Sound City busy was reliant not on technical specifications but on accounts of great music and commercial success. As noted in the book *The Art of Record Production* [37]:

> [R]ecording has a rich tradition of story-telling which has involved its own versions of the canon, its own accounts of authorship and creativity, its own myths of motivation and accomplishment.
>
> (p. 9)

By paying attention to this tradition, we can arrive at an understanding of the studio as a complex site of creativity and production, and also of how 'studio tales' can take the form of a myth. As Philip McIntyre puts it in his 2015 paper 'Tradition and Innovation in Creative Studio Practice' [6]:

> There are not only myths that exist around particular pieces of studio equipment but there are also mythologies that surround the common sense romantic and inspirationist views of creativity that tend to prevail in the studio.

While on the subject of communication, it should be obvious that music production is more of an art than a hard science, but we should also consider the shifting power balance in the industry. As a matter of course, The Beatles' recording sessions took place at EMI Studios and were supervised by the head and A&R manager of Parlophone, the subdivision of EMI with whom they had signed. During their earliest recordings the boundaries were clear: as Paul McCartney remembers in the authoritative book *Recording The Beatles* [38], 'George Martin was the Supreme Producer In The Sky and we wouldn't even dare ask to go in the Control Room'

(p. 351). By contrast, in later decades, artists were picking their own independent studios and freelance producers and engineers. In a recording studio marketplace which has been described as 'emotionally driven' [39], it's likely that musicians' decisions are more affected by fuzzy terms like 'analogue warmth' than technical specifications—so studio owners and personnel might well engage with this type of storytelling. On a somewhat related note, the priorities of recording personnel are different from those of performing musicians, and the typical 1960s music studio was constructed to be an efficient workplace, not to provide an inspirational atmosphere. So when Jimi Hendrix opened Electric Lady Studios in a converted New York nightclub in 1970, his vision was primarily to have a place where he felt comfortable as an artist, which included a lighting system capable of producing any colour [40].

12.3.3 What Exactly Are We Missing?

To maximise sound quality in the early days of music production—when a 'live in the studio' approach was common, as tape machines then had only a few tracks—engineers would do live mixes to a master recorder whenever possible. The multi-track tape machine was primarily used as a safety measure: the tape was used only where the balance between instruments (and vocalists) was deemed unsatisfying and a remix (in the original sense of the word) had to be done. This was because the extra electronic circuitry involved—and most certainly the electromagnetic conversions of the multi-track recorder—would degrade the signal quality. The quest for high fidelity was taken to the extreme with 'direct-to-disc' recording, where the music was cut directly to vinyl—requiring the musicians to play an entire LP side live, with ample proficiency, echoing the very beginnings of the recording industry, before tape recorders. A proponent of this movement was Nippon Columbia [41], which 'by the late 1960s was investigating how to improve LP sound quality, and criticism centered on distortions caused by analog tape recorders'. In the pursuit of audio fidelity, its Denon brand became an early developer of digital technology, in 1972 presenting the DN-023R, an 8-track recorder of 13-bit resolution and 47.25 kHz sampling rate (pp. 2–3). An odd word length may seem curious in a binary digital system, but it is usually a result of the remaining bit(s) being required for 'packaging' the data for the recording medium or transmitting channel. MIDI 1.0's use of 7-bit data can be seen as a later example of this, and I am aware of one earlier instance, when the Research Department of the BBC in 1967 considered 11-bit 32 kHz digital transmission systems in compliance with Post Office standards [42].

It should be noted that many early digital systems during the 1970s and early '80s were far from ideal and, to quote Keith Spencer-Allen, were not received with much 'empathy' [43]:

> perfection (it is not), accurate (it may be, but I don't like the sound of it), and tomorrow's technology (it's not working yet!).
>
> (pp. 267–8)

In general, limitations of storage capacity and computational power meant a less than optimal bit depth and/or sampling frequency of early digital products. Reducing the number of bits per sample increases noise and distortion, while reducing the number of samples per second limits the bandwidth. At 8-bit resolution, distortion and noise is clearly audible with most playback systems, while at 12 bits it just might go unnoticed, at least with less than optimal listening conditions. Technically inclined readers are encouraged to try this for themselves (I suggest using Avid's LoFi-plugin as it *actually* outputs a correspondingly reduced number of bits). This is in agreement with the findings of Bromham et al., who through listening tests were able to correlate 8-bit resolution with a perception of a 'bright' sounding signal, whereas 12 bits were linked to the term 'warm' [44]. In the former case the distortion provided 'brightness' through added harmonics, while in the latter case the missing high frequencies resulting from the lower sampling frequency were more apparent. A limitation of the test is that the inner workings of the plugins used are not known, as the authors also note—in fact, they produce outputs of 32-bit resolution, which a bitscope such as Stillwell's Bitter clearly shows, so it's a case of producing the side effects of early digital systems, rather than the actual 'vintage' bitrate.

The introduction of the compact disc in 1982 provided consumers with a 'mature' 16-bit 44.1 kHz digital audio medium, which should satisfy all but the most critical listeners. However, it was initially limited by D/A-converter performance, more specifically the challenges of designing a low-pass analogue filter that would provide 20kHz bandwidth at the relatively low sampling frequency of 44.1 kHz. Much better quality had just been made possible through the development of oversampling converters with digital filters, by for instance Philips [45], but seemingly, manufacturers did not employ them instantly: At the AES Convention in 1984, Lagadec and Stockham noted how 'passband ripple, steep roll-off and non-linear phase response' were 'typical' problems with digital audio of the day, leading to a 'smearing' effect [46]. While these issues were rapidly eliminated from most quality equipment by the end of the decade, the notion of digital sound as something 'harsh' lingered on.

By the 1970s, for genres such as pop and rock, record production no longer meant the documentation of a musical work, but rather the *creation* of one, involving an increasing number of studio effects and 'tricks'. A classic—and, to this day, impressive—example is that of British rock band 10cc, who in 1975 recorded a total of 624 voices for the a capella backing of their hit single 'I'm Not in Love'. As band member and producer Eric Stewart described in 2005 [47]:

> Each note of a chromatic scale was sung 16 times, so we got 16 tracks of three people . . . standing around a valve Neumann U67 in the studio, singing 'Aahhh' for around three weeks. . . . I mixed down 48 voices . . . to make a loop of each separate note, and then I bounced back these loops. . . . We moved the faders up and down and changed the chords. . . . [S]ound degradation caused by all the bouncing didn't matter at all because . . . we had a chromatic scale sizzling underneath the track. . . . We actually created 'hiss' on the track, when we would normally have been fighting to get rid of hiss!

Two decades later, when digital recording technology as pioneered by Denon and others had developed further and was rapidly taking over, the sound of analogue was being missed, as describer earlier in this chapter. While tape issues are responsible for a large part of 'analogue warmth', there are, as Hugh Robjohns put it in 2010 [13], 'several factors that make up "analogue warmth" and our ears almost certainly require a combination'. The signal flow of music recording changed considerably in the 40 to 50 years since AOR meant 'album-oriented rock', going, broadly speaking, from microphone->mixing console->tape recorder to microphone->preamp->A/D converter. Recent recording techniques often include just one or two stages of amplification, before the conversion to digital happens (as in the RME Fireface 802) [48] (p. 104), whereas in the 'classic' scenario, a signal typically passes through five to seven stages on its way from mic to tape, as described by the manufacturer AwTAC [49]. This fact may seem of little consequence if one imagines the process of amplification to be rather perfect—perhaps akin to optical magnification. In reality, the amplification of an electrical signal is the creation of a *new*, larger output signal, made by modulating a more powerful supply of electricity based on the fluctuations of the input. The challenges of this large-scale recreation result in different amplification circuits being perceived as having different tonal and dynamic qualities. Such implications have been described by Wally Wilson, founder of Sphere Electronics [39], and the basic design concept is neatly illustrated in the 'Amplifiers' chapter of *Modern Recording Techniques* [50]. While it can be argued that the microphone preamplifier adds the most 'character'—as it provides the most gain, and thus has the more difficult job—the effect of added amplifier stages cannot be disregarded. This phenomenon has been noted by engineers such as Jim Scott, who states [5]: 'when you patch something through a Neve equaliser and you don't even engage the EQ, it sounds better. It's just a combination of the amps' (p. 15). This also applies to more humble equipment: I remember noticing early in my career how a Soundcraft k3 Theatre mixing console sounded subjectively better with the EQ section engaged, though flat—a difference that was remarkable even on an older PA system in a small venue with less-than-optimal acoustics. It should be mentioned that AwTAC also blames the proliferation of parametric equalisers in the 1980s for producing modern recordings that 'don't sound like anything at all' [49], as opposed to the more idiosyncratic sounds that passed through 1970s EQ designs, with more limited possibilities for 'correction', but this is somewhat of a tangent.

With the accuracy of digital technology, how can companies such as Universal Audio claim to incorporate the sound and warmth of analogue [26]? The function of any signal processor can be modelled digitally; indeed, an analogue circuit is an electronic realisation of a process, designed to provide suitable results for the given task. And, broadly speaking, the function of audio equipment comes from the *arrangement* of the electronic components (i.e. the design), not the individual components per se. However, many practitioners argue that digital emulations do not *sound* identical to the originals, and this is where we come back to distortion. I argue, with Robjohns and others, that the 'analogue sound' comes from imperfections in components and circuitry. Specifically, non-linearity results in various forms of distortion, and these effects are not integral to the intended function of the circuit, as they were mostly unintended and not a part of the original design plan. While some manufacturers modify their modelling products to sound more 'original'—or even include 'repeatable randomness', like Brainworx's patented 'Tolerance Modeling Technology' which claims to 'replicate variations in the values of . . . individual electronic components' [51]—digital *models* have yet to completely match the most 'colourful' side effects of their analogue origins.

An alternative technique used for emulation is more akin to sampling. By obtaining the impulse response of a device, signals can, via the mathematical process of 'convolution', virtually pass though the device. This yields identical results to an actual pass-through, including any distortion products. A problem with this approach is that a single impulse response cannot capture the 'sound' of a processor in its totality, as the produced distortion is level-dependent—such is the nature of non-linearity. A convincing 'multisampling' of a device requires many impulse responses, which results in significant amounts of data and a rather heavy computational process of selecting and switching between impulses based on the level of a constantly fluctuating signal [52]. Early realisations of this process were produced with dedicated hardware by Sintefex Audio, established in 1997, and later incorporated into Focusrite's Liquid product series [53]. In 2005, Acustica Audio made a similar process available in software [54]. As computational power increases, I expect more manufacturers and users to take this route in their quest for vintage and analogue sounds.

The fact that distortion no longer comes part and parcel with the recording process but can be selectively introduced and designed gives rise to a previously unheard-of flexibility in the manipulation of sound. In his thorough, pedagogical book on getting the mix right in less-than-perfect environments, *Mixing Secrets for the Small Studio*, Mike Senior explains how the traditional method of using equalisation to 'sculpt' the spectrum of a sound is limited by the frequencies present in the signal [55]. For instance, no amount of 12 kHz boost (unless it's so wide that the frequency becomes irrelevant) will brighten a sound which has no frequency components present above 2 kHz. Applying distortion to the signal will, however, produce new frequencies—meaning more sonic material to sculpt and, fittingly, it's the very first tool mentioned in the chapter 'Beyond EQ', where 'A-list mixing engineers' Rich Costey and Tchad Blake are noted as regular users of distortion on instruments and vocals. Incidentally, the use of this technique corresponds with my personal experience of hearing the plugin AmpFarm (introduced by Line6 in 1998 [56]), in use on most mixes by audio engineer Jørgen Knub, whom I worked for as an assistant engineer in Denmark in the early 2000s. Around this same time, producer and engineer Marco Manieri developed his approach to 'creative distortion' by deliberately overloading TDM plugins in his Pro Tools system at GULA Studion in Malmö, Sweden. He mostly did this by boosting high-mid frequencies 'well into the red', until the resulting hard 'clipping' provided a very audible dose of harmonic distortion, and then used a subsequent plugin to reduce the signal level to be within bounds, thereby avoiding further distortion. Arguably it's a no-no for a trained engineer to overload a (fixed-point) digital system, and Manieri had attended the well-known School of Audio Engineering in London [57]. But his 'lo-fi deluxe' results were deemed pleasing, and so he was quite in demand at the time, including teaching mixing classes for the Rhythmic Music Conservatory in Copenhagen, where I was a music technology student.

While distortion was traditionally achieved through 'creative misuse' of analogue gear—the 'all-buttons-in' mode of the 1176 compressor being a famous example [58], or perhaps the application of a guitar stomp pedal—dedicated devices have since been introduced, apart from vintage emulations and recreations. One early example is the Culture Vulture distortion unit, which Thermionic Culture has produced since the company's founding in 1998 [59], even introducing a mastering version in 2011 [60]. Likewise, Looptrotter has, since 2012, received attention for a range of products offering *saturation*, which seems to be the preferred marketing term for products claiming to provide the revered 'warmth' (somewhat akin to limiters being sold as 'maximisers') [61]. In 2015, The provider of electronic kits DIYRE launched the so-called Colour format, which provides DIYers with a 'palette' of electronic circuitry ranging from subtle filtering to full-on distortion [62]. Further examples, in the form of software plugins, are offered by Soundtoys [63] and FabFilter [64], among others, and selecting—perhaps even *designing*—a suitable type of distortion (by any other name) has become part of recording, mixing and even mastering in many circumstances.

Having now arrived at the current situation, I'd like to take a brief look back to 1975, when Aphex introduced the Aural Exciter. While supposedly bringing clarity and, well, 'excitement' to the signal being processed, it relied on adding a (high-passed and) distorted version of the input. One of its uses was to combat the detrimental effects of cassette tape duplication [65]—a 'fight fire with fire' scenario. Seeing how distortion came to be deemed usable, if not necessary, in the heyday of 'classic' studio recording, it's no wonder that we also see its widespread use in today's thoroughly digital era.

Since the notion of distortion as a creative effect was a novelty around the turn of the millennium, it is simply absent from earlier literature. For instance, Paul White makes no mention of distortion *as effect* in his 1989 book *Effects and Processors* for the Creative Recording series [66], nor does Richard Elen in his chapter 'Sound Processing' in *Sound Recording Practice* (4th ed. published in 1994 for the Association of Professional Recording Services) [67].

12.4 Conclusion

With the change from analogue to digital recording and a shorter and more precise signal path, the impact of recording equipment on the audio signal was minimised, and since the 1990s, this 'sonic fingerprint' is increasingly being missed. This is the case not only for genres such as classic rock, which can claim a more or less direct lineage to a 1970s recording style; rather, 'retromania' is found within many popular genres where production forms part of the sonic aesthetic, as I have shown to be the case with EDM-oriented music. A possible motivation for this is that many new acts wish to be seen as 'established', and a 'classic'-sounding production might conjure such an impression, somewhat subconsciously, in the listener. It seems probable that many music critics would react more favourably towards 'tried-and-true' production values [6], since 'the music press in particular pushes an idealised past', as Philip McIntyre argues (with Becky Shepard) while also reminding us that, on a general level, '[w]hat one sees as quality work is always in relation to what one already knows. Quality might be derived in part from the technical but it is by and large a culturally defined attribute.' In this sense, it seems likely that 'vintage sounds' are typically employed to reference recording history and tradition—rather than to signal a nonconformist approach to the advances of (especially digital) technology, as David Mellor, in 1997, envisaged 'truly creative musicians' would take [11].

Notes

1 https://askkaereby.bandcamp.com/track/a-saturated-market-audio-example-1
2 https://askkaereby.bandcamp.com/track/a-saturated-market-audio-example-2

References

[1] Orkin, D.: *Fender and the CBS Takeover*, https://reverb.com/news/fender-and-the-cbs-takeover
[2] Styles, A.: *Story of the AKG C414*, www.saturn-sound.com/Curio's/story%20of%20the%20akg%20c414.htm
[3] White, P.: *AKG C414B XLS & XLII* (2004), www.soundonsound.com/reviews/akg-c414b-xls-xlii
[4] Coletti, J.: *Curing Condenser Confusion: An Audio History of the AKG C414—Page 2 of 2* (2016), https://sonicscoop.com/2016/10/27/curing-condenser-confusion-an-audio-history-of-the-akg-c-414-2/
[5] *Universal Audio, Inc.: LA-2A User's Guide* (2000), https://media.uaudio.com/assetlibrary/l/a/la-2a_manual.pdf
[6] McIntyre, P.: Tradition and Innovation in Creative Studio Practice: The Use of Older Gear, Processes and Ideas in Conjunction with Digital Technologies. *J. Art Rec. Prod.* (2015), https://www.arpjournal.com/asarpwp/content/issue-9/; https://www.arpjournal.com/asarpwp/tradition-and-innovation-in-creative-studio-practice-the-use-of-older-gear-processes-and-ideas-in-conjunction-with-digital-technologies/
[7] *Sound On Sound: 25 Products That Changed Recording* (2010), www.soundonsound.com/reviews/25-products-changed-recording
[8] Howard, D.M., Angus, J.: *Acoustics and Psychoacoustics*. Routledge, New York; London (2017)
[9] Mozart, M.: *Compression Part 1—the Royal Harmonics Orchestra!*, https://mixedbymarcmozart.com/2014/10/23/compression-think-part-1-music-tube-compressor-royal-harmonics/
[10] McLuhan, M.: *The Medium is the Massage: An Inventory of Effects*. Bantam Press (1967)
[11] Mellor, D.: *Does Going Retro Spell the End for Future Tech?* (1997), https://web.archive.org/web/20150607060905/www.soundonsound.com/sos/1997_articles/apr97/retromyth.html
[12] White, P.: *Vintage Is a State of Mind* (2010), www.soundonsound.com/people/vintage-state-mind
[13] Robjohns, H.: *Analogue Warmth*, www.soundonsound.com/techniques/analogue-warmth
[14] Styles, A.: Crafting the Signal. *Resolution Magazine* (May/June 2005), pp. 68–69, https://www.resolutionmag.com/wp-content/uploads/2016/02/Crafting-the-signal.pdf
[15] Aisher, B.: *What Does Analogue Gear Do to Sounds?* (2015), www.attackmagazine.com/technique/tutorials/lessons-analogue-recording-techniques/
[16] *Computer Music: 12 Sure-fire Ways to Achieve a Vintage Sound* (March 07), www.musicradar.com/how-to/12-sure-fire-ways-to-achieve-a-vintage-sound

17. Vintage King: Our Story—Vintage King, https://vintageking.com/about
18. Funky Junk: About Us, www.proaudioeurope.com/info/about
19. Guitar Center: Company Information | Guitar Center, www.guitarcenter.com/pages/company-information
20. Guitar Center: Vintage Gear | Guitar Center, www.guitarcenter.com/Vintage/
21. Blackbird Audio Rentals: Rent Pro Recording Equipment | Blackbird Audio Rentals | Nashville, www.blackbirdaudiorentals.com
22. Wood, T.: Vintage Audio Rentals in Hamburg, www.vintageaudiorentals.com/
23. NEUMANN: M 149 Tube (230 V EU), https://en-de.neumann.com/m-149-tube
24. FAQ Microphones, https://en-de.neumann.com/faq-microphones
25. The TEC Awards 1997 Winners, http://legacy.tecawards.org/tec/1997.html
26. Universal Audio, Inc.: About | Universal Audio, www.uaudio.com/about/
27. Lydkraft: The LYDKRAFT Story, www.tube-tech.com/the-lydkraft-story/
28. Vintec Audio: VINTECH AUDIO, http://vintech-audio.com/
29. Vintec Audio: VINTECH AUDIO: New Graphs, http://vintech-audio.com/new%20graphs.html
30. Rupert Neve Designs, LLC:—History, www.rupertneve.com/company/history/
31. Groupdiy.com: GroupDIY—Index, https://groupdiy.com/
32. DIY-Audio, www.pcbgrinder.com
33. Warm Audio: History, https://warmaudio.com/history/
34. Warm Audio: Our Mission, https://warmaudio.com/our-mission/
35. Welcome to 1979: Welcome to 1979—A Recording Studio Like No Other, https://welcometo1979.com/
36. Grohl, D.: Sound City. Therapy Content, Roswell Films (2013)
37. Zagorski-Thomas, S., Frith, S. eds.: The Art of Record Production : An Introductory Reader for a New Academic Field. Routledge, London (2012)
38. Ryan, K., Kehew, B.: Recording the Beatles: The Studio Equipment and Techniques Used to Create Their Classic Albums. Curvebender, Houston, TX (2006)
39. Wilson, W.: Sphere Electronics: A Brief History of this Console Company (2018), https://tapeop.com/interviews/126/sphere-electronics/
40. Jimi Hendrix | Electric Lady Studios—Web Exclusives | American Masters | PBS, www.pbs.org/wnet/americanmasters/jimi-hendrix-electric-lady-studios/2735/
41. Fine, T.: The Dawn of Commercial Digital Recording. Assoc. Rec. Sound Collect. J. Volume 39 (2008)
42. Howorth, D., Shorter, D.E.L.: Pulse-code Modulation for High-Quality Sound-Signal Distribution. The British Broadcasting Corporation Engineering Division (1967)
43. Borwick, J.: Sound Recording Practice. Oxford University Press, Oxford; New York (2001)
44. Bromham, G., Moffat, D., Barthet, M., Danielsen, A., Fazekas, G.: The Impact of Audio Effects Processing on the Perception of Brightness and Warmth. In: Proceedings of the 14th International Audio Mostly Conference: A Journey in Sound, pp. 183–190. ACM, Nottingham United Kingdom (2019)
45. van de Plassche, R.J., Dijkmans, E.C.: A Monolithic 16-Bit D/A Conversion System for Digital Audio. In: Presented at the Audio Engineering Society 1st International Conference: Digital Audio June 1, Rye, New York (1982)
46. Lagadec, R., Stockham, Jr., T.G.: Dispersive Models for a-to-d and d-to-a Conversion Systems. Presented at the 75th Audio Engineering Society Convention, Paris (March 27, 1984)
47. Buskin, R.: CLASSIC TRACKS: 10cc "I'm Not In Love" (2005), www.soundonsound.com/techniques/classic-tracks-10cc-not-love
48. RME: User's Guide Fireface 802, www.rme-audio.de/downloads/fface_802_e.pdf
49. AwTAC: Philosophy, http://awtac.com/philosophy
50. Huber, D.M., Runstein, R.E.: Modern Recording Techniques. Focal Press, New York (2013)
51. Brainworx bx_console N, www.plugin-alliance.com/en/products/bx_console_n.html
52. Robjohns, H.: Sintefex FX2000 & CX2000, www.soundonsound.com/reviews/sintefex-fx2000-cx2000
53. Sinetefex Audio Lda: Sintefex Audio Products Using Our Technology, www.sintefex.com/?targ=licensedproducts&src=products/

54. Acusticaaudio s.r.l.: *Analog Sampled Plugins for Mixing and Mastering—Acustica Audio*, www.acustica-audio.com/pages/company
55. Senior, M.: Beyond EQ. In: *Mixing Secrets for the Small Studio*, pp. 191–202. Focal Press, New York and London (2011)
56. *Line 6: Amp Farm 4.0*, //line6.com/amp-farm/
57. *Marco Manieri* (2020), https://sv.wikipedia.org/w/index.php?title=Marco_Manieri&oldid=47675651
58. Moore, A.: All Buttons in: An Investigation into the Use of the 1176 FET Compressor in Popular Music Production. *J. Art Rec. Prod.* (2012), https://www.arpjournal.com/asarpwp/content/issue-6/; https://www.arpjournal.com/asarpwp/all-buttons-in-an-investigation-into-the-use-of-the-1176-fet-compressor-in-popular-music-production/
59. *Thermionic Culture Limited: Thermionic Culture: The Culture Vulture*, www.thermionicculture.com/index.php/products/the-culture-vulture-1-17-192012-03-20-11-02-02-detail
60. *Thermionic Culture Limited: Thermionic Culture: The Culture Vulture (Mastering Plus)*, www.thermionicculture.com/index.php/products/hand-shovel-12012-03-20-06-10-04-182012-03-20-11-02-02-detail
61. *Softube: Weiss MM-1 Mastering Maximizer*, www.softube.com/mm-1
62. Hougton, M.: *DIYRE Colour* (2015), www.soundonsound.com/reviews/diyre-colour
63. *Decapitator*, www.soundtoys.com/product/decapitator/
64. *FabFilter: FabFilter Saturn 2—Saturation and Distortion Plug-In*, www.fabfilter.com/products/saturn-2-multiband-distortion-saturation-plug-in
65. Robjohns, H.: *Enhancers* (2000), https://web.archive.org/web/20150609072853/www.soundonsound.com/sos/jan00/articles/enhancer.htm
66. White, P.: *Creative Recording Effects and Processors*. Music Maker Books, Cambridgeshire (1989)
67. Elen, R.: Sound Processing. In: Borwick, J. (ed.) *Sound Recording Practice*, pp. 204–229. Oxford University Press, Oxford; New York (2001)

Part IV
Musicology of Distortion

13 The Studio's Function in Creating Distortion Related Compositional Structures in Hard Rock and Heavy Metal

Ciro Scotto

13.1 Introduction

Distortion, specifically the distortion produced by guitar amplifiers and stomp boxes, plays an essential role as a genre identifier and source of compositional structure for many hard rock and heavy metal bands. While lyrical content often earned bands the moniker of hard rock/heavy metal in the early years of the genre, loud distorted guitars soon became another of the genre's defining characteristics and the defining characteristic of many of its sub-genres. For example, the sound of a Boss HM-2 Distortion Pedal with all its controls set to maximum plugged into a 100-hundred-watt tube or solid-state amplifier defines the sound of Swedish Death Metal bands, such as Entombed (*Left Hand Path*), Dark Tranquility (*The Gallery*), and At The Gates (*Slaughter of the Souls*). Moreover, as the genre developed, the quest for heavier and more intense distortion (i.e., more gain) by many bands initiated an amplifier arms race by manufacturers to meet the demand for increasingly heavier sounds. The early Marshall JMPs of the 1970s morphed into the high-gain JCM 800 in the 1980s replaced by the JCM 900 with even more gain in the 1990s and the JCM 2000 and JVM series in the new millennium. Mesa Boogie, whose founder Randal Smith invented the cascading gain stage circuit, countered the British ordnance with the Mark series amps, such as the IIC+, which is synonymous with the sound of Thrash Metal and its biggest icon Metallica, and the dual and triple rectifiers that are synonymous with the sound of Nu Metal and its biggest icon Korn. Metal musicians often sought out the different distortion characteristics produced by the various amplifiers to serve as the sonic foundation for the various sub-genres of metal, as this quote by James Hetfield from *Guitar Player* magazine's now famous 1992 issue devoted to distortion demonstrates [1]:

> Distortion always starts with the amp . . . you can recognize Marshall distortion in an instant; that's why I shied away from that and went with MESA/Boogies. I basically use the Boogie's distortion with a non-programmable studio quality Aphex parametric EQ to fine tune certain frequencies, dipping out some of the midrange, though my tone isn't quite as 'scooped' as it used to be. On my old records there's some serious low-end chunk. . . . I've been trying to get more clarity but still have the chunk, so I've been adding more mids. Just adding low end isn't enough; you can have lots of it without getting any crunch.
>
> (p. 46)

Hetfield's quote suggests that while distortion may start with the amp, studio equipment also plays a key role in shaping its sound and characteristics. Moreover, Hetfield references Metallica's records, not his or the band's live sound, so by extension we can assume that the process and techniques of recording in a studio can and will further shape the distortion characteristics sought by the musicians. Of course, as the distortion wars continued and continue to escalate, recording engineers must develop new equipment and new techniques to realize and achieve the musician's goal. Some techniques contributing to the production of ever heavier distortion sounds are shifting location in the stereo field, double tracking guitars, microphone placement, microphone choice, EQ, compression, limiters, levels in a mix, and down tuning or using down tuned seven- and eight-string guitars.

While the goal of achieving heavier distortion sounds on records may have fostered the development of innovative recording techniques [2, 3], it also produced new approaches to compositional structures, such as form, movement, and progression in hard rock and heavy metal compositions [4]. For example, The Pixies used heavily distorted guitars in the choruses and clean guitars in the verses of their compositions to structurally distinguish these formal elements [5]. Kurt Cobain and Nirvana adopted this technique in their song "Smells Like Teen Spirit," which Cobain says was "a rip off of The Pixies" [6]. However, the changes in distortion levels do not have to always follow The Pixies soft-loud format. The changes in distortion structure can be more gradual, for example, articulating many different levels of structure or the changes may generate an overall progression to a climax. In my article "The Structural Role of Distortion in Hard Rock and Heavy Metal," I developed an analytical tool called Dist-space for mapping gradual changes in distortion to demonstrate how distortion can create form, movement, and progressions in compositions. For example, the Dist-space tool maps the gradual changes in the song "Nothing Else Matters" by Metallica, demonstrating how the changes in distortion lead to a climactic section creating a unified progression from the modular verse-chorus paradigm. My analysis, however, only tangentially focused on the contribution made by studio recording and production techniques that enhanced the gradual changes in distortion that created the progression leading to a climax, a progression that might be more difficult if not impossible to create outside the studio environment.

This chapter presents a more complete picture of the roles distortion and studio recording techniques play in creating and shaping compositional structures in hard rock and heavy metal compositions by coupling the Dist-space analytical tool with Moore's concept of the sound-box, which is the rectangular space mix that recording engineers use to locate instruments within the sound stage. Moore developed the sound-box concept to trace textural changes and their effect on form in rock/pop music. I will adapt the sound-box space to be an analog of Dist-space for demonstrating how the many different levels of distortion that fill the space can contribute to the creation of compositional structures. For example, many heavy metal bands will have a wall of Marshall stacks (amplifier head plus two 4 x 12 cabinets) as a backdrop in their live shows to visually represent the wall of sound their music produces. Similarly, saturating the sound-box by double tracking distorted guitars and panning them left and right in the sound-box enhances the perceived level of distortion, creating a wall of sound that invokes the visual image of a stage filled with a wall of Marshall stacks. More specifically, I will adapt the contour theory analytical foundation supporting the Dist-space analytical tool to the sound-box concept to demonstrate that used in conjunction both tools present a more complete picture of how distortion can create form, movement, and progression in many heavy metal and hard rock compositions. The information regarding the formulation of the Dist-space analytical tool in the next two sections, Contour Theory and Dist-space, summarizes the more complete formalization and in-depth discussion that appears in my article, "The Structural Role of Distortion in Hard Rock and Heavy Metal" [4].

13.2 Contour Theory

Contour pitches or c-pitches model the general shape of a melodic line without specifying the exact distance (in semi-tones) between one c-pitch and another. In other words, C-pitches simply indicate the relative highness or lowness of pitches without specifying the exact intervallic distance between them. Contour space or c-space contains two dimensions, an x-axis indicating sequential time and a y-axis indicating the relative highness or lowness of a c-pitch, so plotting the c-pitches of a melodic line in c-space produces a contour segment or c-seg that models the shape of a melodic line. Comparing c-segs facilitates demonstrating a common underlying structure between melodic lines generated by very different pitch systems, for example [7]. The first four pitches of the theme from *Webern's Symphonie, Op. 21, mov. 2* generate the c-seg <1 3 0 2> by assigning the lowest pitch, G, the c-pitch 0 and assigning the remaining pitches, F, F#, and Ab, the c-pitches 1, 2, and 3, respectively based on their ordering from low to high (see Figure 13.1). Similarly, the four pitches, D, Bb, C#, and G, from the theme of the Gigue from *Suite No. 2* for 'cello by Bach, also generate c-seg <1 3 0 2> by assigning the lowest pitch, C#, the c-pitch 0 and assigning the remaining pitches, D, G, and Bb, the c-pitches 1, 2, and 3, respectively based on their ordering

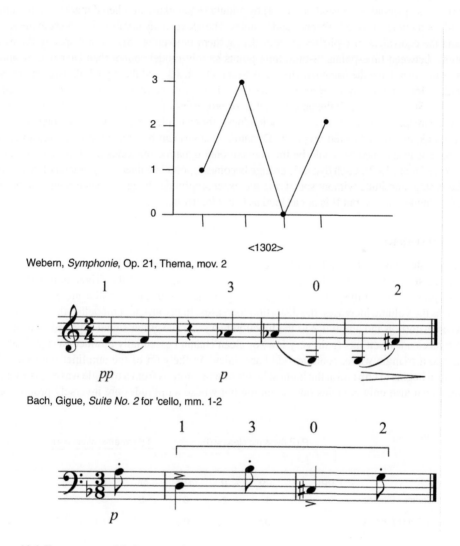

Figure 13.1 Two c-segs and their c-space interpretation.

(reprinted from Scotto, "The Structural Function of Distortion in Hard Rock and Heavy Metal," *Music Theory Spectrum*, 2016)

from low to high. Plotting the c-seg in c-space demonstrates that both melodic fragments have identical contours despite their vastly different pitch-class languages and intervallic distances.

While contour analysis most often models and demonstrates relationships among ordered pitch segments (i.e., melodic lines) by means of c-seg transformations and equivalence, Robert Morris's formalization and generalization of contour theory extends c-space to dimensions other than pitch [8]. For example, many analysts have successfully adapted c-space to model rhythmic structures [9]. According to Morris's generalization, the dimension being modelled with contour theory must be sequential, so any musical dimension whose objects (or states) permits the objects to be linearly ordered can be heard as sequential. Therefore, the objects (or states) of the sequential dimension become the equivalent of c-pitches that do not specify the exact distance (or change) between the objects. For example, the linear time axis in a digital audio workstation (DAW) becomes sequential time (i.e., a sequential dimension or s-time) by adding sequentially numbered markers to the timeline indicating time passage in seconds and minutes. The sequentially numbered markers in the series become the equivalent of c-pitches by considering them time-points that do not specify the exact durations between time-points. S-time time-points simply model sooner than/later than relationships. Since they have the same structure, the s-time dimension and the c-pitch dimension are isomorphic. Moreover, the isomorphism allows time to order or be ordered by any other sequential dimension, such as timbre, if the objects, states, or properties of the dimension can be sequentially ordered. Perceptions of dynamic change or loudness, for example, are relative even though changes in amplitude can be measured precisely. Dynamic changes can be sequentially ordered to create dyn-space whose c-pitches would be the conventional dynamic markings used in many types of music (see Figure 13.2). Each dynamic change becomes a point in a linear progression of increasing or decreasing amplitude without specifying the exact amplitude change between points, so amplitude C is louder than A, but B is not as loud as C but louder than A.

13.3 Dist-space

The dyn-space model of amplitude changes serves as a foundation for Dist-space since the structure of the spaces are nearly identical. Guitar amplifiers, especially tube-driven non-master volume amplifiers, distort the guitar signal gradually as the gain of the amplifier increases. In other words, as the volume increases, the distortion increases. In my article, I modelled the relationship between distortion and amplitude as the function $F(A)=D$. The guitar signal supplied to an amplifier on its cleanest settings (gain very low) or clean channel essentially produces a signal, which similar to a triangle wave, contains odd harmonics. As the gain of the amplifier increases, the waveform begins to clip and at the highest levels of gain morphs from a triangle wave into a square wave, which also only contains odd harmonic but whose amplitudes are inversely proportional

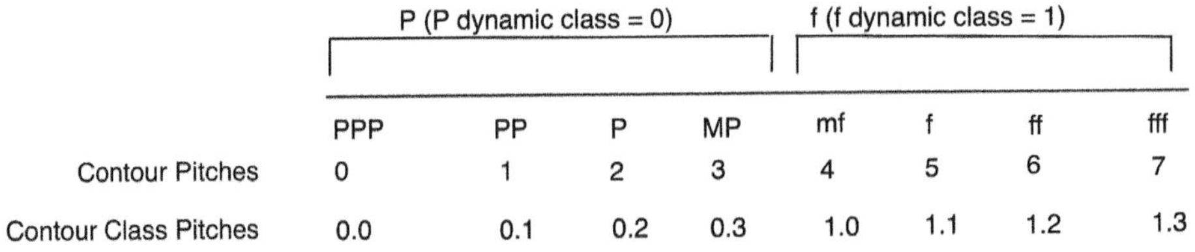

	P (P dynamic class = 0)				f (f dynamic class = 1)			
	PPP	PP	P	MP	mf	f	ff	fff
Contour Pitches	0	1	2	3	4	5	6	7
Contour Class Pitches	0.0	0.1	0.2	0.3	1.0	1.1	1.2	1.3

Figure 13.2 Sequential dimension of dynamics and its c-pitches.

(reprinted from Scotto, "The Structural Function of Distortion in Hard Rock and Heavy Metal", *Music Theory Spectrum*, 2016)

to the harmonic number. If 0 represents a clean, undistorted guitar signal producing a triangle wave and 1 represents a totally saturated, distorted guitar signal producing a square wave, then any point along the 0 to 1 continuum would indicate the amount of change in the waveform as it morphs from triangle to square wave. Spectral analysis would be able to pinpoint a distorted sound's location on the continuum and therefore precisely measure the degree of change from a previous point as the distortion of the sound progresses toward complete saturation.

While the distortion function indicates a sound's location on the distortion continuum, and spectral analysis can precisely measure the change in distortion between any two points on the continuum, the perception of changes in distortion levels is relative. That is, we can hear C as more distorted than A and B as less distorted than C but more distorted than A. Dist-space converts the quantitative distortion function into a sequential dimension that measures perceptual changes in the levels of distortion without precisely indicating the interval of change. Transforming an undistorted guitar sound into a heavily distorted sound is perceptually analogous to hearing the difference between consonant and dissonant intervals. The upper harmonics produce frequency ratios associated with dissonant interval, and their perception increases as gain (i.e., distortion) increases. A single guitar pitch, analogous to a consonant interval, will sound incrementally more distorted (i.e., dissonant) as gain increases and the sound morphs from triangle to square wave, so the ordering of distortion sounds qualifies as a sequential dimension since it is linearly ordered and heard as linearly ordered without specifying the exact intervallic change between the sounds [10].

The distortion sequential dimension segments into distortion regions or categories when it is linearly partitioned. Each segmental partitioning or category becomes a c-pitch in Dist-space analogous to Hindemith's Series 2 formation of a perceptual linear ranking of all intervals from most consonant to most dissonant (see Figure 13.3) [11]. The five

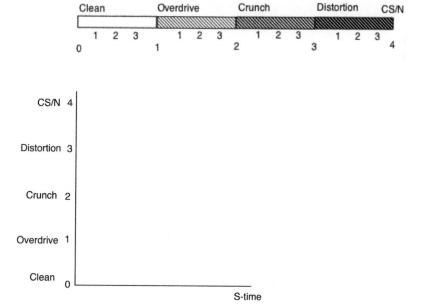

Figure 13.3 Dist-space.

(reprinted from Scotto, "The Structural Function of Distortion in Hard Rock and Heavy Metal", *Music Theory Spectrum*, 2016)

partitions of the distortion continuum create a sequence of c-pitches, or more precisely dist-pitches, each representing a different perceptual level of distortion. According to the dist-pitches, a heavily clipped sound (2) will be heard as more distorted than a mildly clipped sound (1) while a mildly clipped sound will be heard as less distorted than a heavily clipped sound and more distorted than an undistorted sound (0). Coupling s-time (sequential time—sooner than/later than) to the collections of dist-pitches creates a tool that can trace distortion changes in a composition revealing structural patterns. Associating each dist-pitch with an appropriate term from the distortion lexicon of rock musicians creates a familiar, precise, and analytically useful taxonomy. Dist-pitch 4, the terminus of the distortion continuum, receives the label CS/N denoting that the total saturation of the pulse wave or noise produced by dist-pitch (4). Dist-pitches like their dyn-space counterparts form classes since perceivable changes in distortion levels can occur without changing the larger category of distortion or dist-pitch. Rock musicians often refer to gradations within a category, such as overdrive, as light, medium, or heavy, so each dist-pitch class contains the members .1 = light, .2 = medium, and .3 = heavy. Figure 13.4 contains a graph of the changes in distortion for the song "Nothing Else Matters" by Metallica. The graph illustrates the gradual increase in distortion that culminates with the guitar solo reaching nearly the limit of Dist-space. Basically, the entire piece is a distortion crescendo, an essential structural formation that is independent of the work's harmonic progressions and pitch structures. The distortion peaks also demonstrate an essentially continuous three-part form overlaid on the more modular verse-chorus song structure. Even though verses and chorus may repeat the same harmonic and pitch material, they don't sound repetitious because distortion levels constantly change.

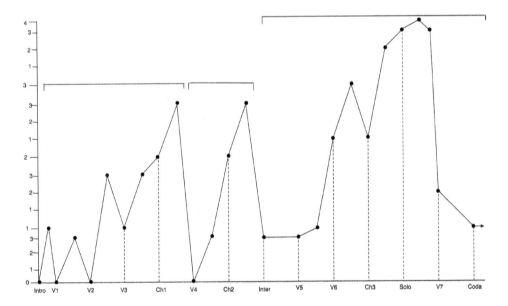

Figure 13.4 Dist-space analysis of "Nothing Else Matters" by Metallica

(reprinted from Scotto, "The Structural Function of Distortion in Hard Rock and Heavy Metal", Music Theory Spectrum, 2016)

13.4 Dist-space and the Sound-box

Allan F. Moore and Remy Martin conceptualized the sound-box as an analytical tool for examining how bands structured spatial positioning and texture on their recordings [12]. Of course, the sound-box is isomorphic to the space a mix engineer works in, but Moore wanted the analytical perspective to reflect listening rather than production. An analysis of production techniques, however, can add depth to an analysis focused on listening. My conceptualization of the sound-box will not draw a sharp line between listening and production because production techniques can also become as important as chord progressions or motivic development to the structure of a song/composition. Nevertheless, analyzing the specific production technique responsible for a particular effect on recordings from listening can often be analogous to trying to determine the inner workings of a black box that generates an output different from the input. Even so, focused listening of particular effects often becomes the first step in determining how to recreate those effects.

Multi-tracking recorders and mixing desks, for example, with their ability to pan sounds right, left, or center serve as the source of production techniques that easily corresponds to the listening experience. Moreover, multi-tracking underlies Moore's conceptualization of the sound-box since, as he states, multi-tracking and panning facilitate structuring compositionally both texture and spatial location. A cube of finite dimensions best represents his visualization of the sound-box since a cube has the additional dimension of depth, which allows the texture of the sound-box to have a foreground, middle ground, and background as well as the dimensions of center, left, right, high, and low. The importance of this development for Moore resides in the control the technology provides over spatial location, which elevates texture to a separate dimension on par with pitch and rhythm. Texture, as he defines it, becomes "the presence of and relationships between identifiable strands of sound in a music" [4]. Additionally, control over these parameters gives rise to the virtual concepts of time and space, since recordings have the ability to create the aural illusion that the generation of the sounds occurred as the recording presents them even though multi-tracking layering techniques in the studio produced them. He also defines a textural strand for rock music as instrumental timbre. I build on this definition by including different levels of guitar distortion and distortion in general as identifiable strands.

While the sound-box potentially elevates texture to the level of an essential compositional structure, Moore's concept of density in the sound-box coupled to the Dist-space tool becomes an even more important concept for the analysis of distortion in production. Equating increasing the density of a space to increasing levels of distortion extends Dist-space to the sound-box, which potentially elevates how distortion fills the space to the level of an essential compositional structure. That is, simply filling the space with more instruments increases the density of the space, but it does not necessarily increase the perceived level of distortion. However, filling the space with distorted instruments, specifically how distortion fills the space, becomes an extension of Dist-space that can be analyzed with the Dist-space tool demonstrating how the sound-box Dist-space contributes to compositional structure.

For example, density and texture are interrelated since density becomes a measure of the texture, and holes in the texture become compositional opportunities. Moore considers Jethro Tull's "Songs from the Wood" an advance in treating texture as a compositional structure, especially with regards to density, since at the time of the album's recording engineers attained much more control over the spatial placement of elements in the sound-box. He notes that over the duration of the song the texture, which starts in the center of the sound-box, gradually spreads to fill the entire space. In other words, the texture becomes saturated, and saturation also applies to distortion as a description of the effect. The textures filling the space in "Songs from the Wood," however, are not themselves distorted, so the saturation of the space does not necessarily reflect

208 *Ciro Scotto*

or become analogous to the saturation of the instruments or an extension of instrument saturation. Heavy metal bands, on the other hand, often use the saturation of the sound-box or the sound-box Dist-space as an extension of the guitar distortion that identifies the genre to enhance or increase the distortion effects of the guitars and underscore the structural role distortion plays in their compositions. A single heavily distorted guitar, for example, panned left does not sound as distorted as two distorted guitars panned right, left, and center. The perceived level of distortion increases as the sound-box becomes saturated, even if the two guitars exhibit identical levels of distortion in Dist-space.

13.5 The Sound-box Dist-space

Many heavy metal bands have the aesthetic goal of producing a wall of distortion on their recordings analogous to the wall of amplifier heads plus 4 x 12 cabinets used as a backdrop in their live shows. The wall of distortion aesthetic represents both the band's heaviness and its power. The wall of distortion also represents the highest achievable level of saturation in the sound-box Dist-space. Moreover, like Jethro Tull's gradually filling the sound-box on "Songs from the Wood," many heavy metal bands gradually build the wall of distortion one distortion brick at a time. A detailed analysis of the structure of a composition that gradually fills the sound box with a wall of distortion, therefore, necessitates analyzing the incremental changes in saturation of the sound-box Dist-space, which further necessitates creating an analytical tool that uniquely represents incremental changes in the saturation of the Dist-space sound-box.

A rectangle divided into three columns and three rows where each of the nine regions becomes a c-pitch adequately represents the sound-box Dist-space, and the c-pitches plotted on an s-time graph (sequential time-line) represent the level of saturation present in the sound-box Dist-space at each s-time timepoint (see Figure 13.5a). The numerals on the left vertical axis of Figure 13.5b translate the regions of the sound-box Dist-space into a density/distortion scale. The integer 0 indicates an empty sound-box, which might occur if the band suddenly stops playing for a dramatic pause. The integer 4 at the other extreme indicates the total saturation of the sound-box (analogous to the total saturations of a waveform) where the sound pressure levels appear to push against and bend the walls of the sound-box. The integers 1, 2, and 3 to the left of the decimal indicate the frequency ranges, low, middle, and high, while the integers 1, 2, and 3 to the right of the decimal indicate density within a given frequency range. For example, the decimal .1 signifies that only a single sonic event either to the left, right, or centre of the space occurs at a given time point in the frequency range, but .2 signifies a second sonic event sounding simultaneously with the .1 event. Saturation of a particular frequency range occurs with the entrance of a .3 sonic event against the .1 and .2 events. The numerical scale prioritizes increasing density or distortion of the sound-box from low to high frequencies, and the frequency scale affords the dimension its sequential structure.

Of course, the numerical scale could just as easily indicate increasing density from high to low frequencies since the choice of direction is, to a degree, arbitrary. In fact, even the left-to-right or right-to-left columns in the sound-box could function as the primary indicator of increasing density. The sound-box Dist-space c-pitches 1.1 through 3.3 in the left-to-right or right-to-left orientation would then represent the low, middle, and high frequencies, respectively, that increase the density within the left, centre, or right columns of the sound-box in Figure 13.5a that eventually saturate it. The symmetry of the sound-box allows reconfiguring its c-pitches to represent many aspects of compositional structure. For example, the left-to-right or right-to-left orientations could also facilitate analyzing location changes with c-pitches within the sound-box.

In spite of the sound-box's many possible c-pitch arrangements, increasing density from low to high does for the most part match intuitions (and experience) of how boxes are filled. Even if one pours sand into a box from the top, the box fills from the bottom up. Moreover, most heavy metal bands prioritize the lower register as the foundation upon which they will lay distortion/density bricks as they build a wall of sound, even if the first sonic brick occurs in the centre-high region of the space. A frequent compositional strategy, in fact, for saturating the sound-box begins the process with a high-centre event (analogous to sand being poured in the box) followed by saturation of the low-frequency region and a gradual filling in of the sound-box to connect with the original centre-high sonic event.

Analyzing movement around the sound-box with c-pitches would require associating and fixing each location to a particular c-pitch, 1.1 = left, 1.2 = centre, and 1.3 = right, for example. Fixing c-pitches to a location facilitates generating c-segs that can be analyzed for their operational relationships (transposition, inversion, etc.). Analyzing increases and decreases in density/distortion in the sound-box, however, requires a more flexible approach to associating location to c-pitches, such as associating the first sonic event that occurs in a composition with the .1 c-pitch in the frequency region of the event. As Figure 13.5c illustrates, if the first sonic event occurs in the low-frequency region on the left of the sound-box, it becomes c-pitch 1.1. If sonic events occur in locations C and R in the low region at subsequent time-points, then those sonic events become c-pitches 1.2 and 1.3, respectively, and with occurrence of c-pitch 1.3, c-pitches 1.1, 1.2, and 1.3 would saturate the low frequency. That is, with the entrance of c-pitch 1.3, the low-frequency region achieves its highest level of saturation or density and, therefore, its highest level of distortion. In Figure 13.5d, since the first sonic event occurs in the centre (C) of the sound-box in the low-frequency region, it becomes c-pitch 1.1, and the subsequent sonic events at locations L and R become c-pitches 1.2 and 1.3 respectively. Once again, with the entrance of c-pitch 1.3, the low-frequency region achieves its highest level of density distortion. The first three sonic events in Figure 13.5e replicate the gradual increase in density of Figures 13.5c and d, but c-pitch 1.1 represents both the sonic events L and R occurring at time-point 4 and 5 indicating a low-density level since only single c-pitches occur at each time-point. Since the first sonic events in Figure 13.5f occur simultaneously at the same time-point, either the L or the R event becomes c-pitch 1.1 and the other event becomes c-pitch 1.2. Finally, the sonic event at time-point 3 represents total saturation of the sound-box Dist-space since all c-pitches occur at the same time point.

Moore's original conceptualization of the sound-box as a cube included the dimension of depth, which the sound-box Dist-space model with its two dimensions does not represent. Of course, reconfiguring the sound-box Dist-space model as a cube would add the depth dimension, but the representation of c-segs in a cubic space becomes more awkward. Adding a minus sign, however, to any c-pitches would elegantly indicate the depth dimension while maintaining the simpler two-dimensional representation of the space. Perceptions of depth in Moore's sound-box translate to hearing the instrument producing the sound as either closer to or farther away from the listener at the front of the cube. The perception of depth, the relative nearness or distance of a sound to a listener, in the world often depends on the amplitude and frequency content of sound. For example, a hot rod or motorcycle with a loud engine sounds loud when a person stands next to it. As a person increases the distance between themselves and the loud hot rod or motorcycle, the amplitude of the engine, of course, decreases. A loud sound heard with low amplitude, therefore, translates to greater distance between the listener and the sound. Similarly, amplitude changes or relative amplitude of a sound often creates the perception of depth in the sound-box in the studio. For example, a loud distorted guitar sounds farther away from the listener when its amplitude is low in the mix. Increasing the amplitude of the distorted guitar creates the impression of moving the instrument closer to the front of the sound-box and

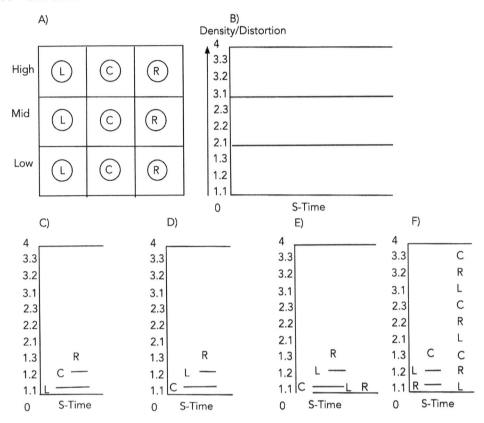

Figure 13.5 Sound-box Dist-space.

therefore closer to the listener. Adding a minus sign to any c-pitch, such as −2.1, indicates a low amplitude for a normally loud sound and, therefore, greater distance from the front of the sound-box and the listener. The successive c-pitches −2.1 and 2.1 would indicate a change in amplitude creating the impression of moving from the back of the sound-box to the front and consequently closer to the listener.

13.6   Analysis

The song "Blackened" from the *And Justice for All* (1988) album by Metallica exhibits many of the sound-box Dist-space features outlined in the previous section. The guitar distortion in "Blackened" mostly moves within the highest region of Dist-space between distortion c-pitches 2.2 and 4.0, as my article demonstrates. The main guitar in the first slow interlude, for example, that begins at 2′ 23″ presents two distortion levels. The guitar repeatedly plays two pitch sequences, E-A-Bb-E and A-Bb-F-E, solidly in the 3.1 or 3.2 distortion range of Dist-space. Between the two pitch sequences the guitar chugs a palm-muted open E on the sixth string whose very high transient content (almost all transient and no pitch) produces the Dist-space c-pitch 4, complete saturation or noise. The differing distortion levels between palm-muted chugging E and the lower distortion level pitch sequences essentially create distortion polyphony or a compound distortion melody.

Distortion Related Compositional Structures 211

The distortion levels used by the guitars in "Blackened" always move within a very limited range of the upper regions of Dist-space, between the distortion c-pitches 2.2 and 4.0. Studio techniques, however, that take advantage of sound-box Dist-space's compositional possibilities, such as double tracking, multiple guitar tracks panned left and right, and movement from negative to positive sound-box Dist-space c-pitches, can create the perception of a wider range of distortion levels and greater movement through the regions of Dist-space. The analysis will demonstrate how the initial expansion of the guitar at the first slow interlude (2' 33") from its more constrained space and location in the introduction, first and second verses, and choruses into other regions of the sound-box creates the perception of increased distortion levels. As the process of expansion continues, the nearly complete saturation of the sound-box occurs at the beginning of the guitar solos starting at 4' 07", creating another perceptual increase in distortion levels.

As stated in the previous section, in the studio where amplitude changes are fairly simple to implement either by riding a fader or automating the fader motion to increase the volume, amplitude changes or the relative amplitudes of sounds often create the perception of depth in the sound-box. The Dist-space level or distortion perception level of a loud distorted guitar sound heard low in the mix and therefore farther away from the listener increases when the loud distorted guitar's amplitude rises in the mix and the sound moves closer to the front of the sound-box and closer to the listener.

The studio recording of "Blackened" begins with a 35" fade-in employing the rising amplitude technique. The fade-in creates the impression of traversing the entire range of dist-space, and it creates a crescendo that leads to the introduction's main guitar riff at 37", whose Dist-space c-pitch is in the highest region (3.2) of Dist-space (see Figure 13.6a). The fade-in features a duet of distorted guitars in their highest pitch register in the 2.3 to 3.0 region of Dist-space located

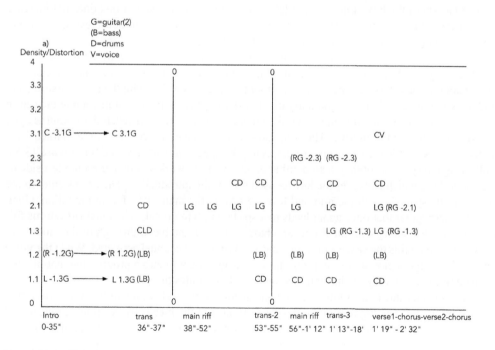

Figure 13.6a "Blackened" sound-box Dist-space graph.

in the centre of the sound-box Dist-space. Of course, the guitars do not sound very distorted at the start of the fade-in because, as their −3.1 sound-box Dist-space c-pitch indicates, their low amplitude creates the perception of barely hearing something that is far away. As the sound-box c-pitch gradually changes from −3.1 to 3.1, the volume increase moves the guitars from the back of the sound-box to its front, creating the perception of increasing distortion levels.

The fade-in includes several guitars located on the left of the sound-box in the low register, but their volume does not increase in a 1:1 ratio with the high guitars in the centre of the sound-box. In fact, the low guitars enter almost as a canon with their crescendo starting several seconds after the high guitars. Figure 13.6a indicates that the bass guitar enters simultaneously with low left guitars, but as the parentheses suggest, the mix levels reduce its prominence. The mixing on *And Justice for All* notoriously reduces the presence of the bass guitar in the mix almost to the point of eliminating it from the ensemble. Nevertheless, the sound of the low guitars appears enhanced by the resonance of the bass guitar since the bass lines in many pieces by Metallica duplicates the lowest notes or the pedal notes performed by the rhythm guitar. When the bass guitar appears to be adding resonance to the guitars, I will include it in the analysis enclosed in parentheses.

The parentheses for the low guitars located to the right of the sound-box c-pitch 1.2, like the bass guitar, feature less prominently in the mix but for a different reason. The sonic events occurring to the right of the sound-box in the introduction up to the slow interlude almost appear to be the result of one channel bleeding into the other rather than purposefully placing the event on the right side of the sound-box. In other words, the events on the right side of the sound-box almost sound as if they are echoes or reflections of the events on the left. The echo technique keeps the right channel activated but reduces its prominence in the mix, yielding a very left-leaning mix. This strategy appears to be a compositional choice that creates anticipation for future events in the right channel.

The fade-in of the low guitars on the left and right sides of the sound-box does not increase in a 1:1 ratio with the high guitars in the centre, creating a canon with their crescendo starting several seconds after the high guitars. The gradual increase in volume of the low left and right guitars creates the impression of gradually filling the sound-box space from top to bottom that mirrors the gradual increase in distortion produced by the volume increase. Filling the sound-box creates a space between the top and bottom that events starting with the first transition and ending with the verse 2 chorus gradually fill. The drums enter in the first transition in the centre mid region replacing the high guitars while guitars in the low register reach their volume apex marked by a new pitch motive. The transition motive repeats twice, each time marked by a sudden silence that is the equivalent of emptying the space, indicated by the 0 in Figure 13.6a. Following the space emptying, a solo guitar enters with the work's main riff on the left side of the sound-box, which enhances the left-focused mix of the introduction. The drums enter in the centre of the space in both the mid and low regions on the main riff's fourth repetition, along with the bass guitar that once again leads to a space-emptying break. The main riff returns following the break along with the drums and bass guitar, but the bleed-through of the guitar into the right channel fills the space a bit more and keeps the right channel activated. When the voice enters in the high region in the centre of the sound-box, it completes a progression that fills the frequency gap created by the introduction and nearly saturates the sound-box.

The slow interlude that follows the first and second verse and chorus fulfills the expectation created by the left-leaning mix in the first two minutes of the piece. The section begins, as did the main riff section at 38", with a solo guitar solidly in the 3.1 or 3.2 distortion range of Dist-space playing the E-A-Bb-E and A-Bb-F-E motives and the drums in the centre with bass guitar on the left. A new element, however, appears in the right channel. A high E harmonic sounds on the right side of the sound-box, and despite its low volume in the mix, it becomes the first event in the right channel that does not sound like bleed-through or an echo. In fact, the high harmonic functions as preparatory gesture for the entrance of a second guitar (most likely the result of multi-tracking) in the repetition of the eight-bar slow interlude E-A-Bb-E and A-Bb-F-E motive at 2' 52". The second guitar on the right of the sound-box signals its entrance with a guitar slide that starts from a pitch region of the high harmonic that leads directly to E-A-Bb-E and A-Bb-F-E motives and full activation of the right side of the sound-box, as indicated in Figure 13.6b. The second guitar's entrance duplicates the events of the left channel, saturating the sound-box in both the low and mid regions. Moreover, the entrance of a second guitar in the right channel not only saturates the low and mid regions of the sound-box, but it also increases the perception of the distortion level of the guitars. The sudden filling of the space makes the guitars sound massive and even more distorted.

The high region begins to add to saturation of the sound-box with the entrance of the voice in the centre of the sound-box space in the second chorus ("opposition") at 3' 07". The full saturation of the sound-box (see Figure 13.6c) occurs with the beginning of the interlude-duet at 4' 07" with return of the high guitars from the introduction in the centre of the sound-box but also panned slightly left and right. The sound-box saturation level really does not decrease once achieving the highest level of saturation. The final section of the work, however, returns to the left-centric mix of the introduction with the guitar main riff solidly in the left channel, signalling the work's close and recalling the progression through the sound-box Dist-space of the entire work.

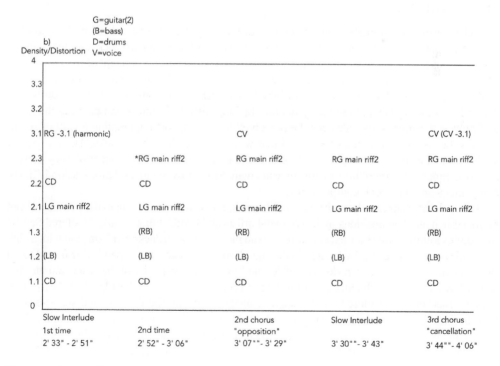

Figure 13.6b "Blackened" sound-box Dist-space graph.

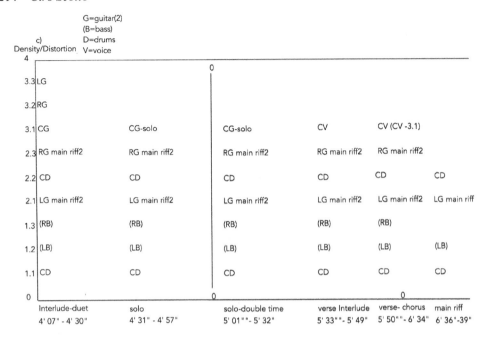

Figure 13.6c "Blackened" sound-box Dist-space graph.

13.7 Conclusions

This chapter presents a more complete picture of the roles that distortion and studio recording techniques play in creating and shaping compositional structures in hard rock and heavy metal compositions by coupling the Dist-space analytical tool with Moore's concept of the sound-box. The analysis using the sound-box Dist-space tool demonstrated how the many different levels of distortion that fill the space can contribute to the creation of compositional structures. For example, saturating the sound-box by double-tracking distorted guitars and panning them left and right in the sound-box enhances the perceived level of distortion, creating a wall of sound that invokes the visual image of a stage filled with a wall of Marshall stacks. The Dist-space analytical tool also demonstrates that used in conjunction with the Dist-space tool, they present a more complete picture of how distortion can create form, movement, and progression in many heavy metal and hard rock compositions.

Although the chapter only analyzed a single work, the techniques and analytical outcomes can be found in many compositions by heavy metal artists and bands. For example, "Colored Sands" by Gorguts follows almost the exact compositional plan found in "Blackened," starting with a high harmonics in a guitar centrally located in the sound-box Dist-space that expands downward to fill the box with multiple guitar tracks panned left and right. The remainder of the work saturates the sound-box Dist-space with guitars that live and never move from the highest levels of Dist-space. Unlike "Blackened," "Colored Sands" appears intent on testing the limits of the sound-box by hard-limiting the entire mix, increasing the pressure on the walls of the box to the point of bursting.

Finally, while the chapter demonstrated the analytical application of sound-box Dist-space, I hope the ideas presented in this chapter will also spark creativity, leading to new and more imaginative ways of making the studio and recording process an integral part of the compositional process.

References

[1] The GP Staff. *Distortion Tips from the Loud & Mighty*, Guitar Player, October (1992).
[2] Moylan, William. *Recording Analysis: How the Record Shapes the Song*, Focal Press, New York (2020), 594 pp.
[3] Zak, Albin J. *The Poetics of Rock: Cutting Tracks, Making Records*, University of California Press, Berkeley (2001), 276 pp.
[4] Scotto, Ciro. The Structural Rock of Distortion in Hard Rock and Heavy Metal, *Music Theory Spectrum*, Vol. 38, No. 2, Fall (2016), pp. 178–199.
[5] *loudQUIETloud: A Film about the Pixies*. Cactus Three, Stick Figure Productions (2006).
[6] Songfacts. *Smells Like Teen Spirit by Nirvana*. Accessed January 2020 from www.songfacts.com/facts/nirvana/smells-like-teen-spirit
[7] Marvin, Elizabeth West, and Laprade, Paul A. Relating Musical Contours: Extensions of a Theory for Contour, *Journal of Music Theory*, Vol. 31, No. 2 (1987), pp. 225–267.
[8] Morris, Robert D. *Composition with Pitch-Classes: A Theory of Compositional Design*, Yale University Press, New Haven (1987), p. 281.
[9] Marvin, Elizabeth West. *A Generalization of Contour Theory to Diverse Musical Spaces: Analytical Applications to the Music of Dallapiccola and Stockhausen, Concert Music, Rock, and Jazz Since 1945*, ed. Elizabeth West Marvin and Richard Hermann, University of Rochester Press, Rochester (1995), pp. 135–171.
[10] Herbst, Jan-Peter. Distortion and Rock Guitar Harmony: The Influence of Distortion Level and Structural Coplexity on Acoustic Features and Perceived Pleasantness of Guitar Chords, *Music Perception*, Vol. 36, No. 4, April (2019), pp. 335–352.
[11] Hindemith, Paul. *Craft of Musical Composition*, Vol. 1, Schott, London (1942), p. 82.
[12] Moore, Allan F., Martin, Remy. *Rock: The Primary Text*, 3rd edition, Routledge, Oxon (2019), pp. 95, 149–153, 270.

14 The Distortion of Space in Music Production

Matthew Barnard

14.1 Introduction

On the observed shift away from the acoustic characteristic of the 'cathedral of the symphony' in early recording ventures, Gould highlights that [1]:

> The more intimate terms of our experience with recordings have since suggested to us an acoustic with a direct and impartial presence, one with which we can live in our homes on rather casual terms.
>
> (p. 46)

Although referencing in part the informality of the domestic presentation of music recordings, these 'casual terms' are a pertinent concept to digest before we proceed further. The ingestion of spatial information by the listener is apparently a painless, fluid process through which countless varieties of conceivable acoustic situations can be transferred. Music production practitioners are spatially intrepid, sometimes cavalier. A modern listener of music productions is psychoacoustically unfussed by variety and complexity; spatial situations that in visual terms could be striking, confusing or fascinating, could in auditory terms be commonplace and familiar, unquestioned. These casual terms compose what Doyle might call the 'spatial contract' [2], which for our purposes we will draw towards the music production context, which is agreed by creator and listener tacitly, with no formalised moment to speak of. These spatial contracts are an understanding that a recorded music production, whatever its nature and characteristics, is primarily 'intended to impress the listener rather than depict reality' [3], utilising—to extremes—an artistic license in its articulation of the spatial image. Indeed, this contract is seemingly reflected by some practitioners through an attitude of 'doing everything for the listener' [4], in a quest that entails 'building musical images' [5] when crafting a music production. A primacy of the ear doesn't preclude a seemingly complex spatial practice. In reference to Ferguson's study of design engineers and their identification of the need for intuitive, non-verbal *visual* thinking in the 'mind's eye', Schmidt Horning usefully draws analogy for our purposes [6]:

> I believe an analogous *aural* thinking takes place in the 'mind's ear' of recording engineers, becoming an essential component of their craft that was crucial to developing skills in editing, mixing and envisioning sound. I use the phrase 'aural architecture' to describe their work because stereo and multi-tracking enabled them to 'build' and 'design' sound.
>
> (p. 264)

DOI: 10.4324/9780429356841-18

This 'aural architecture' is full of adventure and abandon. When introducing the creation of the 'stereo image', Geluso proclaims that 'surely, the possibilities are endless and a great degree of artistic license and creative potential exists that should be explored' [7]. Indeed it does, indeed it has. The appraisal of the spatial image in music production is an adventure in fantasy, paradox and hyper-reality where the surfaces of these 'architectural' sonic creations intersect, superpose and warp in simultaneity in front of our ears, and figures present in caricature, surreal scale, even in *multiperspectivity*. Already, we have adopted visualist language in this exploration and the potential problems of this are conceded, but it is idiomatic of the topic area to do so and is exploited for appropriate analogy.

Let's establish what we mean when we speak of the 'distortion of space' in music production. Firstly, when referring to space in this discussion, we will be considering both the craft and the experience of sonic spatial image; that tendency for sound to evoke a sense of spatial form of some description, in full acknowledgement of its varied intended form by the creator(s) and its varied interpretation by the listener. So, we'll be considering this in both its creation and in its reception. We are not considering further interpretations of space in relation to music production, such as pitch-space, narrative- and symbolic-space, or the broader social and cultural spaces of such endeavours.

Secondly, an interpretation of distortion needs establishing: to distort is to alter form. This interpretation is broad, but it should emerge at a number of junctures why it is useful to adopt this position. The processes that accumulate in the realisation of a music production are numerous and nuanced, varying in the apparent degrees of alteration to the signals that contribute to the whole. It is contended that creating a music production is inherently a distortion of space to varying degrees; simple acts of microphone deployment, juxtaposition of sources in the crafting of a composite sound image and more apparent signal processing, provide a basic gradient of instances. We could even extend this towards more nuanced aspects of music production presentation and include subtleties such as the distortion that a format imbues upon the whole spatial image, something Zagorski-Thomas dubs 'media-based staging' [8], as if we are peering through dirty lenses—the aural architecture obscured by artefact. The vinyl format is exemplar here, imparting audible noise floor, crackle and occasionally boom and hum. The concept of 'media-based staging' distorts the aural architecture to an extent—to distort is to alter form after all—so it is worthy of acknowledgement, if not focus, in this discussion. It is, however, indicative of the level of nuance we could inspect when comprehending the distortion of space in music production.

Some circumscription is clearly necessary, otherwise this discussion will be a succession of minor alterations to signals declared 'distortion!', with nothing substantial ventured or new ideas postulated. Therefore, the scope of this exploration is: (1) to outline some of the burgeoning theorising of space in music production that can relate to the idea of distortion in select contexts; (2) to consider how spatial distortions are so casually permitted by the listener; (3) to propose the idea of spatial sampling in relation to microphone deployment; (4) to grapple with the implications of multi-microphone capture on the spatial image; and (5) to comprehend the spatial implications of amplitude dynamics processing in music production.

14.2 Music Production's 'Spatial Turn'

A discussion of the distortions of space in music production can sit within the burgeoning writing on the topic of space in sound and music more widely. Recent discussions have included a greater focus upon the role and nature of space in its creation and reception. By drawing upon terminology used to describe a trend for greater contemplation of space in 'philosophy, cultural studies, geography, art history, and literary studies', Vad argues a need to 're-imagine the attention to space in musicology and analysis of space in music as part of the spatial turn' [9]. A range of sustained investigations have attempted to comprehend collectively an ontology of space across the auditory experience from varying idiomatic pivots. This theorising is essential for a practice that is inherently, inescapably spatially informed and engaged. This chapter utilises some foundations that the following investigations have provided.

The interpretation of music production space through reception is the primary angle investigated by: Moylan [10] through the proposed 'sound stage' and 'perceived performance environment'; Dockwray and Moore [11] with the 'sound-box' concept and its resulting mix arrangement taxonomy; Brøvig-Hanssen and Danielsen's 'virtual sound room' [12]; the analysis of interpretations by Lacasse [13]; and the incision provided by Zagorski-Thomas [8]. Indeed, Zagorski-Thomas' concept of the 'sonic cartoon' is of particular service to this discussion, utilising a relation of Lakoff and Johnson's 'conceptual blending' and Gibson's 'image schema'. A 'sonic cartoon' is akin to the visual concept: 'schematic' in form, relying upon 'mapping of a limited number of features or properties' [8] for interpretation in some abstract, conceptual space. It's a neat circumscription for how the reduced form of a sound, represented through a recorded form, can be powerfully meaningful. Additionally, Zagorski-Thomas' concept of 'staging'—essentially the act of crafting the aural architecture of a music production—is a developed overview of the many permutations of spatial arrangement, its dynamic and its peculiarity in the virtual, musical space.

Additionally, conceptualisations of the nature and mechanics of interpretation of 'recorded' musical space are provided via: the comprehensive contemplation of sound as form by Zak [14]; Doyle's comprehensive probing of the 'fabrication' of space in popular music up to 1960 [15]; Clarke's ecological interpretation of 'musical meaning' [16] and developments [17]; Brøvig-Hanssen and Danielsen's popular music case-study approaches [12]; Born's comprehensive survey that encapsulates such themes [18]; and finally, overviews provided by Vad [9] and Kraugerud [19].

Finally, in the domain of electroacoustic and acousmatic music, Wishart identified key concepts relating to the spatial language of sound and how it is reconfigured in an acousmatic context. One particularly useful concept is that of the 'virtual acoustic space' [20]: the artificial, unbridled sonic landscape and the associated freedoms it affords. Additionally, Smalley [21] has outlined a significant, resonating contribution in the shape of 'space-form', exhaustively detailing a taxonomy of spatial arrangements that can be suggested in acousmatic practice, as a broader phenomenology of the aural spatial experience. Both of these texts have been drawn upon within developments in the popular music domain (Smalley's effectively by Brøvig-Hanssen and Danielsen [12] and Zagorski-Thomas [8]).

14.3 The Distortion of Space

'STOP! Don't read any further if you are afraid to find out how popular recordings are really made!' is Somer's 1966 warning to the reader seeking insight into 'popular recording, or the sound that never was', a gentle exposé on the 'aural frontier' of the burgeoning popular recorded music movement [22]. But, a virtual musical space is full of sonic freedoms that needs no introduction, contextualisation or warning before being experienced. Although we are outlining pronounced alterations and interventions in the representation of space in sounds, it is appropriate to consider the lack of cognitive resistance that it excites in the listener and subsequently how spatial distortions are permitted liberally in the 'spatial contract'.

There exists an overwhelmingly object-based language that surrounds discussion of musical sound with a tendency to objectify 'source' in recorded sound. Chion highlights Bailblé's interpretation of this tendency to objectify as an autonomous outcome of our aural perceptual mechanisms that 'work at a distance', characterised through the tendency of 'reprojection toward the source of sensory excitation' [23] as we grasp for localisation and identification. Smalley has termed this relation of sound to source and situation as 'source-bonding' [21], which can be of varying accuracy, based upon familiarity or otherwise. It's worth acknowledging the limits of these bonds, as Chion contends that [23]:

> reprojection onto the source in the case of sound is only partial and that sound is not anchored to the source as solidly as in the case of visual reprojection.
>
> (p. 107)

This could be considered a particularly reliable interpretation in the context of virtual musical sources, where we can be presented with the familiar or relatable, to the unfamiliar. Brøvig-Hanssen and Danielsen, channelling Smalley, suggest that a sound is inextricably linked to space—it is spatial in nature—and we can't avoid an automatic tendency to 'recognise these different acoustic reflection patterns from our previous experiences' allowing us to 'imagine specific actual spaces' [24]. We can confidently slink away from any suggestion that we are Schaefer's 'causal listeners', striving to identify a sound's source, and instead gravitate towards an acceptance that we are Chion's 'figurative listeners' [23], concerned with what the sound *appears* to represent, in a sort of poststructuralism where sounds are 'signifiers' with multiple, ambiguous meanings [25]. The intent of the creator is already decoupling from the reception in the listener's 'mind's ear'.

Concerning space specifically, then, an ear tuned to figurative interpretation can encounter potentially dense, even contradictory signifiers in a virtual musical space. These various arrangements of signifiers can invite relation to artificial, surreal, unnatural and even impossible situations. In contemplation of synthesising the recognisable, relatable with the unreal in the visual world of M.C. Escher, Hofstader comments that [26]:

> one can . . . manufacture hypothetical worlds, in which Escherian events can happen . . . but in such worlds, the laws of biology, physics, mathematics, or even logic will be violated on one level, while simultaneously being obeyed on another, which makes them extremely weird worlds.
>
> (p. 99)

As the 'laws' of the aural experience of reality are being obeyed on one level, they can be violated on another without consequence because, as Zagorski-Thomas states, when contemplating how this visualised 'artificiality and realism' relates to the aural 'sonic cartoon', 'it is still something meaningful. We might call it a form of truth rather than a form of reality' [8]. It is clear that we can uncouple the aural architecture of the virtual, musical space from any notion of reality without hesitation, because as Katz neatly highlights, it is 'a musical space unique to the work . . . with no physical counterpart' [27]. Indeed, in a call-back to the casual terms of the 'spatial contract', Warner suggests that 'in recorded music there is no need for any kind of "natural" diegesis' to tune into, claiming that there is acceptance by the listener of the 'implied environment of a specific timbre' that's encapsulated in the broader 'musical statement about that instrument or voice' [3]. Even if the listener deigns to consider the unrealism of the presented virtual space, there's the inescapable, automatic tendency to relate stimuli to experience to reckon with first. In further considering the digestion of M.C. Escher's visual works, Hofstadter explains the futility of attempting to disentangle the paradoxical [26]:

> which rely heavily upon the recognition of basic forms, which are then put together in nonstandard ways; and by the time the observer sees the paradox on a high level, it is too late—he can't go back and change his mind about how to interpret the lower-level objects.
>
> (p. 98)

So, there's apparently no easy reversal of interpretation. Somer's 'sound that never was', at the time of popular music's technologically innovative boom, draws parallels with cinema and its decoupling from the 'limitations and restrictions of live theater'. Through use of different perspectives, juxtapositions and framings that can 'hardly be said to be representations of "reality", but ... are indubitably successful artistically and dramatically', the cinema expands 'visual horizons', much like the 'popular recording industry ... can be as effective in expanding aural horizons' [22]. Indeed, as highlighted in Born [18], Connor identifies the primary peculiarity of the auditory experience being 'its capacity to ... reconfigure space' [28]. In a reference to the change in production trend that Somer exposed—one that shifted gradually away from attempts to emulate the spatial qualities of a musical performance 'the "illusion of reality" (mimetic space)' towards 'a virtual world in which everything is possible' or 'the "reality of illusion"' [29]—Moorefield handily provides carte blanche for spatial distortions: *everything is possible ... hearing is believing.*

14.3.1 Permutations of Distortion

As Prokofiev places a bassoon close to the microphone and the trombone at a distance for a recording, he recognised that the studio construct engenders that which 'would have been impossible in compositions for symphonic orchestra' alone, if it were not represented in the form of a recorded music production [30]. Even this simple act of managing the relative proximity of source from microphone is concluded as a heavily modified spatial representation. As a production is a succession of processes, there is a multitude of junctures at which the spatial qualities of a given sonic situation can be distorted. The exploitation of monophonic, stereophonic and extended 'multichannel' formats is the first act of reconfiguration, where figurative sources are positioned or 'staged' in varying affordances of area, from point-source to plane to periphony. The affordances of the two most explored presentation formats—monophony and stereophony—invites some consideration in this context, alongside some select permutations of music production praxis that are illustrative of distortive practice.

Initial and evolving explorations of monophonic presentation exhibit complex spatial creations that defy what's possible in reality. Consider the works of Phil Spector and his 'Wall of Sound' technique that involved cooking a sort of 'musical soup in the studio' [14]. It involved multiples of instruments in the arrangements (including drum kits) to blend into 'kind of a mulch' [29]. These instrumental sources are necessarily scattered across the studio floor, captured with different microphone sources (or 'spatial samples', as is argued later) with bleed between them, to be balanced in the control room. Add to this the tendency that 'the entire mix was then fed to a loudspeaker in one of [the studio] Gold Star's ambient chambers, located in a converted bathroom' [14]. And then, in the final act of spatial trauma, the whole mix—a composite of these elements—is extruded through a single loudspeaker when reproduced in monophony. It's said that Spector's aural vision was slow to emerge in a session, being exploratory in nature as 'its only existence was in his imagination' [14] prior to realisation. It certainly has no analogue, no equivalence in reality. In this 'reality of illusion', figures are superimposed onto a vanishingly small origin, with a sense of proximity to the perimeter of the loudspeaker, but no extent otherwise: a truly distortive system of capture and presentation.

The nature of monophonic space is quite incomprehensible in figurative terms, presented as we are with a vanishingly small keyhole through which our 'mind's ear' must interpret the virtual musical space. In any case, the tendency is at no moment to be troubled by this issue upon audition. In stereo praxis, the space-less origin of monophony is given planar substance, generalised as a continuum across two points. This produces the regularly exploited affordance of positioning sources along the continuum, often in an arrangement without any pretence of Moorefield's memetic space. The stereo image affords a more robustly defined sense of proximity, with triangulation of source position and a more richly expressed sense of figurative acoustic situation. That luxury affords further spatial adventures, be they attempts at mimesis to represent in some approximation the staging and situation of a proposed or actual musical performance, or the 'virtual space' of imagination, where the logic of a relatable situation is silent casualty. Typically across all spatial formats, sources are from different times and spaces, or even the same apparent source across different times, such as Stevie Wonder and his nearly one-man band (as heard on 1972's *Music of My Mind*, amongst others). This synthesis of chronologically and spatially unrelated source produces a 'composite image of an apparently unitary musical performance' [14] that reifies in the 'mind's ear'. The stereo continuum can also be exploited for simplistic spatial distortions that involve positioning a source in one suggested location and having a second, processed iteration in another, almost contrapuntally. An example of this can be heard on Yussef Dayes' 'Love Is the Message (Live at Abbey Road) (feat. Rocco Palladino, Alfa Mist and Mansur Brown)', as during the first half, Brown's guitar sound is anchored towards the left of the continuum, but we also hear a reverberated iteration of the guitar sound anchored to the right. The spatial situation of the guitar has apparently been cleaved apart, with the direct component divorced from the indirect in location. There is one apparent 'source' event for these counterpointed spatial figurations.

The recorded signals that compose any music production, particularly those from acoustic sources, can have varying prominence of 'internal space' [23] that betrays the acoustic origin. The use of secondary, typically artificial reverberation and other psychoacoustic cues that are suggestive of space adds a second-order space to the 'internal space' of the signal. This is potentially distortive, as it is not necessarily an 'exposed' mediation, and the 'internal space' can be jeopardised by obscuration: imagine hearing the acoustic of a bedroom *inside* the acoustic of a cathedral. Let's go even further and add additional layers of reverberation atop the first-order 'internal space' and second-order space to produce third-order spaces, or more. Smalley would term this space-inside-a-space as 'spatial nesting' [21]. This nesting is variously inflected but is pronounced in its alteration of a sense of spatial form. Further, this affordance of applying secondary reverberation is a catalyst for 'a broad range of spatialities . . . to coexist in a single, possibly crowded multi-track recording'. As Case continues, its accepted to be typical 'in popular and rock music' where 'simple productions commonly run three or four different reverb units at once' with 'a globally applied single space' being the exception [31]. A mix can be acoustically busy, then. There is a potential nuance to this use of multiple secondary reverberations, as illustrated by Zagorski-Thomas, as a drum kit can entail different component instruments (the snare, for example) having a particular reverberation characteristic applied to it, with other component parts of the kit having only their 'internal space' or even a contrasting one [32]. In Smalley's lexicon of space-form, this is termed 'spatial simultaneity' [33], when one is aware of the simultaneity. Reflecting on the implications of this situation to the listener, Brøvig-Hanssen and Danielsen explain that we don't try to relate such a suggested situation to a relatable space and are instead 'forced to attempt an awkward synthesis of a number of such spaces' in juxtaposition, possibly superimposition. Going further, it's explained that [12]:

> As we project these previous 'real-world' experiences onto a single virtual environment, we hear the music in question as unnatural or surreal and opaque, because it clearly signifies a spatial environment that could never occur in a real, physical, technologically unmediated environment.
>
> (p. 33)

So, very easily we step into Hofstadter's realm of partial illogic, forced to confront the weird of the technological mediation. This mediation and its 'opacity' are perhaps the analogue of the distortive qualities of the spatial interventions variously made.

The continuity of space is a further attribute that is occasionally distorted. The reversal of time and the speeding up and slowing down of the playback rate of a sound are overt distortions, as is the interruption of the expected, organic decay of a reverberant signal. Zak [14] and Brøvig-Hanssen and Danielsen [12] have in particular given regard for the use of 'gated reverb' across works by Peter Gabriel, and Prince and Kate Bush respectively. In Brøvig-Hanssen and Danielsen's analysis of 'Get Out of My House' by Kate Bush, the use of a gated reverberation produces a spatial distortion where 'instead of the sound fading into a big space, the space disappears altogether' [12], as if the virtual, musical space appears and disappears in a succession of momentary illuminations. The interruption of space and other selected distortive approaches and attributes are part of an integrated whole in the shape of a music production that feels more cartoonish at each juncture.

14.3.2 Microphones as Spatial Samples

Where there is no sound, there is no acoustic reveal of space. A silent space is a 'dark' space. Sounds, then, are illuminations in the acoustic darkness, which gives exposure to the physical topology of the environment, delimiting feature or expressing free-field. Only this is in opposition: sound is sculpted into form by the environment, given volume and occupancy through containment and architectural coercion. A figurative impression of sound is the spatial positive to the environmental spatial negative [34]. LaBelle illustrates a related interpretation with a clap of hands, which 'describes the space' as it propagates and reveals a 'multiplicity of perspectives and locations'. It is not merely a sound with source but instead a 'spatial event' [35] that illuminates more than the simple idea of source, spilling around the environmental delimitations. So, as sound isn't object, but instead a signifier and essentially a communication of space, it is pertinent to turn our attention towards the capture of sound through the deployment of the microphone. The capture of the 'sound of a thing' is not object feature extraction, but instead a sampling process: a sampling of space. These remarks by Kostelanetz are a reflection of when he auditioned the control room interpretation of a studio performance, as conductor for the CBS orchestra when making the Chesterfield radio programme [36]:

> I realised that equally important with the seating of musicians and with carefully marking scores for bowing and other subtleties of interpretation was the placement of the microphone. Some instruments were virtually lost; others came through several sizes too big, so to speak. One of the first corrections was to hang the mike above the whole group of musicians, not favouring any section.
>
> (p. 70)

This is an early comprehension of the nature of sound capture in spatial terms. As Prokofiev ventured to rearrange the orchestra to sample the space as desired, placing some instruments in close proximity to the microphone and others at a distance, Kostelanetz describes a rearrangement of the spatial sampling point. They are orchestrating space, crafting 'tonal "camera angles"' as Eisenstein dubbed Prokofiev's outcomes [37]. The positioning of a microphone becomes a sampling point in a volume of acoustic space. These samples can be singular or multiple, offering greater or lesser 'spatial resolution' of sampling through a given volume. Weidenaar's poetic turn feels apt to deploy, as 'just like the camera, the recorder never lies', it's just that it can only reveal a 'truth' permitted to it [38] or perhaps Zagorski-Thomas' 'form of truth'. A greater number of spatial samples yields a higher-resolution representation of that space in raw terms, any potential deployment of those samples notwithstanding. The nature of the spatial sample is of course dictated by the characteristic of the microphone itself, with its directional pattern of sensitivity being the most relevant to this concept. Different directional sensitivities (omnidirectional, unidirectional and bidirectional) become sample qualities, as if naked blub or focused torchlight is revealing the sound space. We dip our toes into the acoustic pool when we deploy a microphone: Spector's 'musical soup' is fluid, it turns out.

14.3.3 De Facto Cubists: Multi-Microphones as Multiperspectivity

As exemplified through the juxtaposition of figurative sources in a virtual, musical space, spatial samples can regularly be deployed in combination in a given recording situation. Particularly relevant to much modern practice, there exists a spatial peculiarity in the use of multiple microphones on what would be considered source objects, typically instruments such as a piano or drum kit. Single microphone techniques are not uncommon, but there is a trend for multiples to be adopted, from simple, fairly 'mimetic' stereo framing that attempt to represent the characteristic scale and form of the source from a logical, single perspective, to more spatially maverick, intuitively distributed arrays. These more adventurous combinations can include 'mimetic' stereo framing, alongside focused, close-perspective sampling to combinations with no spatial 'logic' related to the reality of the sampled situation. The configuration of the deployed microphones is employed with some sonic result in mind from each spatial sample that, according to Zagorski-Thomas, 'exaggerates or distorts particular features', typically to 'enhance the attack transients' [8], drawing focus to those defining sonic characteristics that help shape the sonic cartoon. These cartoons are the essence of the exercise, achieving the 'record-sounding drum kit' according to Reznor [39], not some sonic transportation to the *sound of that drum kit in that space in that moment* as if you were there. We will now consider that this practice is 'aural architecture' in a cubist tradition.

The multi-microphone arrangement produces contemporaneous spatial samples, focused in varying fashions on a source to be identified as a 'totalisable' image [23], as if there is a single, figurative drum kit, or piano, or whatever it is in the representation. The cubists of the visual domain explored the representation of form through the employment of multiple perspectives and spatial interpretations. Cézanne's *Still Life with a Basket of Apples* is a study that encapsulates subtle parallax of a scene, as the perspective alters across the work, as if we are moving across the figurative space as we scan along the work, with the two sides seemingly having a non-linear relationship. A more pronounced articulation of the aesthetic is seen in Picasso's *Les Demoiselles d'Avignon*, where figures present in apparent contortion and distortion, as we see limbs and features in differing apparent orientation to our relative position. We are seeing many 'angles' from one. A final indulgence in visual contextualisation is found in Popova's *The Pianist*, as an overt expression of the principle. The pianist and piano are fractured into multiple, mosaic perspectives and features that are juxtaposed into a single visual frame. We're afforded both a glut of spatial information in variety *and* a sparsity, as the 'totalised' form isn't immediately apparent. In the visual domain, the spatial stimuli are indispensable and stable as fragments, whereas in the aural domain, these spatial stimuli are quickly dispensed with to produce a more coherent synthesis as a 'totalised' image. The usefulness of the visual analogy may now be exhausted, but the principle fits.

Borrowing a term adopted to characterise certain narrative form and historical process that encapsulates multiple voices, sources or *perspectives* to describe or represent a narrative, 'multiperspectivity' can be considered, according to Pfister, in a binary configuration of being either 'open' or 'closed' [40]. An 'open' multiperspectivity is in essence composed of dissonant, contradictory perspectives in various combination, relying upon contrast for triangulation or enrichment. A 'closed' multiperspectivity is composed of multiple perspectives which instead coalesce into a coherent sense of form through consonance. Reznor's 'recording-sounding drum kit', for example, can be a complex of spatial samples that provide 'perspective' from a variety of contrasting positions in the original environment. In Zagorski-Thomas' note, we can suppose this isn't intended to represent those prominent markers that help to define the drum kit in the visual domain, but is instead intended to produce appealing sonic results whilst accommodating the performer. This has the curious result that, in most situations, such an instrument is presented in a 'totalised' image that entails what we might term complex perspective, where we are auditioning from more than one position in the same space, simultaneously. Take, for example, the standard procedure of positioning a microphone both above and below the snare drum: these two spatial samples are typically presented in the same spatial position in the mix, combining to a curious perspectival paradox. Continue to combine further spatial samples of the instrument, and the form becomes increasingly 'logically' problematic in its construction as a 'causal complex', but of course not its reception. This multi-microphone multiperspectivity is regulated by typical spatial sampling concerns, that rightly put primacy in the final form, controlled by concerns such as time vis-à-vis phase coherence. This can entail either managed spatial sampling that ensures coherence, or after-the-fact time alignments that involve adjusting the relative position of different spatial samples in time. Whatever the method, a successfully coherent virtual, musical source is a 'closed' multiperspectivity that 'totalises' into a sense of a 'record-sounding drum kit'. The aural architecture of a multi-microphone capture is cubist in essence, the originators of these forms are cubists de facto.

14.3.4 Distortion through Amplitude Dynamics Compression

There are conceptions and subsequent considerations of the spatial image that are focused upon virtual displacement of sources, apparent acoustic situations and the dynamic thereof. Amplitude contours of signals are regularly altered, or distorted, when the aural architecture of the mix is being constructed. There is a widely considered identification of the changing characteristics of space across a given musical work, particularly those changes that inform or amplify the musical structural language, but attention is now drawn to the short-term changes that signals can undergo.

Compression of amplitude dynamics is a common process to adopt across a range of signals in a mix. The aim is to reduce the dynamic range, making the *quiet bits louder* by making the *loud bits quieter*. This is a simple treatment to comprehend: a signal's amplitude is more consistent, with fewer fluctuations in level, smoother contours. The spatial ramifications are easily overlooked, but are, in Zak's words 'conceptually problematic' [14], as we are dealing with LaBelle's 'spatial events' after all. Just like the apparent scale and displacement of a figurative source can be defined through treatment and juxtaposition, its internal dynamics or spectromorphology [33] can be altered or distorted through compression processes, resulting in curious spatial morphologies and troublesome analogy. What seems to be happening to the aural architecture or figurative source, when a dynamics process is applied and the quieter moments of the signal are made louder? Does its apparent scale increase? Does it move closer to the listener? Does the apparent acoustic environment contract? Is the 'exposure' of the acoustic increased as the acoustic torchlight or lightbulb shines more brightly? Perhaps the resonance of LaBelle's hand clap is dragged into the foreground? In reference to Peter Gabriel's 'The

Intruder', Zak assesses the heavily compressed and gated drum sound, remarking that the liberal use of compression was 'in effect squeezing its energy into a tighter space and thus increasing its explosive character and elongating its decay' [14]. This description provides some analogy for us to grasp onto in considering the ramifications of spatial compression. Zak's suggestion that the energy of the perceived event being 'squeezed into a tighter space' identifies the perspectival modifications that ensue, with three possible interpretations: (1) the apparent figurative space is contracted; (2) the perceived event is energised towards its 'explosive' qualities; or (3) both occur. The quality that Zak outlines was crafted by Hugh Padgham at The Townhouse in studio 2's stone room, who utilised it perhaps most famously on Phil Collin's 'In the Air Tonight'. The compression was created in part from the 'listen-mic' function on an SSL console, which applied extreme compression processing to an incoming signal, producing 'the most unbelievable sound' [41]. Unbelievable indeed, with no equivalence in reality.

Doyle has proposed a spatial quality of being 'concave' to describe the perception that a production appears within the virtual, musical space, at a distance from the perimeter of the loudspeaker. The opposition to this is that of a 'convex' quality, where the production appears to exist in part at the perimeter of the loudspeaker but seems to extend into the listening environment, usually as a result of the 'dry', more direct-sounding figurative sources [15]. Considering a further example, the degree of compression exhibited upon materials on Vaetxh's 'Cuntpresser' is so pronounced as to iron out the amplitude features of many materials into textures, with some moments of reverberation 'ringing out' enough for the digital noise-floor to be dragged into stark audibility by the compression amount. The space is being actively contracted, crushed into oblivion before our ears. Perhaps the effect of compression on this space is a quivering oscillation between concavity and convexity? Perhaps there is a concave/convex spatial morphology across compressed (and 'expanded') materials as a principle. The modern utilisation of 'side-chain' compression—where the compression of a given signal is informed by another—fits satisfyingly into this concept: a prominent example of the technique is heard in Stardust's 'Music Sounds Better With You'. In this work, the bass drum occurrences are informing the compression of all other elements in the mix, resulting in a characteristic 'pumping' sound when the bass drum sounds. The spatial quality of this 'pumping' could be characterised as a momentary convexity, as the non-bass drum materials seem to contract back from the perimeter of the loudspeaker.

The use of amplitude dynamics processing is of sometimes quite pronounced consequence for the apparent spatial characteristics or situation of figurative sounds. As Zak declared, these consequences are 'conceptually problematic', representing a challenge to find appropriate description of the figurative forms that are produced.

14.4 Conclusions

Produced fixed-media works are an awkward complex, a fundamentally distorted function of sound signal capture, generation and juxtaposition in presentation. As fixed-media music productions inescapably involve use of sound via some technological mediation, they are always representative, 'schematic' formulations [8], using sounds that are 'always a sonic cartoon' [19]. A full survey and consideration of all the possible iterations of the distortion of space has been outside of the scope of this discussion, but we have traversed a contemporary production soundscape that is home to severely altered, *distorted* forms of sound that all seem to have survived the experience, sitting comfortably in combinations that can produce orders of distortive character, always 'intended to impress the listener rather than depict reality'. The modern listener condones these distortive tendencies through silent acceptance of the 'spatial contract', listening beyond the illogic, reconfiguring the aural architecture into totalised, relatable forms, or even suspending any need for figurative meaning, defeated by the challenge of making sense of it. The spatially intrepid music production practitioners can continue to craft illogic uninhibited by these observations.

As highlighted in Born [18], Connor has detailed the shift in auditory experience that resulted from the development of modern technological means as a drift from the 'Cartesian grid' of 'visualist imagination' has given way 'to a more fluid, mobile and voluminous conception of space' [28]. This 'more fluid, mobile and voluminous' comprehension of space is a satisfying fit for the peculiarly pliant and potentially complex 'virtual space' that is conjured in the music production process, with its selective spatial sampling, multiperspectivity, superimpositions, paradoxical acoustics, physics-defying dynamics, in tandem with its inescapable potential to leverage architectural, reified figures in the 'mind's ear', manufacturing a 'reality of illusion' that works to the 'spatial contract' that probably reads, simply: *anything goes*.

References

[1] Gould, G., The Prospects of Recording. *High Fidelity*, 16(4) (1966), p. 46
[2] Doyle, P., Ghosts of Electricity: Amplification, in Bennet, A. & Waksman, S. (eds.), *The SAGE Handbook of Popular Music*, New York: SAGE (2015)
[3] Warner, T., *Pop Music—Technology and Creativity: Trevor Horn and the Digital Revolution*, Aldershot: Ashgate (2003)
[4] Mitch Miller 1999 Interview in Schmidt Horning, S., *Chasing Sound*, Baltimore: JHU Press (2016)
[5] Martin, G. *All You Need Is Ears*, New York: Macmillan (1979), p. 35
[6] Schmidt Horning, S., *Chasing Sound*, Baltimore: JHU Press (2016)
[7] Geluso, P., Stereo, in Roginska, A. & Geluso, P. (eds.), *Immersive Sound: The Art and Science of Binaural and Multi-Channel Audio*, Abingdon: Routledge (2017), p. 72
[8] Zagorski-Thomas, S., *The Musicology of Music Production*, Cambridge: CUP (2014)
[9] Vad, M., Perspectives From the Spatial Turn on the Analysis of Space in Recorded Music, *Journal on the Art of Record Production*, 11 (2017)
[10] Moylan, W., *Understanding and Crafting the Mix: The Art of Recording*, 2nd edition, Boston: Focal Press (2007)
[11] Dockwray, R. & Moore, A., Configuring the Sound-Box 1965–1972, *Popular Music*, 29(2) (2010), pp. 181–197. http://doi.org/10.1017/S0261143010000024
[12] Brøvig-Hanssen, R. & Danielsen, A., *Digital Signatures*, Cambridge, MA: MIT Press (2016)
[13] Lacasse, S., Interpretation of Vocal Staging by Popular Music Listeners: A Reception Test, *Psychomusicology: A Journal of Research in Music Cognition*, 17(1–2) (2001), pp. 56–76
[14] Zak, A., *The Poetics of Rock: Cutting Tracks, Making Records*, Berkeley, CA: UCP (2001)
[15] Doyle, P., *Echo & Reverb: Fabricating Space in Popular Music Recording, 1900–1960*, Middletown, CT: WUP (2005)
[16] Clarke, E., *Ways of Listening: An Ecological Approach to the Perception of Musical Meaning*, Oxford: OUP (2005)
[17] Clarke, E., Music, Space and Subjectivity, in Born, G. (ed.), *Music, Sound & Space*, Cambridge: CUP (2013), pp. 90–110
[18] Born, G., Introduction—Music, Sound and Space: Transformations of Public and Private Experience, in Born, G. (ed.), *Music, Sound & Space*, Cambridge: CUP (2013), pp. 1–69
[19] Kraugerud, E., Meanings of Spatial Formation in Recorded Sound, *Journal on the Art of Record Production*, 11 (2017)
[20] Wishart, T., *On Sonic Art*. Emmerson, S. (ed.), Amsterdam: Harwood Academic Publishers (1996), p. 134
[21] Smalley, D., Space-Form & the Acousmatic Image, *Organised Sound*, 12(1) (2007), pp. 35–58

[22] Somer, J., Popular Recording, or the Sound That Never Was, *HiFi/Stereo Review*, 16 (1966), pp. 54–58
[23] Chion, M., *Sound: An Acoulogical Treatise*, 2nd Edition, Durham, NC: Duke University Press (2015)
[24] Brøvig-Hanssen, R. & Danielsen, A., The Naturalised and the Surreal: Changes in the Perception of Popular Music Sound, *Organised Sound*, 18(1) (2013), p. 74
[25] Ireland, M. S., *The Art of the Subject: Between Necessary Illusion and Speakable Desire in the Analytic Encounter*, New York: Other Press (2003), p. 13
[26] Hofstadter, D. R., *Gödel, Escher, Bach: An Eternal Golden Braid*, 20th Anniversary Edition, New York: Penguin (2000)
[27] Katz, M., *Capturing Sound: How Technology Has Changed Music*, 2nd Edition, Berkeley, CA: UCP, (2010), p. 42
[28] Connor, S., The Modern Auditory I, in Porter, R. (ed.), *Rewriting the Self: Histories from the Renaissance to the Present*, London: Routledge (1996), pp. 203–223
[29] Moorefield, V., *The Producer as Composer: Shaping the Sounds of Popular Music*, Cambridge, MA: MIT Press (2010)
[30] Prokofiev in Eisenberg, E., *The Recording Angel: Music Records and Culture from Aristotle to Zappa*, 2nd Edition, London: Yale University Press (2005)
[31] Case, A., *Sound FX: Unlocking The Creative Potential Of Recording Studio Effects*, Boston: Focal Press (2007), p. 314
[32] Zagorski-Thomas, S., The Stadium in Your Bedroom: Functional Staging, Authenticity and the Audience-Led Aesthetic in Record Production. *Popular Music*, 29(2) (2010) pp. 251–266
[33] Smalley, D., Spectromorphology: Explaining Sound-Shapes. *Organised Sound*, 2(2) (1997), p. 124
[34] Barnard, M., The Studio, in Schulze, H. (ed.), *The Bloomsbury Handbook of the Anthropology of Sound*, New York: Bloomsbury (2020), pp. 369–384
[35] LaBelle, B., *Background Noise: Perspectives on Sound Art*, New York: Bloomsbury (2016), p. x
[36] Kostelanetz, A., *Echoes: Memoirs of André Kostelanetz*, San Diego: Harcourt (1981)
[37] Eisenstein in Eisenberg, E., *The Recording Angel: Music Records and Culture from Aristotle to Zappa*, 2nd Edition, London: Yale University Press (2005)
[38] Weidenaar, R., Composing with the Soundscape of Jones Street, *Organised Sound*, 7(1) (2002), p. 66
[39] Rule, G., Trent Reznor, *Keyboard Magazine*, 20(3), no. 215 (1994), p. 5
[40] Pfister, M., *The Theory and Analysis of Drama*, 2nd Edition, Cambridge: CUP (1988)
[41] Flans, R., Phil Collins "In The Air Tonight", *Mix*, 29(5) (2005)

15 Distorting Jazz Guitar
Distortion as Effect, Creative Tool and Extension of the Instrument

Tom Williams

15.1 Introduction

The following research explores a microcosm of what Dawe calls the 'guitarscape' (Dawe, 2010). For many, the imagery conjured by the phrase 'jazz guitar' is tied closely to the traditions that encapsulated jazz 60 years ago. Jazz guitar, however, is no longer defined by the archtop-playing acolytes of Wes Montgomery and the quest for a purity of sound (which encapsulated early developments in electric jazz guitar). As Les Paul discusses, 'all I really want to do is to reproduce sound as it really is, and with depth perception so you can *see* the performer' (Waksman, 2001, p. 61). Contemporary jazz guitarists are known for their abandonment of such imagery, in the pursuit of a wider timbral palette not available in the 'traditional' model.

'Neo-trad' (Hersch, 2008, p. 8) artists such as Jonathan Kreisberg, Kurt Rosenwinkel and Gilad Hekselman are testament to the continued timbral development of 'preservationist' forms of jazz guitar. Conversely, Bill Frisell, John Scofield, Wayne Krantz, Marc Ribot, Allan Holdsworth, Scott Henderson and John McLaughlin are just a few who remain in the pantheon of jazz guitarists while eschewing the traditional almost entirely. Regardless of allegiance to tradition or modernity, all of the previously mentioned guitarists have used (and continue to use) distortion in a range of ways.

The way in which the guitar community interacts is fascinating. The electric guitar is arguably the most ubiquitous of all popular music instruments (Waksman, 2001); its presence seen at both the extreme ends of idiomatic and nuanced style (Dawe, 2010), and also more generally as the lynchpin of the 'trunk' of popular music development (Gracyk, 1996). As such, the guitar continues to be at the forefront of technological and timbral development in popular music practice (Lähdeoja et al., 2010). Most relevant here is the way in which various factions of the guitar world have differing views to these technologies and timbres, and the multitude of ways in which we experience the guitar, as performers, listeners and producers of music.

While the perspective of the performer is frequently dissociated from that of the producer and the production itself, this research aims to explore the intersection between the performative and the production. More specifically, the way in which distortion—contextualised within the 'guitarscape'—has different uses, which affect not only the production but also the tactility, identity, definition and history of the instrument.

This study will map the evolving landscape of jazz guitar as viewed through the prism of distortion, production and wider notions of the emergence of a post-modern jazz 'guitarscape', which embraces developments in technology (and by extension distortion) as part of its identity. Examples will help build a narrative across the development of jazz guitar to demonstrate the many ways distortion has been used both as an aesthetic effect and as a creative tool in and of itself, how creatives may adopt such practices, and finally how embracing a modernist approach is crucial in enabling the progressive and developmental elements which define jazz guitar culture today.

DOI: 10.4324/9780429356841-19

Crucially, the author invites the reader to consider the guitarist here not only as performer but as curator of sound, not dissimilar to the role we traditionally associate with a producer.

15.2 Definitions and Scope

15.2.1 Distortion

While a short study can do no justice to the multiple perspectives that identify distortion, some initial perspectives can help un-mire what is a 'complex phenomenon that resists simple categorisation' (Scotto, 2017, p. 179). Scotto states: 'distortion is any change between the input and output of a signal in a system' (2017, p. 179). Poss labels distortion as 'The departure from some presumed ideal linearity' (1998, p. 46).

Although often met with disdain from traditionalists, distortion is now one of many devices which have enabled traditional performative/compositional processes to be expanded, extending the creative possibilities for jazz guitar improvisers and composers. Discourse surrounding the guitar and distortion is not new, but it usually takes place in the context of rock music. The term *distortion* can be considered pejorative in nature within the jazz community, but this research intends to show, as Poss states, that 'distortion and purity are not antithetical' (1998, p. 45).

Distortion is discussed in a range of ways in the guitar community. Overdrive, drive and distortion are often used interchangeably. There is, however, no academic or professional consensus on the difference (Herbst, 2019a, p. 8). Other terms are used regularly to describe the sound in analogous terms: 'crunch', 'dirt', 'boost', 'fuzz' (although this is mostly differentiated). These are sometimes mistakenly used interchangeably, although they are distinct (Ahmadi, 2012). On a continuum of differences, the terms 'overdrive', 'distortion' and 'fuzz' are place markers for various levels of saturation (see Scotto, 2017). As will be discussed, certain stages of distortion have associations which are identified aurally and inform the tonal choices in an electric guitarist's distortion toolbox. Overtone production, for example—a side effect of signal compression—is paramount to the development of spectrally rich and unique timbral identities.

15.2.2 Jazz

The following section discusses the contemporary jazz guitarscape, and will provide useful boundaries, definitions and ways of aligning the context of this research. It is worth briefly mentioning at the onset that the terms 'jazz' and 'jazz guitar' are used here with more of an associative approach. More objectively, the examples, references and case studies of guitarists have been considered by a range of associations, including but not limited to those that: work(ed) with bandleaders widely accepted as in the jazz canon; have notable use of musical elements associated with jazz (harmony, vocabulary, rhythm, style); use processes integral to the craft (for example improvisation or ensemble interaction); interact with the canon of repertoire in various jazz styles; and have been referred to secondarily as part of the jazz canon.

15.3 Research Field Overview

While discussion surrounding jazz guitar and distortion is limited, there are, however, a range of fields which intersect around jazz guitar and distortion.

15.3.1 Timbre

Study of guitar timbre is often localised to rock discourse. While more recently, timbral discourse surrounding jazz has developed, Western pitch and harmony study still dominates the field. As Scotto writes, 'timbre only appears analytically relevant when it highlights, supports, or reinforces pitch-class relationships' (Scotto, 2017, p. 178). Scotto's assertion here is aimed at rock music discourse; for jazz-related music, the precedence of timbral importance is yet more removed.

The effect of timbre on perceived consonance and dissonance is not without study either. Authors such as Terhardt (1984), Sethares (1998) and McDermott et al. (2010) explore the ways in which consonance is perceived, how it can be altered and how this differs in various settings. Herbst's text *Distortion and Rock Guitar Harmony* (2019b) provides a useful meta study.

15.3.2 Properties of Distortion

Discourse surrounding the hard properties of distortion, its production, control and the myriad subfields which occupy this space, deserve a dedicated meta study of their own. A balanced approach is used here to enable the research to be suitably informed, whilst not creating a quagmire in which the focus of this research—jazz guitar and distortion—is lost. At this juncture I will, however, refer to Bloch (1953), Dutilleux and Zölzer (2002) and Rossing et al. (2014).

15.3.3 Jazz/Electric Guitar Studies

While there are no in-depth specific studies on jazz guitar holistically, it is important to note the excellent work of Waksman (2001), Millard (2004) and Ingram (2010) in positioning the guitar's development. Perhaps most significantly, Dawe's seminal concept of the guitarscape (2010) is redefining the way in which we view guitar from a range of perspectives. Dawe's model is here adapted to conceptually frame what I call the contemporary jazz guitarscape. As the guitar is often at the epicentre of popular music developments, it is important to note the authors who draw attention to a range of facets of the guitar, such as Middleton (1990), Frith (1996), Walser (1993) and Gracyk (1996).

The field of organology is also useful at this juncture. Research such as De Souza (2017), Lähdeoja et al. (2010) and Kartomi (1990) help us to consider that the guitar is an ever-changing instrument and should be considered in light of the various 'scapes' (Dawe, 2010) that surround it. Dawe reminds us, however, that 'organologists have mainly focused upon the investigation of the material and acoustic properties of musical instruments, the description of probable methods of construction, tuning systems, timbre, the techniques required to produce sounds from them, and an analysis of their repertoire' (Dawe, 2010, p. 51). An organological approach must then be tempered with the emerging understanding of the augmentation of the guitar—that which is not the guitar itself, but is every bit as important to creating the sound of the instrument (the stomp boxes, the amplifiers, the microphones, the post-production, the picks, the occupation of 'space' within a mix are part of this construct).

15.3.4 Distortion and Guitar

In the past decade, research has begun to unpick the relationship between guitar and distortion, albeit mainly in a rock music context. Herbst's work, for example, considers the effect distortion has on perception of chords; aesthetics, history and production of guitar distortion; perceptions of 'heaviness'; and a range of other interrelated subjects concerning distortion and rock/metal guitar practice (Herbst, 2017a-b, 2018a-b, 2019a-b). Herbst is one of many scholars now exploring the relationship between distortion and guitar (see Hanada et al., 2007; Scotto, 2017).

My intention here, as a popular musicologist, sitting on the fringes of all these subjects, is to create a platform to spark jazz guitar and distortion discourse.

15.4 Dismantling the Western Scaffolding Surrounding Jazz and Popular Music

Popular music discourse and analysis still suffers from an over-reliance on Western pitch-class relationships. While rhythmic analysis has somewhat developed, representation of timbre in jazz studies is marginal at best. Part of this is systemic, considering the history of Western music analysis, but until more recently, analysts have simply not had the tools to discuss timbre in an objective way.

Analyses methodologies have now developed to include in-depth discussion of timbre, acoustics and the perception of sound, often through readings of spectrograms, detailed frequency analysis and structural mapping of specific timbral qualities (see Herbst, 2019b). These are often supported by surveying and mapping perception to hard acoustic correlations of timbre and sound (Herbst, 2019a).

Comparative analysis of distortion is becoming more commonplace and with this additional tools and perspectives. Pakarinen (2010) uses a software tool to compare distortion when created using differing audio systems (e.g. fx/stompboxes) and can map frequency responses that inform the *identity* of certain systems. Herbst's article 'Put It Up to Eleven' (2017a) is exemplary in its comparative methodology. Using clean and distorted reproductions of canonical distorted pieces, the research gives direct insight into the effect distortion has on playability (or tactile augmentation) of the guitar and wider musical expression afforded to the player. Indirectly, Herbst's research gives objectivity to many theoretical discussions in the canon of guitar and popular music research (i.e. Walser, 1993; Gracyk, 1996; Théberge, 1997; Waksman, 2001).

Larger-scale analytical methodologies also exist. Williams' (2014) model uses a psychoacoustic approach to discuss timbral changes in metal productions over time. Through comparative analysis, it highlights trends in production—in this instance, to demonstrate changing levels of perceived loudness—and the nuance of individual tracks. No such comparative chronological timbral analysis currently exists for the jazz guitar. Additionally, Scotto's 'Dist-space' model (2017) uses a range of perspectives to measure the structure of distortion use in relation to the composition, creating a way of viewing a composition or production through the lens of distortion entirely.

While a number of researchers have used perceptual scales to identify and categorise timbral qualities (Williams, 2014, p. 42), no study looks at the perception of a jazz guitarist specifically, or indeed the differences between those in the rock domain and those in jazz. If jazz scholars are to have any chance of rebalancing the lack of timbral analysis, more work should be done to apply the methodologies used by rock scholars.

15.5 The Jazz Guitarscape

Dawe (2010) uses the term 'guitarscape' to illuminate the landscape of the developing guitar community, from the perspective of the instruments, individuals, music, social, technological and cultural aspects of the guitar. Dawe's work discusses the many interrelating, and amorphous, cultures of guitar playing and the affect and effect caused by broader musical, social and cultural developments.

Most poignant in relation to jazz guitar is Dawe's emphasis of heterogeneity over homogeneity. As with many cultural developments before the 1960s, the parameters defining the edges of jazz cultures were more heavily demarcated. During the swing era, Charlie Christian, whose importance to the jazz guitarscape cannot be understated, was the first to give the guitar a unique voice and an 'authentic existence in jazz' (Mongan, 1983, p. 84).

Django Reinhardt also had significant impact on many guitarists who would go on to adopt the bebop style, including Barney Kessel, Herb Ellis and Tal Farlow. At this point the role of the archtop jazz guitar had been fully established. Its role, while emblematic, was also rather homogenously defined. The tone was warm and thick, with clarity being of utmost importance. This was furthered with the advent of the solid-body electric guitar. Les Paul championed the solid-body electric guitar and solid-state amplifier, advocating for a sound that was pure/clean and free from interference/noise (Waksman, 2001, p. 60).

The contemporary jazz guitarscape has a richly interconnecting and diverse topography, with many micro- and macro-scapes. It is not an artefact, nor should it be viewed as such. Dawe's observations encourage us to reconsider guitars not merely as familiar instruments of wood and metal, but rather as 'sites of meaning construction' (2010), whose effects permeate social, cultural, political and economic domains. What then *is* a guitar—technology and timbre are so important to the identity of an instrument. One could argue then that distortion is one of many reactive extensions to the instrument and as much a part of the instrument as the wood and metal themselves.

Consider, for example, what we might call the Contemporary New York Jazz Guitarscape (NYJGS), a localised post-traditional style popularised in the early 2000s. Led by guitarists such as Jonathan Kriesberg, Kurt Rosenwinkel, Miles Okazaki, Ben Monder, Lage Lund and Julian Lage, to name but a few, this culture of jazz guitar is built on a solid foundation of traditional jazz elements such as the use of jazz standards and a common repertoire, functional (albeit developed) harmonic principles, the centricity of the 'line' and of course the archtop or semi-acoustic jazz guitar. It is relatively homogenous in construct, not dissimilar to the earlier homogenous jazz styles and cultures. While these traditional elements give a binding and can authenticate the musicians within the wider jazz community, aesthetically, musically, tonally and stylistically, many innovations are distinctly anachronistic to a preservationist's view of jazz culture. The tonal characteristics of the NYJGS, for instance, unify the traditional, with new technologies such as the use of ring modulator, distortion pedal and Electroharmonix 'freeze' pedal.

Micro-cultures can exist under the larger body of work or stylistic aesthetic of a band leader or composer. John Zorn's stylistic blend of the avant-garde, jazz, Jewish traditional music, rock, surf and other musical styles can be seen in the multistylistic guitar work of those on his own projects or those on his label Tzadik, such as Bill Frisell and Marc Ribot. Traditional jazz lineage is less of a concern for these players, instead focusing on extended techniques, a variety of tonal configurations and a much greater use of timbre as an improvisational device itself is part of the Zorn's aesthetic.

15.5.1 *Prime Movers—Early Jazz Guitar and Acoustic 'Distortion'*

To understand the role of distortion, its adoption and utility within the jazz guitarscape, we must acknowledge the complex cultural structures which create the scaffolding of meaning. The following section will provide a range of examples and analyses to demonstrate the varying ways in which jazz guitarists interact with distortion.

We tend to associate the sound of guitar distortion with amplification, stompboxes and potentially complex signal chains; it is important though to remind ourselves that the guitar itself can produce a rich spectrum of timbres without any additional equipment. In the case of early jazz guitar, the many factors that inform the 'acoustic fingerprint of a performance' (Williams, 2014) may perhaps be experienced perceptively as analogous to the staged distortion models described in recent literature (see Scotto, 2017). For instance, the purposefully harsh attack (symptomatic of the right-hand technique) used by Django Reinhardt in the earlier quintette recordings provides instances where the attack is so aggressive, the instrument itself produces what we might perceive as a heavily distorted quality in other settings. It is entirely possible that a model such as Scotto's could be used to analyse the various timbral stages that exist within Reinhardt's playing.

Early guitar recordings, such as those by Lonnie Johnson and Eddie Lang, have an almost distorted quality to them (the former more so). 'Blue Ghost Blues' is a clear example of this. Johnson would eventually use an electric guitar, and as a pioneer crossing over from blues to a more jazz-informed approach, produced many recordings with a guitar timbre that was distinctly ahead of its time ('He's a Jelly Roll Baker' for instance). Even in his work with Duke Ellington, on tracks such as 'The Mooche', Johnson's heavy attack and distinctly different timbral identity separated him from a typical rhythm section guitarist.

It is perhaps interesting that both Johnson and Reinhardt went on to record with electric guitars, perhaps in an effort to have more consistency of dynamic control. Reinhardt's 1947 recording of 'Minor Swing', for instance, is similar in dynamic to the earlier quintet recordings. Where Reinhardt strikes a chord, or utilises octaves, the amplifier sounds on the cusp of 'breaking up' entirely, in stark contrast to the uniformity of the single-note lines that surround it. The compromise is made here between the convenience, uniformity of sound and the decreased requirement for physical effort an electric guitar affords a musician (Herbst, 2019b) vs. a grounding in acoustic playing styles, where very aggressive picking hand attack velocities are required in order to be heard at all. Perhaps then, early jazz guitarists were aware of the value of timbral control and the role of distortion in creative jazz improvisation.

There is a distinction to be made here, of course, given that what we might now perceive as analogous to distortion (as we have come to understand and experience it) was at this point something entirely different. For instance, 'violinists do not describe their instruments' sawtooth waved voices as "distorted", nor do most bass clarinetists probably feel kinship with a fuzzbox' (Poss, 1998, p. 47).

15.5.2 Early Amplified Sounds and Distortion

As distortion started to become more prevalent in rock and pop idioms, around the turn of the 1950s many guitarists in the rock arena were experimenting with new sounds. Mythologies surrounding the creating of these 'new sounds', such as Rocket 88's broken amp distortion, Link Wray poking holes in his speaker cone with a pencil, and the Kinks using razor blades to slash holes in theirs (Herbst, 2018b), helped connect the identity of a guitarist to the equipment used.

The same cannot be said for jazz music. In the late '30s/'40s the beginnings of what we might now call the 'traditional' sound were born. Led by an ideology championed by the likes of Les Paul, the timbral aim was to create a warm, smooth and round sound which was less bright, aggressive and jagged than the guitar's previous associations. While this was in part down to the playing style of individuals, the instruments (large-body archtop jazz guitars such as the Gibson ES-175), pickup technology and amplification had developed significantly enough that achieving a much cleaner signal and tone was possible (Zwicker & Buus, 1998).

The historical pantheon of jazz guitarists is perhaps at its most dense during this period, due to the clear homogenisation and identification. Perhaps spearheaded by Charlie Christian, Wes Montgomery, Tal Farlow, Joe Pass, Herb Ellis, Jim Hall, Kenny Burell, Jimmy Raney and Barney Kessel are among many who steered the guitar from the associative trappings of the roots blues and jazz toward a more serious occupation in the late swing, bop and cool eras. Considering Gracyk's concept of 'record consciousness' (Dawe, 2010, p. 149), it is perhaps this period from which listeners and practitioners create the patriarchal image of the jazz guitar as we know it.

During this period, we see less in the way of timbral exploration than perhaps in earlier jazz styles. Perhaps in part, the stigma that traditionalists have for distortion comes from this era, where distortion was being popularised in what was then diametrically opposed to jazz—rock music. For the most part the sound that began in the late 1930s encapsulated jazz guitar until the 1960s. The dynamic range in a player's approach here is perhaps more attenuated and regulated, in contrast to the earlier styles of acoustic jazz and blues. Charlie Christian is an outlier here in some ways, however, as the novel sound of his electric guitar was perceived as distorted at the time (Waxman, 14), though by today's standards may not affect similar connotations.

15.5.3 The '60s and Miles' Acolytes

Although Miles Davis is perhaps the embodiment of change and reinvention in jazz, the effect he had on the jazz guitar community has never been fully explored. In Davis' output from the mid-1960s until his death, he employed a range of eclectic guitarists each with a unique voice. Davis' guitarists have included John Mclaughlin, Mike Stern and John Scofield. Others who have collaborated with Davis include George Benson, Larry Coryell, Robben Ford, Michael Landau, Bjarne Roupé, Jean Paul Borelly and Pete Cosey. These guitarists are renowned for their use of distortion and effects. One wonders here whether Davis employed these guitar players for their innovative approach to sound or whether that approach was curated as a result of working with Davis.

During this time, the harmonic language of the players changed too. While there were still allusions to former styles, the trappings of bop were cast off and the visceral agency and power of rock was adopted. Timbre here becomes fully central to the understanding of the music, and with that a much wider range of distorted tones. At this point distortion is no longer simply a colour, it is an extension of the instrument itself, with the focus on the improvisation moving towards the creation of detailed and varying fields of sound (see the Miles Davis (1970) Track 'John Mclaughlin').

Davis' effect on the development of jazz guitar runs deeper than this, as we will see by looking at secondary connections created by his other sidemen who make up the who's who of the '70s/'80s fusion movement, and by extension their guitarists.

15.5.4 The Post-Miles Fusion Guitar Landscape

Experimentation with sound developed exponentially during the late '70s and '80s, due to the rapid development of modular technologies that could be placed within the signal chain. In this respect, the space that occupies the '80s to the early 2000s is a rich area of jazz guitar and the place where many historical accounts simply stop. At this point guitarists were becoming more experimental with their use of sound. During this period the boundaries surrounding the definitions of jazz continue to blur, which in turn sparked a number of slip streams of development—some more rooted in the sound popularised in the '60s/'70s and some more traditionally focused. While the scope of this period is enough for a large-scale study of its own, there are a number of seminally important guitarists whose sound collectively represents a vast difference in tonal identity in contrast to earlier homogenous eras of jazz guitar.

Jazz guitar's history, while often presented in a linear fashion, is anything but. Traditionalists (Pat Martino for example) continued to occupy a large space within the jazz guitarscape, along with inspired newcomers such as Jimmy Bruno and Emily Remler. Traditionalism was developing, however. For example, Pat Metheny, who utilises many authentic traits of a traditional jazz guitarist, along with a deep history of engaging with the canon of traditional repertoire, performers and contexts, continually redesigned his sound, creating incredibly complex setups, unique sound worlds and timbral explorations.

This period gave a home to a number of guitarists, who are often separated from traditionalism under the term 'fusion'. (While by no means exhaustive, notable mentions include John Abercrombie, Hiram Bullock, Mike Stern, Frank Gambale, Al Di Meola and John McLaughlin.) Perhaps most interestingly, the guitarists in this period that occupy the jazz guitarscape almost entirely eschew the trappings of earlier styles, considering both musical and sonic vocabularies. We now turn to those who have furthered jazz guitar's evolving relationship with distortion. Guitarists like John Scofield, Bill Frisell and Allan Holdsworth are part of a larger constellation of post-1960s jazz guitar; however, each has a uniquely personal relationship with distortion.

Scofield's sound has evolved across the course of his career. While some may be more familiar with his popular *Blue Matter* (1987) period, which demonstrates a less nuanced approach to timbral control and use of distortion, sound has become a unique identifying quality of Scofield's approach.

Scofield's *The Low Road* (2007) is an acute example of a guitarist using distortion as a creatively expressive device in and of itself. The level of distortion heard here sits on the line between two different levels of saturation and is reactive to picking hand dynamic. This enables Scofield to create and control multiple distorted 'voices', through attack level.

The opening is a series of harmonics, open strings and what might be described as 'noise'. It is visceral in nature and projects a primal aesthetic, not normally associated with jazz guitar. The distorted quality here is built by saturation of frequency, created by progressive blending of harmonics and open strings. Furthermore, the intervallic content adds to the perceived timbral quality here—most notably the contrast between the root Db and the tritone G. Scofield builds through a series of layered frequency textures (whose combined amplitude contribute to the perception of more distortion) and harmonic dissonances. Perceptively this is perhaps experienced as more distorted or 'noisy' due to the irregular and jarring rhythm used.

The ostinato which follows (0:22) presents an interesting arrangement of differing distorted 'voices' separated and controlled by their harmonic content and the attack. The first measure, consisting of a low drone note, followed by a chord is representative of voice A. Scofield here uses a harsh attack to drive his signal into a further distortion. Perceptively, this presents as more distorted, given the tension in his chosen close voicing of the dominant 7#9 chord, leaving the major 3rd and raised 9th in the same register, creating a minor 2nd dissonance. This is further emphasised with the use of open strings within the voicing, providing a richer range sonority of sound (Stewardson, 1992, pp. 20, 23). The second measure, featuring the subsequent voice, has a different timbral quality. Scofield here plays a low-register riff, with significantly less attack, to create a softer marginally 'broken-up' sound.

The following section, including the main melodic figure, is again a composite of voices, built initially on a single-line melody which is supported using dyads spaced equally through each measure. The dyads allow Scofield to present a mildly distorted or *driven* voice. The attack here is softer, and so a third softer, vocal-like voice is presented. This is placed in contrast to the reappearing chordal figure from the pre-verse ostinato (now also voiced by the horn section). While the introductory sections are enough to demonstrate his unique command of varying distorted 'voices', a close reading of the solo shows a wide range of attack, technique and dynamic control, which provide an interesting and developmental schema of distortion timbres.

Scofield's sound is often characterised by the timbral qualities heard in this track. Most interesting to this study is how Scofield appears to have learned how to use his guitar to control levels of distortion, by mapping certain techniques, levels of attack and harmonic dissonances into an approach that feeds. Scofield's acute level of detail here demonstrates that he is interacting with distortion, every bit as much as he is with the harmony, rhythm and phrase structure performatively. Thus, distortion becomes an interactive part of Scofield's improvisatory toolbox, as important as the lines and chords themselves. As Lähdeoja writes, 'the electric guitar becomes a source of a wide variety of musical materials, such as multi-temporal sonic masses, noise-oriented sonorities, and syntaxes and vocabularies stemming from the instrument's extended playing techniques' (Lähdeoja et al., 2010, p. 53).

Bill Frisell is perhaps one of the most well-known explorers of contemporary jazz guitar timbre. Frisell uses distortion in a range of ways and embodies all of the rationales presented in the following section. While Frisell's use of effects and adventures into novel timbral settings for *jazz* guitar merits an independent study, it is important he is mentioned here. Frisell is perhaps the furthest out on the spectrum here, and while many would argue Bill is firmly outside of that categorisation, it is this term that Frisell personally identifies with, albeit more abstractly (Steinberg, 2019). Ron Carter discussed Frisell's enormously wide approach, which takes from a myriad of American and European musical traditions, as analogous to a 'cloud' (Randall, 2019). The spatial density, breadth and malleability of approach allow Frisell a unique relationship with distortion and timbre.

The track *Throughout* (1983) is a clear example of Frisell's relationship with distortion. The piece begins with an acoustic guitar accompaniment whose harmony is spectrally enhanced by an electric guitar with a lightly distorted and heavily affected sound. Frisell utilises swells and feedback in this section to project an almost pedal steel–like quality. The solo which follows is exemplary in demonstrating the applications for distortion in jazz guitar. Attention is paid to the dynamics throughout to create a compelling timbral narrative, continuing with the use of volume swell, feedback and use of dyads (and careful attention to note duration/crossover) to create various intermodulation effects. There are also times when you would be forgiven to think you are hearing a violin.

As much as improvisatory vocabulary can change and shape the identity of a performer over the course of their career, so too can development of a timbral identity. Allan Holdsworth, for example, spent a lifetime meticulously honing and tweaking modular elements of his instrument, in order to produce the iconic sound he is celebrated for. Chang (2017c) suggests that Holdsworth used distortion to ultimately remove the 'twang' of the guitar and impart a 'bowed quality' to the tone. Compare two ends of his career spectrum, for example: 'Proto Cosmos' as played in the Tony Williams Lifetime band (1975) and 'The Drums Were Yellow' (2000), on Holdsworth's final studio album *Sixteen Men of Tain*. The sound is starkly different:

> [with the synthaxe] I didn't have to deal with distortion and shaping a distorted guitar sound into something musical, which is a real challenge. It's been one of the problems I have all of the time with the guitar—I want to make it sound more like a horn. But at the same time, the fact that you have to use any sort of distortion to get sustain is a kind of a Catch-22.
>
> (Chang, 2017a)

The guitar playing on 'Proto Cosmos' is still distinctly Holdsworthian and exhibits many of the characteristics of his later developments, from a technical and harmonic perspective, but sits almost entirely somewhere else tonally. The solo (0:24 onwards) is far more representative of the rock/fusion crossover sound of the time. There is a harshness to this tone, which Holdsworth would go on to eliminate, at least in part. As such, there are moments when harsh attack and string noise combine, providing clear allusion to the genus of the instrument being played. Holdsworth's picking hand is more 'present' in the performance. Holdsworth would go on to change this, however, through a combination of new technologies and available distortion tones, his picking technique and phrasing approach more generally. In this earlier phase, he was more beholden to the sound as colouration, and as the prior quote suggests, which at times was unwieldly.

Holdsworth's lead tone in 'The Drums Were Yellow' (2:27 onwards) is perhaps the closest realisation of his horn-like timbral development. Certain passages could easily be mistaken for a saxophone (Chang, 2017c). Holdsworth's technique supports this too. A typical legato phrase on the guitar usually still has an *impetus*, created from the first pick stroke. Holdsworth developed his technique to be around his fretting hand and regularly used the percussion of his left hand alone to instigate the first note of a sustained phrase.

> On the one hand, I had to use distortion—quite unnatural to a percussive instrument like a guitar—to get the kind of sustain and vocal quality I wanted from my instrument. At the same time, I'm left with the 'schzzhhhh' of it all. I find that I leave a lot more 'holes' and pauses in my playing with the SynthAxe, whereas with the guitar's sustain, there's always some kind of note hanging on.
>
> (Chang, 2017b)

To do so, Holdsworth would have had to refine his distorted tone, to allow for a significant amount of sustain, shaping of frequencies to disguise string noise, and also provide the 'colour' of the wind instruments he sought to emulate. In this sense, the tone produced through a simple engagement of left-hand percussion (hammer on) gives rise to the breath-like quality of Holdsworth's sound. Holdworth also discusses the issues with this, most notably, a difficulty in adding rests or, as he calls them, 'holes'. It would seem then that Holdsworth actively learned to *play* the distortion, as opposed to simply using it as colour.

15.5.5   21st-Century Jazz Guitar

A curious scene started to emerge in the mid-'90s in New York. At the head of this was guitarist Kurt Rosenwinkel. The sound that Rosenwinkel produced was entirely unique and finely straddled the traditional and the modern. His sound consisted of an archtop electric guitar, Electroharmonix POG, reverbs, delays and most important, his distortion of choice—The ProCo Rat pedal (a distortion used by many seminal jazz guitarists).

What he managed to do here was recreate the tight homogeneity of earlier eras of traditional jazz, and in doing so an identifiable sound community. This was a sound that is at once traditional and modern, with a timbre as equally considered as the harmony and rhythmic choices. Rosenwinkel's use of distortion can be heard most effectively on the live album *The Remedy* (2008), where he manages to imbue a foundation of traditionalism with a distorted timbre that clearly challenges the traditionalists' tonal expectations.

The boundaries surrounding jazz guitar have all but dissipated in the 21st century. While there are still strands of hardcore innovators and traditionalists, who are sometimes at odds, there are many who sit comfortably within the jazz domain (perhaps due to the growing appreciation of other communities which have adopted jazz practice, and have thus been positioned within the jazz guitarscape, irrespective of the refute of purists) who use distortion expressively, including Wayne Krantz, Oz Noy, Nir Felder and Marc Ribot.

Wayne Krantz explores timbre as an improvisatory strategy, combining idiomatic guitar techniques (such as frequent use of open strings), novel harmonic vocabulary and a distorted tone to build textural structures and narratives in his approach (see Williams, 2017). In doing so he often disregards the traditional pitch hierarchy almost entirely to create unique (and idiomatic) timbral textures, which can act in a conversational way within an ensemble (see 'Whippersnapper' (1995) 4:16–4:22).

This is by no means an exhaustive set of examples and explorations. It has, however, demonstrated that there is a much deeper significance to distortion, effects and wider timbral considerations in the historic and evolving jazz guitarscape. The pursuit of a tonal identity has become every bit as important to the pilgrimage of a jazz guitar player as harmonic vocabulary is. It is often a 'fetish of the most serious electric guitarists' (Poss, 1998), and as such should be given adequate consideration.

15.6 Distortion and Guitar: Evolving Relationships

Considering the intersecting research domains, primary research (through interview, survey and the author's own correspondence and experience within the jazz guitar community) and historical account, we are now able to begin to unpick the question: *Why do jazz guitarists use distortion?*

Utilitarian purposes (sustain, technique)—At its simplest, amplified distortion affords the guitar more sustain, giving the player more command of the length of their notes. In doing so, it makes the touch required to produce given notes easier, which can be a positive or negative. Equally it gives rise to certain techniques which when played 'cleanly' may require a more articulate and careful approach. Distortion can hide slips in clarity that would otherwise be heard. As Walser states, the 'sustain of the electric guitar . . . increases its potential as a virtuoso instrument' (1993, p. 50). Furthermore, research has demonstrated that high amounts of distortion decreases the need for physical effort playing the instrument (Herbst, 2017b), thus making it easier to navigate and realise creative ideas. Of course, there must be a compromise here, and the player must balance this affordance with clarity and sonority of tone (Mynett, 2012).

15.6.1 *Authentication through Association*

As the marketplace grows and more distortion types become available, each with subtle nuances and finely tuned characteristics, it becomes easier to curate a distinctive sound. As a player or set of players begin to gather around a specific sound, that sound becomes attached to the domain of which they are part. Guitarists may wish to recreate the sounds of those they wish to emulate in part to authenticate themselves within the same domain. Timbral 'vocabulary' or vernacular is analogous in this sense to harmonic, rhythmic and melodic vocabulary and in some domains every bit as, if not more, important. The market structure of instrument, amplifier and effects pedals sales is testament to this, growing year on year.

Of course, what might appear as authentic to one person could be entirely inauthentic to another, depending on their position on the continuum, of what Charles Hersch (2008, p. 8) calls 'neo-trads' and 'anti-trads', the former more likely to favour the tones of the homogenised eras of jazz between the '30s and '60s. An authentic tone can emerge, however, even in a localised setting. The Contemporary New York Scene has a tonal identity of its own, of which the ProCo RAT distortion pedal is an immutable artefact.

Considering Theberge's assertions here (1997, p. 198), that instruments are connected to genres and styles, it is reasonable to consider that they may hold authenticating properties, and not just in the physical, but also the augmented instrument signal chain, which as demonstrated, often contains distortion in its foundation.

15.6.2 Timbral Creativity

Just as a guitarist can learn new scales to improvise with, they can curate new timbres to do the same: 'The manner in which you play an instrument can transform both the instrument itself and the nature of musical sounds produced' (Théberge, 1997, p. 166). Distortion is one of many categories of timbre manipulation, and it can be used to create dynamic contours of timbre, which over the course of a singular piece of music, an album or the entire output of a jazz guitarist gives a narrative of sound every bit as important as the notes themselves. Considering the control of timbre in this way realigns its position within the hierarchy of jazz practice. Choice of, use and control of distortion becomes not simply a sound colouration, but an improvisatory device in and of itself.

15.6.3 Extension/Augmentation of the Instrument

As Théberge discusses, musical instruments are developed and not fixed at the point of design. In fact, the musicians who play them define the range of parameters which go on to define them (1997, p. 160). Instrument augmentation is a process where technology is used to expand the sonic palette available to its user, whilst still maintaining an authentic ergonomic, aesthetic and tactility (Miranda & Wanderley, 2006).

The developments in portable audio manipulation devices—stompboxes—allows a guitarist to have the full spectrum of analog and digital sound production spectrums at the tap of a foot, when considered alongside the development in amplification, the myriad of devices that change the input signal (EBow for instance), synthesisers and other augmentations. As Lähdeoja writes: 'the electric guitar appears as a solidly anchored augmented (and augmentable) instrument. Its hybrid instrumental praxis is widely integrated by guitarists who routinely work with the acoustic, electric and digital parts of their instrument' (2010, p. 44). Foulon et al. (2013) discuss 'modes' of playing guitar, but they only consider the physical input—i.e. hands on instrument. Perhaps distortion (or indeed other effects) become part of the instrument and in some ways we play the distortion, every bit as much as we play the guitar.

Jazz guitarists can use distortion (and other effects such as chorus, delay, ring modulation, harmonisers, to name but a few) to extend their instrument by means of modular augmentation, as Lähdeoja et al. support.

> [Distortion is] almost an instrument in itself . . . . There are many beautiful tones and effects you can get from it, as all the little details of your playing, like pick scrapes and overtones come through stronger than if you were to play with a clean sound.
>
> (Herbst, 2017b, p. 231)

The electro-acoustic chain which begins with the player and his/her instrument was once relatively simple in design for the jazz guitarist, and now offers more combinational potential than ever before. This chain can be seen oriented around specific focuses—distortion being one of them. Guitarists do not think of these augmentations as *other*, but instead as integral to the desired sound production. The guitar in the conventional sense simply becomes a controller for these modular elements. The developments in jazz guitar technology should not be considered less important than the development in harmonic practices.

15.6.4 Instrument Emulation

To an extent this falls under the previous category, but a distinction must be made. The hierarchy in jazz has typically been led by brass players. The guitar was not a solo instrument initially, and subsequent eras and movements tend to have been spearheaded by brass players. As such the tone produced by saxophonists and trumpet players is often coveted by guitarists.

In a traditional sense, the sound of guitar in jazz is often driven by a desire to be recognised as a perhaps more serious, natural and authentic instrument within the jazz domain. The electric guitar carries with it many embedded associations, which could be viewed by traditionalists possibly in a negative way.

Perhaps by means of compromise and to seek authentication, many guitarists will then use distortion (and other effects) to create the illusion of a horn timbre, also to eschew some of the aforementioned associations. As Waksman writes: 'Charlie Christian was lauded for playing guitar like a horn'(2001, p. 38), and as was demonstrated in the previous section, Allan Holdsworth relentlessly pursued the development of a horn-like sound, through his experimentation with distortion. Tsumoto et al. suggest that certain harmonics, produced through the use of distortion, are what may give the distorted guitar a horn-like quality. The 5th harmonic produced is described by its 'brightness, richness and horn-like quality'. The 6th harmonic contains a 'delicate shrillness of nasal quality' (2016, p. 7) and could perhaps be the basis for creating a horn-like timbre.

15.6.5   Sonic Identification

Just as a guitar player can be identified for their vocabulary and harmonic approach, their sound is also an important signifier of identity (Dawe, 2010, p. 70), as the previous case studies have shown. Chaining multiple effects together allows a player to create further distinction. For example, John Scofield separates himself from other players like Mike Stern or Pat Metheny, simply by his use of distortion (although the type of chorus effect and guitar sound are often enough to identify these players).

Scott Henderson's abandonment of a complex modular rig, that popularised his sound during the '80s/'90s while playing with Chick Corea and his own group Tribal Tech, has led to a very sparse setup now, which focuses more on amplifier distortion and a series of boosters which help colour the distortion of the amp. In many ways, this return to traditionalism was essential to his transition to garner acceptance for his roots blues–inspired albums.

Kurt Rosenwinkel's use of the Proco Rat distortion pedal led to an association with his own sonic identity, but also the wider contemporary New York Jazz Guitar Scape, as many of his contemporaries modelled a very similar sound, which in turn created a homogeneity within the community.

15.6.6   Rejection of Homogeneity

As jazz guitar has become more popular, there is an increasing need for artists to distinguish their individual voices (as there is in all other performative arenas). Players with more modern ideologies and outlooks often attempt to move away from these fields, in part as a statement of rejection and also to demonstrate that jazz guitar can exist outside of the confines of the 'archtop dogma'. Bill Frisell comes to mind here, when considering his continual reinvention of soundscape and timbral choice, from bright surf guitar to aggressive distortion and everything in between. Perhaps, like Miles Davis, Frisell remains in the pantheon of jazz, through the processes of reinvention, improvisation and redevelopment. Frisell occupies the jazz landscape perhaps more as an impartial explorer, delicately avoiding being bound by a single sound or domain. In Frisell's words:

> Oh, that's country or that's rock—you sort of take all those styles, and you can call it something based on what style you think it is, but for me, jazz is not a style, it's a way that music works.
>
> (Steinberg, 2019)

Certain microcosms of the jazz guitarscape have fairly rigid margins, for what is permissible and what is not. Gypsy jazz, for instance, builds its image around its totemic figure head—Django Reinhardt. What is most intriguing here, however, is that despite the sound traditionalists expect from this genre, Django's later career is largely ignored, despite himself using an electric guitar and amplification which had a coarse and distorted sound far removed from the rounded tone of Charlie Christian and Eddie Lang, whom he modelled his earlier playing on. Ironically, gypsy jazz royalty Bireli Lagrene's attempt at moving away from the tradition to a distorted electric guitar sound was met with a severe critique not unlike that given to Bob Dylan following his appearance at the 1965 Newport Festival.

15.6.7 A Return to Primitivity

Waksman discusses the dichotomy of white technology vs. black nature (2001, p. 223). During the developments in '60s jazz culture, traditionalists opposed the adoption of solid-body electric guitars, electric basses and synthesisers. It was felt that this eroded the identity of jazz in ways. Perhaps ironically, the primitiveness of distortion, and it's connection to unyielding noise (p. 219), is more aligned with the exploratory nature of jazz and may overcome 'alienation from the body, restoring the body, and hence the self, to a relation of full and easy harmony with nature or the cosmos, as they have been variously conceived' (Torgovnick, 1991).

Perhaps then the use of guitar distortion, albeit when considered outside of its associative trappings and connections to rock music, is a way of returning jazz to a more primitive timbral focus, far more in tune with its African heritage than perhaps the structural trappings of the bebop scale, enclosures and extended harmony ever could be. While of course this point is perhaps flippant and said tongue-in-cheek, given that the purpose of the bebop movement was to elevate black music to a level of musicianship, complexity and 'seriousness' that was normally reserved for white band leaders and the Western canon, it certainly gives perspective and tempers the traditionalist view that jazz in some way ended when the strict homogeneity of sound was broken in the '60s.

15.6.8 Tonal Nostalgia

Many guitar players, through education, adoration or otherwise, may seek to emulate a particular era by reproducing the timbres. I separate this here from associative authentication, although there are links. The difference is that here tonal nostalgia is used to transport the listener to a particular time. This may be coupled with the adoption of a similar emulated approach in terms of musical materials and structures employed, or even ignored to create a kind of anachronistic timbre.

Gracyk's concept of record consciousness becomes useful here (see Dawe, 2010, p. 149). As jazz has developed, the records produced at various points in history become artefacts, by which standards of performance and conventions of genre are judged. The pioneers of jazz guitar therefore have all contributed to a normalisation, or cultural acceptance, of distortion.

Aside from this point, it is also entirely possible that it is simply the ubiquity of distortion-based effects that find jazz guitarists experimenting more with it, something which was not possible in the first half of the 20th century.

15.7 Reframing Distortion

The electric guitar is perhaps the most important instrument precursor and catalyst for technological development in popular music (Carfoot, 2006). Through its many augmentations, both physical and digital, the electric guitar sits at the epicentre of technological instrument extension. The distinction between the connection between the fingers and the strings, that occupy many purist discussions of guitar sound and tone, and the 'other'—amplifier, postproduction and effects—is all but dissolved. A guitarist can 'play' a stompbox every bit as much as they can play the instrument itself. As Lähdeoja et al. write, 'it can be considered an augmented instrument, defined as a network of sound production and processing units, spatially extended and configurable by the player according to the desired sonic results' (2010, p. 37). Further to this point, 'by it's augmented and modular aspects, the electric guitar appears as an emblematic example of an instrument that is intimately connected to the present-day live electronic music praxes' (2010, p. 37).

We should not overlook distortion in jazz guitar as simply part of the uptake of new technology or, worse, being of interest only to the rock canon. The term *distortion* is no longer pejorative, and its function, purpose and use across the entire spectrum of jazz guitar warrants a thorough investigation. The wider implications here are that of technology, timbre and what is really at the heart of the improvisatory and interactive processes that drive jazz. Realigning, or at least extending, our understanding in this manner will further challenge the Western hierarchy (with pitch and structure sitting atop) that is so often used in discourse to describe these processes. As has been demonstrated, sound is as equally important as note choice, and a thorough understanding of jazz guitar's relationship with distortion may help us understand how.

As Sigmund Spaeth writes, jazz itself 'is a distortion of the conventional, a revolt against tradition, a deliberate twisting of established formulas' (1928).

References

Ahmadi, K. (2012). *Initial Technology Review: Distortion*. Sydney: University of Sydney.
Bloch, A. (1953). Measurement of nonlinear distortion. *Journal of the Audio Engineering Society*, *1*(1), 62–67.
Carfoot, G. (2006). Acoustic, electric and virtual noise: The cultural identity of the guitar. *Leonardo Music Journal*, 35–39.
Chang, E. (2017a). *24: Wardenclyffe Tower (1992) and Then! (1990/2003)*. Retrieved 25 August 2020, from https://threadoflunacy.blogspot.com/2017/09/24-wardenclyffe-tower-1992-and-then.html
——— (2017b). *20: Secrets (1988–89)*. Retrieved 25 August 2020, from https://threadoflunacy.blogspot.com/2017/08/20-secrets-1988-89.html
——— (2017c). *The Front Desk*. Retrieved 25 August 2020, from https://threadoflunacy.blogspot.com/2017/11/the-front-desk.html
Dawe, K. (2010). *The New Guitarscape in Critical Theory. Cultural Practice and Musical Performance*. Kent: Ashgate Publishing.
De Souza, J. (2017). *Music at Hand: Instruments, Bodies, and Cognition*. Oxford: Oxford University Press.
Dutilleux, P., & Zölzer, U. (2002). DAFX-Digital Audio Effects, chapter Nonlinear Processing. *John Wiley & Sons*, *93*, 162.
Foulon, R., Roy, P., & Pachet, F. (2013). Automatic classification of guitar playing modes. In *International Symposium on Computer Music Multidisciplinary Research* (pp. 58–71). Cham: Springer.
Frith, S. (1996). *Performing Rites. On the Value of Popular Music*. Cambridge, MA: Harvard University Press.
Gracyk, T. (1996). *Rhythm and Noise: An Aesthetics of Rock*. Durham: Duke University Press.
Hanada, K., Kawai, K., & Ando, Y. (2007). A study on the timbre of an electric guitar sound with distortion. 華南理工大學學報 *(自然科學版)*, *35*(s), 96–99.
Herbst, J. P. (2017a). "Put it up to eleven": An experimental study on distorted solo guitar techniques in sixty years of rock music. *Vox Popular*, *3*, 1–20.
——— (2017b). Shredding, tapping and sweeping: Effects of guitar distortion on playability and expressiveness in rock and metal solos. *Metal Music Studies*, *3*(2), 231–250.

——— (2018a). Heaviness and the electric guitar: Considering the interaction between distortion and harmonic structures. *Metal Music Studies*, *4*(1), 95–113.
——— (2018b). "My setup is pushing about 500 watts—it's all distortion": Emergence, development, aesthetics and intentions of the rock guitar sound. *Vox Popular,* (3).
——— (2019a). Empirical explorations of guitar players' attitudes towards their equipment and the role of distortion in rock music. *Current Musicology, 105*.
——— (2019b). Distortion and rock guitar harmony: The influence of distortion level and structural complexity on acoustic features and perceived pleasantness of guitar chords. *Music Perception: An Interdisciplinary Journal, 36*(4), 335–352.
Hersch, C. (2008). Reconstructing the jazz tradition. *Jazz Research Journal, 2*(1), 7–28.
Ingram, A. (2010). *A Concise History of the Electric Guitar*. Fenton, MO: Mel Bay Publications.
Kartomi, M. J. (1990). *On Concepts and Classifications of Musical Instruments* (p. 108). Chicago: University of Chicago Press.
Lähdeoja, O., Navarret, B., Quintans, S., & Sedes, A. (2010). The electric guitar: An augmented instrument and a tool for musical composition. *Journal of Interdisciplinary Music Studies, 4*(2).
McDermott, J. H., Lehr, A. J., & Oxenham, A. J. (2010). Individual differences reveal the basis of consonance. *Current Biology, 20*(11), 1035–1041.
Middleton, R. (1990). *Studying Popular Music*. New York: McGraw-Hill Education (UK).
Millard, A. (Ed.). (2004). *The Electric Guitar: A History of an American Icon*. Baltimore: JHU Press.
Miranda, E. R., & Wanderley, M. M. (2006). *New Digital Musical Instruments: Control and Interaction Beyond the Keyboard* (Vol. 21). Middleton, WI: AR Editions, Inc.
Mongan, N. (1983). *The History of the Guitar in Jazz*. Delaware: Oak Publications.
Mynett, M. (2012). Achieving intelligibility whilst maintaining heaviness when producing contemporary metal music. *Journal on the Art of Record Production, 6*.
Pakarinen, J. (2010). Distortion analysis toolkit—a software tool for easy analysis of nonlinear audio systems. *EURASIP Journal on Advances in Signal Processing, 2010*, 1–13.
Poss, R. M. (1998). Distortion is truth. *Leonardo Music Journal, 8*(1), 45–48.
Randall, M. (2019). *Bright Moments with Bassist Ron Carter—JazzTimes*. Retrieved 25 August 2020, from https://jazztimes.com/features/interviews/bassist-ron-carter-bright-moments/
Rossing, T. D., Moore, F. R., & Wheeler, P. A. (2014). *The Science of Sound*. London: Pearson.
Scotto, C. (2017). The structural role of distortion in hard rock and heavy metal. *Music Theory Spectrum, 38*(2), 178–199.
Sethares, W. A. (1998). Consonance-based spectral mappings. *Computer Music Journal, 22*(1), 56–72.
Spaeth, S. (1928, August). Jazz is not music. *Forum, 80,* 2.
Steinberg, A. (2019). *Bill Frisell: Nothing to Talk About—JazzTimes*. Retrieved 25 August 2020, from https://jazztimes.com/features/profiles/bill-frisell-nothing-to-talk-about/
Stewardson, R. (1992). Altered guitar tunings in Canadian folk and folk-related music. *MUSICultures*.
Terhardt, E. (1984). The concept of musical consonance: A link between music and psychoacoustics. *Music Perception, 1*(3), 276–295.
Théberge, P. (1997). *Any Sound You Can Imagine: Making Music/Consuming Technology*. Middletown, CT: Wesleyan University Press.
Torgovnick, M. (1991). *Gone Primitive: Savage Intellects, Modern Lives*. Chicago: University of Chicago Press.
Tsumoto, K., Marui, A., & Kamekawa, T. (2016). The effect of harmonic overtones in relation to "sharpness" for perception of brightness of distorted guitar timbre. In *Proceedings of Meetings on Acoustics 172ASA* (Vol. 29, No. 1, p. 035002). Acoustical Society of America.
Waksman, S. (2001). *Instruments of Desire: The Electric Guitar and the Shaping of Musical Experience*. Cambridge, MA: Harvard University Press.
Walser, R. (1993). *Running with the Devil. Power, Gender, and Madness in Heavy Metal Music*. Hanover: Wesleyan University Press.
Williams, D. (2014). Tracking timbral changes in metal productions from 1990 to 2013. *Metal Music Studies, 1*(1), 39–68.

Williams, T. (2017). *Strategy in Contemporary Jazz Improvisation: Theory and Practice* (Doctoral dissertation). University of Surrey.

Zwicker, T., & Buus, S. R. (1998). When bad amplification is good: Distortion as an artistic tool for guitar players. *The Journal of the Acoustical Society of America, 103*(5), 2797–2797.

Audio Reference

Davis, M. (1970). *Bitches Brew* [CD] Columbia.
Frisell, B. (1983). *In Line* [CD] ECM.
Holdsworth, A. (2000). *16 Men of Tain* [CD] Gnarly Geezer.
Johnson, P. (2008). *Rocket 88* [CD] Atlantic.
Krantz, W. (1995). *2 Drink Minimum* [CD] Enja.
Rosenwinkel, K. (2008). *The Remedy* [CD] WOMMUSIC.
Scofield, J. (1987). *Blue Matter* [CD] Gramavision.
——— (2007). *This Meets That* [CD] Universal Records.
Williams, T. (1975). *Believe It* [CD] Columbia.

16 'Got a Flaming Heart'

Vocal Climax in the Music of Led Zeppelin

Aaron Liu-Rosenbaum

16.1 Introduction

I should perhaps begin by explaining what is meant by 'vocal climax' and how it can contribute to a book on distortion in popular music. On an intuitive level, distortion and climax both exemplify a kind of extra effort, but just as there are different kinds of distortion, so too are there different ways of understanding climax. Typically, *climax* is defined as 'the most intense, exciting, or important point of something; the culmination' [1]. However, it can also refer to 'the final stage in a succession' or 'a sequence of . . . ideas in order of increasing importance, force, or effectiveness of expression' [1]. Additionally, *climax* can mean 'a major turning point in the action' [2]. This last meaning, as it turns out, is especially relevant here, as research suggests that contrasting musical features create 'a more intense emotional response' in listeners [3]. *Climax* thus carries a plurality of meanings. Agawu explains that the word is derived from the Greek 'klimax,' meaning staircase or ladder (cited in [4]). Tellingly, the Greek word refers not only to the uppermost rung, but also to the entire structure, adding further nuance to its meaning. In the following discussion, we will therefore consider vocal climax in all its potential significations.

Turning now to distortion, in the impressive webscape of discussions surrounding Robert Plant's voice, descriptions of distortion in its many guises abound. A typical example describes his singing as in a 'high range, [with] an abundance of distortion, loud volume, and emotional excess' [5]. That his vocal technique should be so intimately bound to distortion makes perfect sense, given the band's inspiration from the blues, which has a long tradition of distorted and highly virtuosic (read 'excessive') vocalizing [6]. During the band's first recordings, the 20-year-old singer was initially surprised at the power of his recorded voice: 'That first album was the first time that headphones meant anything to me. What I heard coming back to me over the cans while I was singing . . . had so much weight, so much power, it was devastating' [7]. Plant has mentioned blues singer Howlin' Wolf, in particular, as an influence [8]. Howlin' Wolf, who is credited on Led Zeppelin's 'The Lemon Song,' had a voice that was described as 'stronger than 40 acres of crushed garlic' [9]. Similarly, Willie Dixon, credited on Zeppelin's 'Whole Lotta Love,' 'You Shook Me,' 'Bring It On Home' and 'I Can't Quit You Baby,' sang with a voice worthy of his hefty 6'6" stature.

The use of distorted voices in the blues, whether 'natural' or amplified, wasn't restricted to male voices: Bessie Smith was described as singing with a 'combination of field holler and Jazz Age sophistication' [10]; Billie Holiday supposedly 'didn't think of herself as singing but as blowing a horn' [11]; and Memphis Minnie, whose recording of 'When the Levee Breaks' inspired Led Zeppelin's remake [12], had a sound described by poet Langston Hughes as 'harder than the coins that roll across the counter' at the club where he heard her perform [13]. Hughes was reacting to her already strong natural voice 'made harder and stronger' by the microphone

DOI: 10.4324/9780429356841-20

through which she sang [13]. The relationship between distorted voices and the blues evidently runs deep, but blues singing wasn't merely distorted in a 'non-linear' sense; the blues repertoire comprises diverse vocal delivery effects that have always contributed to its expressive power, including 'forceful, passionate singing and the use of vocal extremes from deep growls to falsetto leaps, head voice, and exaggerated vibrato' [6]. Together, these constitute 'distortions,' as well, in the sense writ large. We will consider, then, this rich interplay between vocal climax and vocal distortion in the following analyses.

While a full discussion would be beyond the scope of this chapter, it would also be remiss to not mention, however briefly, the gendered implications of Plant's vocal persona. It is not for want of metaphor that Fast described his singing as 'trademark banshee wails'—banshee being a *female* spirit in Gaelic folklore [12]. The purpose here is not to reopen the now familiar pandora's box of issues of appropriation, be they cultural, racial, or gender, but rather to acknowledge that one component of Plant's vocal persona is his fluidity of sounded gender, the range of which facilely spans both male and female registers—which only differ, incidentally, by about half an octave in either direction [14]. Such gender-sonic fluidity may also be situated within a long lineage not only of falsetto blues singing, but also of 'gendered traversing of vocal registers,' which has been observed in the gospel singing of men of African descent dating back to at least the antebellum rural American South [15]. Whether one characterises Plant's singing as 'exaggerating the vocal style and expressive palette of blues singers' [5], or merely interpreting and extending that same tradition, this obscuring (read: distorting) of gender and race is an indelible part of his vocal identity in tandem with the other characteristics of his distorted voice, all of which distinguishes vocal climax in the music of Led Zeppelin.

16.2 Theoretical Background and Methods

While there is consensus that no single approach is all-yielding when it comes to analysing popular music, there is less agreement concerning which combination of approaches is most effective. In preparing this chapter, I was struck by the dearth of literature on the subject of climax in pop and rock music. Among the few exceptions is a 2013 article by Brad Osborn that focuses on climax in 'post-millennial' rock music. The author, incidentally, also notes an apparent lack of interest in the subject, echoing an even earlier observation of the same by theorist Kofi Agawu back in 1984 [4]. Osborn is specifically interested in songs that are 'directed toward a single moment of new [climactic] material at the end,' which he labels 'terminally climactic form (TCF)' [4]. He differentiates TCF from Mark Spicer's earlier notion of '(ac)cumulative form'(2004), which focuses on pre-millennial pop-rock music from the late 1960s to 1980s, in which the 'climactic pay-off' similarly occurs at the end of the piece, yet does not constitute new material, as with Osborn [16].

Both authors are thus interested in climaxes at the end of a song, albeit, of different provenance. This distinction is somewhat problematic, however, since it is a matter of interpretation whether and to what degree a climactic ending constitutes new musical ideas versus a development of prior musical ideas. Osborn proposes the term 'sectional climax' over what he describes as 'classically oriented "tension and release" and "moment" approaches, which identify climactic points, rather than identifying entire sections as sustained climaxes' [4]. This is helpful, even if the difference may be more semantic than musical, since classical music climaxes are seldom 'points': one need only consider the chorale in Beethoven's ninth symphony as one amongst numerous examples of climax-as-section. Where the idea of sectional climax becomes especially useful is in songs with a traditional verse-chorus structure, since 'choruses often contain the same melodic hook . . ., it makes more sense to speak of sectional plateaus rather than individual peaks, since the same climactic event . . . will likely repeat throughout a section.' (26)

Osborn anchors his theory in Leonard Meyer's musical semiotics, specifically in Meyer's distinction between 'syntactic climaxes'—that is, climaxes resulting from changes to musical parameters such as melody, rhythm and harmony—and 'statistical climaxes' resulting from changes to quantifiable *sonic* parameters such as dynamics, pitch (as frequency), tempo and timbre [17]. One can see straightaway some potential pitfalls in these categorisations, notably in differentiating between melodic 'syntactical' climaxes and 'statistical' pitch climaxes, or how to quantify timbre in the same way as other 'statistical' musical attributes. Creating a taxonomy of climax is clearly not a straightforward affair.

In a challenge to conventional notions of climax in Western classical music, Austin Patty (2009) discusses contexts in which a climax is not prepared by accelerated harmonic or melodic motion, but rather where 'tension is associated with deceleration and resistance as well as with acceleration and activity' [18]. Patty suggests 'free[ing] acceleration and deceleration from any a priori associations with intensification or abatement,' and considering, instead, all possible pairings of acceleration, deceleration, intensification and abatement to yield what he terms 'pacing scenarios' [18] that can form a musical climax. He cites the final climax in Isolde's 'Transfiguration' (Wagner, *Tristan und Isolde*, act 3, scene 3) as an example of an increase in tension due to a decelerating harmonic motion leading up to the climax. Patty notes that while the harmony decelerates, the melody ascends and a crescendo occurs, thus demonstrating that a combination of elements ultimately creates an intensification, rather than any single musical factor. He concludes (rightly) that 'acceleration does not automatically create intensification and tension and deceleration does not automatically create abatement and relaxation' [18]. These observations will aid in shedding some preconceptions when observing vocal climax in popular music.

In addition to these studies, this chapter will benefit from music psychology and voice studies, since, at the cognitive level, 'a good deal of music perception is likely due to the activity of speech processing mechanisms' [19]. Moreover, it is likely that the voice plays a unique role in contributing to the emotional impact of a song. When assessing music's capacity to communicate emotion, psychologists often refer to 'affect,' a multidimensional construct used to measure music's emotional impact on a listener. In one such study, it was found that the mere presence of vocals in a song enhanced listeners' perceived arousal, or energy level—that is, it made them more excited than the strictly instrumental parts of a song, and it appeared to do so even in the absence of lyrics [21]. There are, in fact, numerous examples of non-verbal singing in Led Zeppelin's music, and this practice resonates with some early descriptions of the blues, in which 'the definitive element . . . is not verbal . . . [and] what counts for most is not verbal precision (which is not to say vocal precision) but . . . musical nuance' [11].

In a similar study, 'voice quality [timbral] variations . . . [were] potent in affect cueing even where loudness cues have been eliminated' [22], which suggests that vocal timbre may play a more central role in our perception of musical climax than intensity, at least under certain circumstances. In short, vocal climax is not simply a function of loudness; timbral qualities also contribute to our perception of climactic musical moments.

I will use spectral analysis to identify 'acoustic cues,' or patterns of acoustic information that have been shown to be associated with different emotions in music [23]. The aim is not to claim a definitive causal relationship between the two—an ongoing challenge for researchers in this area [19, 20]—but rather to open the door to various potential significations that may contribute to this music's emotional power. Vocal climax will thus be identified not only on the basis of these definitions of climax, but also on more objective acoustic criteria that have been shown to be associated with how music conveys emotion.

For this chapter, I used Led Zeppelin's studio recordings from the 1990 CD compilation box set [24]. Given the present scope, I selected three songs that lent themselves to a sonic narrative-based analysis [25]—that is, where the vocal climax could be situated within a song's musical and textual 'story.' Live recordings of these songs and other mixed or mastered studio versions will differ to varying degrees from what is presented here, sometimes significantly, due to the improvisatory nature of the band. For the spectral analyses, the range varied according to the musical context but was generally below 9,000Hz to maintain focus on the voice. Excerpts were compared at similar bandwidths. The spectrogrammes were created using Sonic Visualiser, developed at the Centre for Digital Music at Queen Mary, University of London. Mean pitch and intensity were sampled using Praat, developed by Paul Boersma and David Weenink of the University of Amsterdam. C4 is understood to be middle C.

16.3 The Songs

16.3.1 'Dazed and Confused' (Album: Led Zeppelin, Recorded October 1968, Released January 12, 1969)

For those who came to know Led Zeppelin's music via vinyl, 'Dazed and Confused' was the last song on the A-side of their eponymously titled debut LP and contains perhaps the most sexually suggestive vocal climax in all their catalogue—no doubt helpful in enticing listeners to flip the record to hear what awaited on the B-side. More to the point, it also fixed in the minds of many Plant's vocal persona. The song has become such an icon of Led Zeppelin's style, it is easy to forget that it was actually written by Jake Holmes in 1967 and was even performed by the Yardbirds in 1968 before finding its groove on *Led Zeppelin*.

To my ears, there are two primary climaxes in the song: one essentially instrumental and the other predominantly vocal. The structure of the song—perhaps more than any other in the band's repertoire—was left open to accommodate a highly improvisatory and experimental middle section during which Page could explore his bowed guitar technique in concert [26]. Figure 16.1 shows a timeline of the song.

The instrumental climax occurs during the build-up and arrival on the third chorus; in other words, it does not consist in a single musical event, but rather occurs across a series of changing musical textures that increase in energy until plateauing at the third chorus (5:10), after which the energy gradually dissipates into the fourth and final verse. Since the focus here is on vocal climax, we won't tarry with the instrumental climax; suffice it to say that it is prepared by a

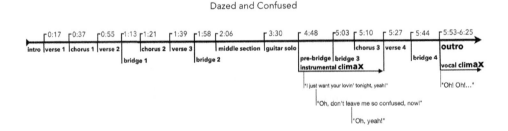

Figure 16.1 Songmap of 'Dazed and Confused'.

singularly occurring power-chord pre-bridge section which nearly buries Plant's screams, 'I just want your lovin' tonight, yeah!', followed by 'Oh, don't leave me so confused now.' Continual drum rolls heighten the intensity during the last two repetitions of the pre-bridge, only to drop off suddenly during the first half of the bridge (5:02). Despite this drop-off of percussive support—the musical equivalent of slamming one's foot on the brake pedal while driving—the tension is maintained during the drop-off and increases when the drums rejoin during the last half of the bridge leading into the third chorus, notably with Plant's distorted scream, 'Oh, yeah!' (5:07) (see Figure 16.1).

On a quantitative level, this third bridge is at an accelerated tempo compared to its earlier two instances (approximately 178 bpm compared to 171 bpm earlier), and when Plant begins the fourth verse, he sings at a higher pitch compared to earlier verses, restricting himself almost exclusively to a single repeated B5 (see Figure 16.2).

The jump to a higher pitch is indicative of an intensified emotional state [19]. Additionally, this kind of restricted vocal 'repertoire' is actually an acoustic signifier characteristic of prohibitive utterances (e.g. 'No, No, No!'), here applied to pitch rather than words, and it occurs in contexts where getting someone's immediate attention supersedes verbal communication in an urgent situation [19]. The highest sung note in the song also occurs during this final verse (E5 at 5:43). Even pitch accuracy becomes secondary: the sung melody at verse 4 is diatonically 'fuzzier' than at verse 1, due to a preponderance of non-diatonic ('out of tune') pitches. It is worth reminding ourselves here that notated pitches are only approximations of heard frequencies. In Figure 16.2, for example, the maximum frequency in verse 1 is 497Hz (slightly higher than the notated B5), whereas in verse 4, the maximum frequency reaches 528Hz (just above

Figure 2

Verse 1 (0:17)
Been dazed and con-fu-sed for so long it's not true

pitch range: 10 semitones
mean pitch: 389.5 Hz (approx. G4)
mean intensity: 73.2 dB

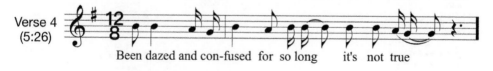

Verse 4 (5:26)
Been dazed and con-fused for so long it's not true

pitch range: 5 semitones
mean pitch: 423.1 Hz (between G#4 / A4)
mean intensity: 80.1 dB

Figure 16.2 Restricted vocal pitch at verse 4 in 'Dazed and Confused'. (Mean pitch refers to recorded pitch.)

C5!). My hearing a 'fuzzy' B5 is a gestalt impression due, in part, to my expectations of B5 (from prior verses). These sonic characteristics of the last verse have the cumulative effect of focusing our attention on the emotional energy and intensity of the communication rather than on the melody or lyrics.

After the instrumental climax has abated, the song does not simply wind down; a coda section becomes a vehicle for the song's vocal climax—a sexually suggestive, synchronised interplay between voice and guitar (at 6:02). The song's 12/8 meter is now suppressed in favor of shorter 3/8 groupings that are articulated by Plant's moans of "Oh!" and accompanied by insistent 16th note octave figures in the guitar. The reduced meter draws the listener into a more fundamental (read: 'primal') listening state, and the heavily distorted and repeated rhythmic accompaniment compounds the sense of urgency. During this last half-minute of the song, we can observe how Plant uses his voice to intensify the final climax through gradually prolonged durations and 'harmonicisation' of his voice—that is, he gradually reduces the roughness in his voice until it sounds harmonic, i.e. non-distorted. Additionally, his moans of 'Oh!', which are only quasi-diatonic, eventually crystallise into fully diatonic descending melodic thirds (see Figure 16.3). This vocal climax thus constitutes a progression from distortion (non-linearity) to harmonicity, culminating in a guitar power-chord surprise ending—'surprise' in that it is struck on the last eighth note in a 3/8 group, transforming a normally weak beat into a strong one and giving the impression of arriving 'prematurely.'

These subtle and fast-moving changes, when taken together, evidence a highly nuanced use of the voice. Moreover, in the context of the song's narrative about the uncertainty of requited affection, the climactic vocal ending that culminates with the 'ejaculatory' final guitar chords suggests the protagonist was able to make good on his avowal, 'Gonna love you baby, here I come again.' Despite the ostensibly 'male perspective' [26] of this song, there is, at the same time, an inherent gender ambiguity in how the vocal-instrumental 'intercourse' plays out during the vocal climax: if the guitar may be understood as a 'technophallus' [27], i.e. representing a

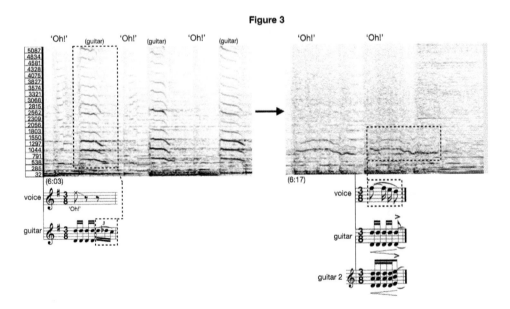

Figure 16.3 Vocal climax in 'Dazed and Confused'.

male persona, then the listener is left to reconcile Plant's voice performing a female persona during this vocal climax. This is further complexified by the fact that guitarist Jimmy Page occasionally performed in women's attire, due to difficulties in finding clothes to fit his 'slight' frame [28]. This vocal climax can therefore also be understood as distorting visible and audible gender roles. As mentioned previously, this chapter does not pretend to do justice to the nuance surrounding gender roles in this music; Fast (1999, 2001) has treated the subject more thoroughly. It should also be noted that the 'ambiguity' only occurs within a perspective that forces personae into a binary opposition; for other listeners, this vocal climax may thus warrant no gender reconciliation.

Ironically, Plant came to have second thoughts about perhaps his most recognisable sonic trademark: 'I took it way too far with that open-throated falsetto. I wish I could get an eraser and go 'round everybody's copy of *Led Zep I* and take out all that "Mmm, mmm, baby, baby" stuff' [29]. Personally, it's difficult to imagine the music without them: sometimes, it takes a cover of a cover to find one's own voice.

16.3.2 *'Communication Breakdown' (Album: Led Zeppelin, Recorded October 1968, Released January 12, 1969)*

Also on *Led Zeppelin*, nestled in the middle of the original vinyl's B-side, is the track 'Communication Breakdown,' a harmonically leaner and rhythmically heavier carnal cousin to The Beatles' 'I Want to Hold Your Hand,' recorded only five years earlier. This time, the protagonist struggles with a 'communication breakdown' when he tries to tell a 'girl . . . that I love you so/I wanna hold you in my arms.' As with 'Dazed and Confused,' this song features two climactic episodes: one instrumental and the other vocal (see Figure 16.4).

While 'Communication Breakdown' is a rather typical alternating verse-chorus form with a solo and an outro, there is more to the song than meets the ear: a sonic narrative can be heard in the vocal part that tells the story of an intensifying 'communication breakdown' that the protagonist tells us 'drives me insane' and which manifests itself primarily across three progressively distorted and prolonged vocal screams on the word 'insane' at the end of each of the song's three choruses, each of which is overdubbed with a second voice and overflows into the following section (see Figures 16.5a-c).

The first episode evidences a movement from distortion toward harmonicity, as the voice audibly becomes less distorted (more harmonic) during the four-second scream. We can also hear a movement in the vocal texture from polyphony to monophony as the overdubbed voice

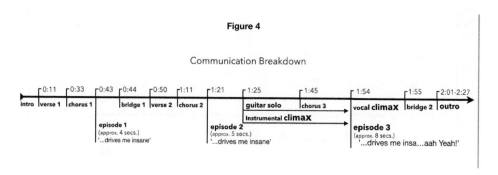

Figure 16.4 Songmap of 'Communication Breakdown'.

252 Aaron Liu-Rosenbaum

Figure 5a

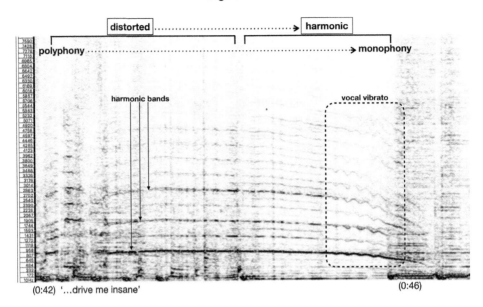

Figure 16.5a First 'insane' episode in 'Communication Breakdown' (0:42).

Figure 5b

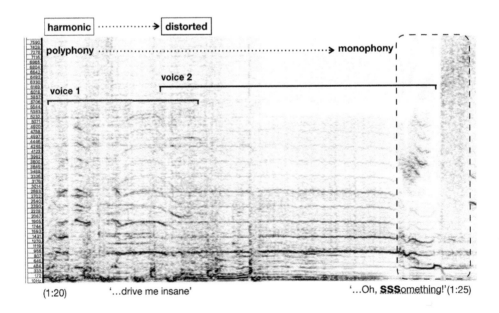

Figure 16.5b Second 'insane' episode in 'Communication Breakdown' (1:20).

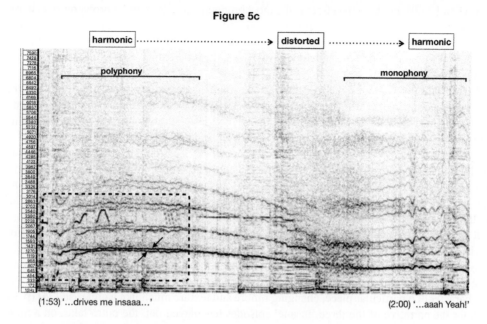

Figure 16.5c Vocal climax at third 'insane' episode in 'Communication Breakdown' (1:53).

is dropped off before Plant finishes singing 'insane.' From the spectrogramme, we can visualise the distortion that we hear: on the left side of the image, there is more diffuse spectral energy surrounding the harmonic bands of the voice, clouding the image somewhat, whereas toward the end of the scream, on the right side, the image becomes less diffuse, i.e. more harmonic with less spectral 'cloudiness.' As his voice becomes clearer, Plant adopts a vibrato effect, indicated in the dashed box. The divergence between the overdubbed vocal parts also becomes particularly visible through the wavy, bifurcated bands of the vibrato.

The second episode, at roughly five seconds, is longer than the prior one and similarly evidences a reduction in voices. However, it reverses the former's progression by moving from harmonicity at the beginning of the word 'insane' toward distortion, via a fade-in of a more distorted voice on top of the prior (see brackets in Figure 16.5b), as well as by Plant's harshly enunciated 's' at the end of the scream ('Oh, something!'), which literalises the theme of a communication breakdown into a signal-to-noise breakdown before leading into the instrumental climax/guitar solo. This enunciated noise is indicated by the dashed box in Figure 16.5b.

For its part, the guitar solo that follows then overflows into the chorus, lending it a higher level of intensity before leading into the vocal climax of the song at the third scream on the word 'insane' (1:53) (see Figure 16.4).

This third 'insane' episode is the most extreme in the song (see Figure 16.5c): it is the longest (eight seconds) and by far the most tumultuous in terms of its trajectory, moving from a harmonic voice to a heavily distorted one before returning to harmonicity on the phrase 'Ah, yeah!' (at 1:59). The diffuse spectral cloud in the middle part of the figure corresponds to the most distorted and raw point in Plant's scream, just before he pulls his voice back into harmonicity (read: 'sanity') at 1:58. The vocal climax is also the densest of the episodes, with what appears—both to eye and ear—to be multiple voices (indicated by arrows in dashed box in Figure 16.5c). As before, the multiple voices are reduced to a single one by the end of the episode (cf 1:53 and 2:00), but here the effect is the most pronounced; it is as if the protagonist's chorus of angst-ridden inner voices were being harnessed back under control. The return to a singular vocal subject can be heard as a retreat from the edge of breaking down, pulling us ultimately back into the moment and into the unified psyche of our protagonist, whose unsettling journey we are witnessing.

Across the three episodes, the overdubbed vocal parts grow in prominence in overtaking the earlier voice, thus highlighting the narrative's drive toward insanity brought about by the communication breakdown. Additionally, the bridge, particularly in this song, serves more as a kind of extension to the chorus than as an independent section; it enables the voice to 'break free' of the syntactical bounds of the chorus. That these 'breaches' occur on the word 'insane' serves to highlight the protagonist's inability to restrain himself within the confines of the song's formal structure. On a cultural level, these musical gestures that disregard a song's formal structure may serve as musical 'metaphors for countercultural values of transgression and openness' [26] (p. 28), drawing attention to the song's historical context.

It is noteworthy that Plant's voice doesn't always proceed from harmonicity to distortion or vice versa; rather, he employs a wide variety of vocal effects, depending on the context, expressing himself through his voice in an artistic way that allows for interpretation based on a song's lyrical and other musical content. As Susan Fast noted, Plant 'mold[s] his voice to suit the character of a particular piece, changing timbre and texture much like a guitarist' [26] (p. 9).

After the narrative of the three 'insane' episodes has played out, the outro takes on a more denouematic, post-climactic feel, during which things start to degrade both timbrally and lyrically. Significantly, the lyrics also support this sonic narrative, for as the communication breaks down during the outro, the sentence structure appears to break down as well: Plant repeats the phrase, 'I want you to love me all night'—by now, a far cry from The Beatles' 'I want to hold your hand.' The phrase becomes fragmented into 'I want' and 'I want you to love me,' which, themselves, are ultimately reduced to 'I want you to love,' almost depriving the lyrics, and our protagonist, of any sense—a veritable communication breakdown.

16.3.3 'Whole Lotta Love' (Album: Led Zeppelin II, Recorded 1969, Released October 22, 1969)

If our first example portrays our protagonist disoriented by a woman, and our second example shows him struggling to communicate his feelings, in this third example, he emerges definitively more self-assured. 'Whole Lotta Love' is the opening track on *Led Zeppelin II*, released just nine months after the band's first album. The song contains some of the group's most sexually explicit lyrics, along with one of its most dramatic and colorful vocal climaxes, instigated by the voice alone, which carries the listener into a climactic spectacle of wet (reverberant) vocal screams, moans both distorted and harmonic (non-distorted), and vocal sound fragments that cascade and rebound in the stereo field during the entire final minute of the song.

Like 'Dazed and Confused,' 'Whole Lotta Love' is essentially a verse-chorus structure with an experimental middle section leading into a guitar solo, followed by a closing section (see Figure 16.6). There are likewise two climactic sections in the song: the guitar solo and the outro that is initiated by one of Plant's characteristic prolonged vocal screams on the word 'love.' Fast writes that this scream 'dissipates' emotion [12], which is understandable, given its descending pitch contour, normally associated with reduced effort and intensity. I would like to suggest, however, that the recording studio aesthetics are designed in such a way as to counter this perception and heighten its intensity in order to connect to the climactic outro, during which the song's accumulated energy is given cathartic voice. As such, Plant's extended scream may be

Vocal Climax in the Music of Led Zeppelin 255

understood as constituting the first 'rung' in a climactic ladder (as in the Greek 'klimax'), as will be shown shortly.

Just before the scream at 4:20, one can hear a distant (pre-)echo of Plant's voice *preceding* the main vocal track. This moment is identified as the 'vocal bridge' in Figure 16.6. Eddie Kramer, engineer on the album, explains:

> It gets to this point in the song where you hear Robert go 'woman', and I'm putting a little bit of reverb on, and I'm hearing another voice. And we [Jimmy Page and Kramer] look at each other and say 'What the hell, Oh!', so track 8 was the main voice and track 7 was another take. And I couldn't get rid of it . . . [so] we just soaked it in reverb, and that's why you hear the 'woman' and in the background you hear this other one . . . the message is leave the damn mistakes in.
>
> [30]

Indeed, this 'mistake' has become an integral part of the character of the vocal climax, and the casualness that its presence suggests belies the complexity of its production: during Plant's scream, as his pitch descends, the level of reverberation is dramatically increased, and consequently fills up the stereo space. By the end of the scream, the single voice has been transformed into a reverberant vocal chorus at 4:27, thus accumulating energy, notwithstanding the descending pitch. In fact, we never hear the last consonant of 'lo(ve),' which becomes merged with the words 'Oh, yeah' just before the outro at 4:28 (see Figure 16.6). This prolonged scream on a word that elides into the phrase 'Oh, yeah' is similar to what we observed before during the vocal climax in 'Communication Breakdown' on the final iteration of the word 'insane.' In contrast to the earlier song example, here Plant's voice progresses from monophonic toward polyphonic. At the same time, the timbre of Plant's voice shifts from distorted to harmonic.

We may thus track several trajectories within this climactic gesture: descending pitch, increasing amplitude, increasing reverberance, a progression from vocal distortion towards harmonicity (compare 4:20 and 4:27) as well as from a monophonic to a (reverb-induced) 'polyphonic' texture. A Praat analysis of this event confirms an overall increase in intensity notwithstanding the descending pitch, albeit not absolutely linear (see Figure 16.7).

In addition, the spectrogramme allows us to visualise these various trajectories: the harmonic bands are more pronounced at the beginning of the scream, albeit with distortion in the form of 'cloudiness' between the bands. Once the reverberation is applied (see dashed box in Figure 16.7), the bands become blurred, even as the voice's growing harmonicity extends those blurred harmonics into the upper register. The lower frequencies likewise become intensified

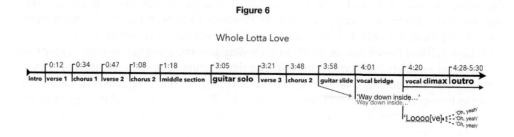

Figure 16.6 Songmap of 'Whole Lotta Love'.

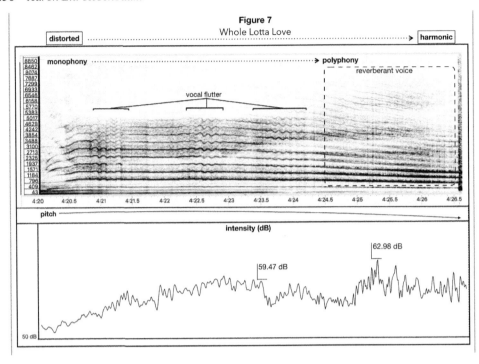

Figure 16.7 Scream initiating vocal climax in 'Whole Lotta Love'.

and blurred (read: 'muddied')—appearing as thickened lines at the bottom right side of the example from around 4:25—again, due to the extreme reverberance. Lastly, we can note three areas of wavy lines, indicating a kind of periodic flutter in the frequency bands around 4:21, 4:22 and 4:23, likely where Plant exerted extra effort to produce and maintain the scream. This kind of vibrato is not only an expressive gesture signifying intensity, but also a performance technique that manifests when a singer tries to sustain a held note [31] (p. 24).

As mentioned previously, the vocal climax steps up the intensity leading into the outro, itself a climactic plateau whose energy is principally maintained through Plant's continual vocal interjections. The first vocal sounds we encounter in the outro are a pair of nearly identical non-verbal phrases, 'Ma!, Ma!, Ah!, Ah!' (at 4:32) (see Figure 16.8). Each phrase comprises four syllables, screamed in a distorted voice. Due to their roughness, the phrases do not readily lend themselves to pitch analysis, but the last syllable of each phrase culminates in a higher pitch than the prior syllables. According to a study by Pfeiderer, phrases ending on high notes signify arousal [32]. Here, not only does each phrase end on a high note, but, as can be seen in Figure 16.8, each syllable constitutes, in itself, an ascending gesture, spanning multiple frequencies. These nested signifiers thus form a kind of hyper-signification of arousal. According to Bryant, arousal states in mammals (both human and non-human) are often triggered with the aim of affecting listeners' behavior in an urgent manner [19]. The non-verbality of these phrases only accentuates their animality and evokes a more 'primitive' level of signification. These sounds may be understood, then, as communicating on both artistic/aesthetic levels as well as on a more basic, instinctive level [19]. One could conceivably hear these sounds, then, as both arousing and, say, frightening. Feeble are the bounds by which we separate music from sound.

Vocal Climax in the Music of Led Zeppelin 257

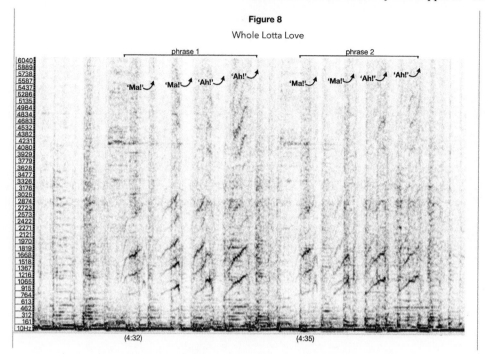

Figure 16.8 Non-verbal signifiers of 'arousal' in 'Whole Lotta Love'.

The voice's importance during the outro is additionally underscored by its heavy dose of treble equalization (which adds presence) and reverb. So resonant is the voice's sibilance, it occasionally dwarfs the presence of John Bonham's drums (hear 'Shake for me girl' at 4:45). The sheer variety of vocal effects used to maintain the climactic energy during the final minute of the song is impressive. Figure 16.9 shows a particularly interesting moment during which Plant vocalises a tense, distorted drone, while gradually transforming his vowel from 'Oooh' to 'Ahhh.' The spectrogramme illustrates the consequent expansion of harmonic partials accompanying the more open 'Ahhh' vowel, the jagged contours of which attest to the sounded tension in his voice. This episode is immediately followed at 5:04 by more harmonic 'Oh!'s, whose smoother harmonic bands evidence the reduced tension in his voice. More than mere gibberish, these vocalisations constitute a 'living and expressive material that speaks for itself and has no need of being put into a [verbal] form' [33].

It should be noted that in between these non-verbal vocalisations are interspersed verbal phrases such as 'Keep it coolin' baby' (5:13), framing these sounds as sounds of carnal pleasure. It is also worth reiterating that this combination of performance technique and production technique is not a one-size-fits-all trick, but rather a sonic narrative particular to this song; in other songs the narrative varies, as we have seen. The use of these techniques, then, 'needs to be understood in relation to the text' [12].

Plant's final vocal gesture is one of the most complex in the song: it begins with a harmonic ascent, distorts at the top, then breaks with a leap down to a lower register, which is fluttered before leaping across a falsetto note to return to harmonicity once again before fading out. The overall form is of a bell shape (see Figure 16.10). These juxtaposed 'voices' are, by now, easy to spot in the figure with their alternating neat, then blurred, parallel frequency bands.

258 Aaron Liu-Rosenbaum

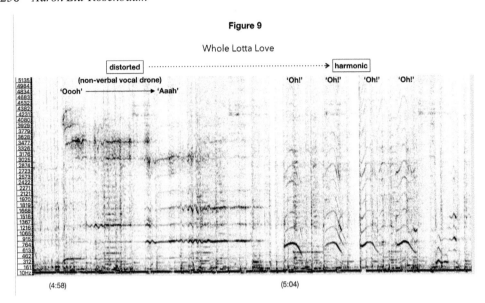

Figure 16.9 Distorted vocal drone followed by harmonic 'Oh!'s in 'Whole Lotta Love'.

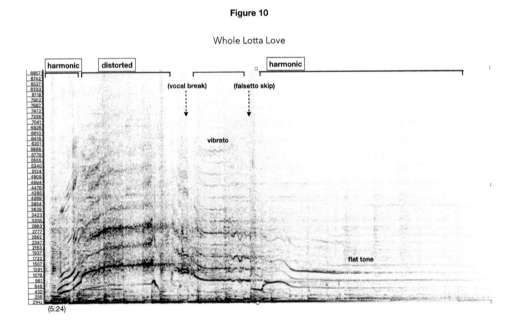

Figure 16.10 Closing vocal gesture in 'Whole Lotta Love'.

The closing to this song is a masterful smörgåsbord of climactic vocal gestures, distorted and 'clean,' from chest-voiced to falsetto, flat tone to vibrato (see Figure 16.10). At the falsetto, we can observe the disappearance of harmonic bands, albeit fleeting, due to the momentary sinusoidal-like tone of Plant's voice (as sine tones comprise energy at only a single frequency band). Returning to questions of gender, the climactic nature of these vocal gestures was enough to inspire one writer to claim with regard to the narrative in 'Whole Lotta Love' that 'Plant virtually has her orgasm for her' (Murray quoted in [26], p. 190). That this gesture occurs even as the song fades out suggests that the narrative is not over: though the curtains be drawn, the 'back door man' still makes his rounds.

16.4 Conclusion

There is, no doubt, much more to say about these pieces than can be covered here. From these few examples early in the group's career, we can already appreciate the crucial and multifaceted role that the voice plays in shaping climactic musical moments in this music, with diverse vocal trajectories between harmonicity and distortion, monophony and polyphony, verbal and non-verbal, and everything in between. Plant's naturally distorted vocal technique certainly benefitted from the variety of recording studio manipulations that further distorted and altered it. Moreover, I hope these examples have demonstrated that it is possible to relate the use of audio effects and manipulations to a song's narrative and thereby enhance our appreciation of it. These varied effects highlight how important recording studio aesthetics are in shaping the recorded sound of Plant's voice and defining the group's sound.

While some have taken up the task, more theorising is needed surrounding musical climax in pop and rock music. Its definition needs to be more nuanced, to take into account all its manifestations. Describing it in superlative terms obscures its comparative nature and how it derives its impact not just from its formal features, but from its musical context as well. Rather than consider it as a singular event, it may be more useful to conceptualise it as a process that unfolds over time, even during the course of an entire section.

A piece of music typically takes us on a journey during which our feelings of intensity ebb and flow, sometimes across multiple climactic moments, both vocal and instrumental. What is hopefully clear is that intensity alone is an inadequate barometer of vocal climax; it doesn't fully account for the diverse phenomena that contribute to our perception of climax, including timbre, lyrical content, apparent effort, as well as other acoustic cues that may trigger emotional responses in us. Thus, we may experience some events as more climactic even though they be quantifiably less intense with respect to signal level, pitch height, etc. If we understand a recorded voice to be the net result of 'affects, emotions and personality traits as well as their stylized vocal expression, of physiological peculiarities and biographical traces as well as of conventionalized cultural patterns' (Bielefeldt, as cited in [32]), then this chapter has only scratched the surface of vocal climax in rock music. A fuller understanding will require integrating and coordinating diverse methodologies and bodies of knowledge from separate fields that only recently are beginning what will hopefully become a long and fruitful discourse.

Lastly, it may be more useful in the long run to think of climax not as a fixed feature of musical structure, but rather as a feature of our own perception—a cumulative cognising of several variables, the intensity of which depends in part on how well each of them resonates with our own musical sensibilities. Such a view poses significant, but not insurmountable, analytical challenges. More importantly, it recognises and validates the uniqueness of each person's musical experience, and points toward a more inclusive analytical practice. Notwithstanding Plant's powerful presence in Led Zeppelin, he has always acknowledged that he is but a part of a larger whole, 'Part of my charisma is . . . reliant on the other three, and the same with everybody else . . . that's the whole secret' [34]. I have only touched on a small sampling of one voice within Led Zeppelin. This chapter is, by design, incomplete; the rest of the story beckons to be told.

References

[1] Climax | Definition of Climax by Oxford Dictionary on Lexico.com also Meaning of Climax, www.lexico.com/definition/climax

[2] Definition of Climax, www.merriam-webster.com/dictionary/climax

[3] Schellenberg, E.G., Corrigall, K.A., Ladinig, O., Huron, D.: Changing the Tune: Listeners Like Music that Expresses a Contrasting Emotion. *Frontiers in Psychology*. 3, 574–574 (2012), https://doi.org/10.3389/fpsyg.2012.00574

[4] Osborn, B.: Subverting the Verse—Chorus Paradigm: Terminally Climactic Forms in Recent Rock Music. *Music Theory Spectrum*. 35, 23–47 (2013), https://doi.org/10.1525/mts.2013.35.1.23

[5] Fast, S.: *Led Zeppelin*, www.britannica.com/topic/Led-Zeppelin

[6] Ellis, W.L.: Mississippi Moan: In: Collins, J. (ed.) *Defining the Delta*, pp. 157–178. University of Arkansas Press (2015)

[7] Crowe, C., Crowe, C.: *The Durable Led Zeppelin* (1975), www.rollingstone.com/music/music-news/the-durable-led-zeppelin-36209/

[8] Page & Plant Wanton Song Later with Jools Holland plus subsequent interview—YouTube, www.youtube.com/watch?v=AtS3F6CHoHs

[9] Howlin' Wolf: *Booming Voice of the Blues*, www.npr.org/templates/story/story.php?storyId=130276817

[10] Forebears: Bessie Smith, *The Empress of the Blues*, www.npr.org/2018/01/05/575422226/forebears-bessie-smith-the-empress-of-the-blues

[11] Murray, A., Devlin, P.: Singing the Blues. In: *Stomping the Blues*, pp. 77–90. University of Minnesota Press (1976)

[12] Fast, S.: Rethinking Issues of Gender and Sexuality in Led Zeppelin: A Woman's View of Pleasure and Power in Hard Rock. *American Music*. 17, 245–299 (1999), https://doi.org/10.2307/3052664

[13] Retman, S.: Memphis Minnie's "Scientific Sound": Afro-Sonic Modernity and the Jukebox Era of the Blues. *American Quarterly*. 72, 75–102 (2020), https://doi.org/10.1353/aq.2020.0004

[14] How do HRT, Sex Reassignment and Other Such Procedures Affect Vocal Production, Particularly the Singing Voice?—Quora, www.quora.com/How-do-HRT-sex-reassignment-and-other-such-procedures-affect-vocal-production-particularly-the-singing-voice

[15] Jones, A.L.: Singing High: Black Countertenors and Gendered Sound in Gospel Performance. In: Eidsheim, N.S. and Meizel, K. (eds.) *The Oxford Handbook of Voice Studies*, pp. 34–51. Oxford University Press (2019)

[16] Spicer, M.: (Ac)cumulative Form in Pop-Rock Music. *Twentieth-Century Music*. 1, 29–64 (2004), https://doi.org/10.1017/S1478572204000052

[17] Meyer, L.B.: Exploiting Limits: Creation, Archetypes, and Style Change. *Daedalus*. 109, 177–205 (1980)

[18] Patty, A.T.: Pacing Scenarios: How Harmonic Rhythm and Melodic Pacing Influence Our Experience of Musical Climax. *Music Theory Spectrum*. 31, 325–367 (2009), https://doi.org/10.1525/mts.2009.31.2.325

[19] Bryant, G.A.: Animal Signals and Emotion in Music: Coordinating Affect Across Groups. *Frontiers in Psychology*. 4, 990–990 (2013), https://doi.org/10.3389/fpsyg.2013.00990

[20] Loui, P., Bachorik, J.P., Li, H.C., Schlaug, G.: Effects of Voice on Emotional Arousal. *Frontiers in Psychology*. 4 (2013), https://doi.org/10.3389/fpsyg.2013.00675

[21] Yanushevskaya, I., Gobl, C., Ní Chasaide, A.: Voice Quality in Affect Cueing: Does Loudness Matter? *Frontiers in Psychology*. 4, (2013), https://doi.org/10.3389/fpsyg.2013.00335

[22] Juslin, P., Laukka, P.: Communication of Emotions in Vocal Expression and Music Performance: Different Channels, Same Code? *Psychological Bulletin*. 129, 770–814 (2003), https://doi.org/10.1037/0033-2909.129.5.770

[23] Bryant, G.A.: The Evolution of Human Vocal Emotion. *Emotion Review*. 175407392093079 (2020), https://doi.org/10.1177/1754073920930791
[24] Led Zeppelin: Led Zeppelin—Box. *Atlantic* (1990)
[25] Journal on the Art of Record Production: *The Meaning in the Mix: Tracing a Sonic Narrative in 'When the Levee Breaks',* www.arpjournal.com/asarpwp/the-meaning-in-the-mix-tracing-a-sonic-narrative-in-%E2%80%98when-the-levee-breaks%E2%80%99/
[26] Fast, S.: *In the Houses of the Holy: Led Zeppelin and the Power of Rock Music.* Oxford University Press, Incorporated (2001)
[27] Waksman, S.: *Instruments of Desire: The Electric Guitar and the Shaping of Musical Experience.* Harvard University Press (2001)
[28] Jimmy Page – Talks about Whole Lotta Love Riff & His Picturebook Autobiography—Radio Broadcast 2010 (2020)—YouTube, https://www.youtube.com/watch?v=_uyrD1BTFto
[29] *The Story Behind Led Zeppelin's "Dazed and Confused"*, www.covermesongs.com/2019/01/the-story-behind-led-zeppelins-dazed-and-confused.html
[30] Eddie Kramer Tells the Story of Led Zeppelin's Iconic Sound on "Whole Lotta Love"—Cosmo Music (2017)—YouTube, https://www.youtube.com/watch?v=HReL_3fCvKI
[31] Ribaldini, P.: Heavy Metal Vocal Technique Terminology Compendium: *A Poietic Perspective*. 108 [Licentiate thesis] University of Helsinki (2019)
[32] Pfleiderer, M.: Vocal Pop Pleasures. Theoretical, Analytical and Empirical Approaches to Voice and Singing in Popular Music. *IJ.* 1, 1–16 (2011), https://doi.org/10.5429/2079-3871(2010)v1i1.7en
[33] Appel, N.: "Ga, ga, ooh-la-la": The Childlike Use of Language in Pop-Rock Music. *Popular Music.* 33, 91–108 (2014)
[34] Led Zeppelin's Robert Plant 1975 Complete Interview—YouTube, www.youtube.com/watch?v=qLQ0e6RA6R8

17 The Aesthetics of Distortion

Toby Young

17.1 Introduction

Popular music has an extraordinary capacity to turn one group's aesthetic rejects into another group's aesthetic ideal. The complex re-appropriation of sonic hallmarks traditionally and politically associated with beauty and ugliness, such as the glitch of Auto-Tune [1] or the reclamation of the scratch of vinyl [2], opens up a complex landscape of transformation from the 'damaged' and 'imperfect' to the artistically authentic and valuable. Sitting somewhere between interference and blemish, distortion has become an important sonic quality for producers in their search for auditory identity. Whether heard as an acoustic feature of live performance or an audio technique deployed in the studio, distortion inherently brings the listener towards an encounter with sonic negation and re-signification; a multi-layered and complex experience which this chapter will attempt to untangle and demystify from the standpoint of aesthetic theory.

Before doing so, however, it is necessary to define some of the key terms and frameworks at play here. Firstly, given that aesthetics is a discipline concerned with mediating the 'perception, appreciation, and production of art' [3], the fact that term 'distortion' itself means a number of different things to a number of different people is a challenge. In its most basic form, distortion—whether acoustic or electronic—is the alteration of something so that it no longer resembles its original state. Colloquially, it often might be taken to mean some sort of destruction, for example, that of an electrical signal resulting in buzzes and hisses infiltrating a sonic object (for example, the hum or a faulty cable or malfunctioning equipment), either to disruptively negative or subversively positive effect. On the other hand, and as many of the other chapters in this book will attest, for the mix engineer or audiophile, distortion is a much more specific act of sonic manipulation in which there is a 'change in the waveform or harmonic content of an original signal as it passes through a device' [4]. These manipulations can be broadly categorised as techniques of sonic excitement [5] through which new harmonics are introduced to an original signal in some way in order to make sounds warmer, denser and even brighter. At its most extreme end, this can produce aggressive, harsh or gritty results far closer to the colloquial definition—for example, the rough fuzz associated with overdriving a guitar signal (or indeed the numerous pedals now commonly used to recreate the signature 'distortion guitar' sound)—but used sparingly it can introduce far more subtle and appealing timbral changes to a sound, such as adding warmth from the gentle clipping of an analogue preamp or compressor, or even the richness and depth from adding subtle saturation to, say, a vocal track [see 6].

The tension here between negatively and positively valanced perceptions and uses of techniques of distortion poses something of a challenge to this chapter. There are a number of ways to try and smooth this difference through a process of aesthetic categorisation, for instance, by dividing occurrences of distortion into those that are musically pleasing vs unpleasing, sonically perceptible vs imperceptible, or intentional vs unwanted (for example, when distortion's presence has a negative impact on mix quality). If we want to avoid being tied down in messy and complex discussions of value—as Alan Goldman [7] urges us against—a less error-focused system of taxonomy might be to focus on the difference between distortion-as-process (i.e. the direct interference or manipulation of a signal path) and distortion-as-product (the effect of that manipulation), though even this grouping is epistemologically limited by the constantly recurring question of intent subsuming any meaningful focus on sonic affordances, not to mention the complication between distortion-as-process being a binary conception (distortion is either added or not) whilst distortion-as-product as a gradable one (i.e. 'that sounds very distorted' or 'it's only a little bit distorted'). It is equally important to consider technical vs colloquial uses of language here as reliable vehicles to report testimonies of taste, though we do not have time to unpack this here. As an example, whilst a mix engineer might have several shared technical words for the subtle gradations of distortions easily to hand (e.g. saturation, overdriving), a layperson might exhibit a more distinctive, evaluator-relative perspective by employing 'distorted' for a number of different uses based on their listening habits and other frames of reference.

Whilst I do not propose taking a fully taxonomic approach to this chapter, it does seem useful to the ensuing discussion to be able to delineate occurrences of distortion in a way which concatenates these distinctions to some extent. One such convenient division is that between distortion on a micro and macro scale. Micro-distortion, for the purpose of this study, can be defined as distortion that maintains the relationship between subject and object by altering only a small element or parameter *within* the sound object, rather than affecting the perceived logic of the whole. These affective 'microdisruptions' necessarily occur at the level of material manipulation, stemming from within the medium in such a way as to make their presence 'felt' (to a greater or lesser degree) whilst still emphasizing their independence from a constitutive listening subject; for example, through the subtle introduction of valve saturation to gently colour a sound without altering its role within a mix and other forms of soft clipping. On the other hand, in the case of macro-distortion, the subject-position of the listener is fundamentally altered by a manipulation that reconstitutes—rather than alters—the sound object; for example, through the inclusion of white noise in a mix, the harsh overtones produced by hard clipping, or the buzz of a faulty piece of equipment. On the surface this second form of distortion seems drastically different from the first, as the subject experience of the object becomes altered in such a way as to render it a fundamentally different and indeed disruptive force, but as we will see later, both types of distortion have the potential to invoke comparable aesthetic properties.

One further consideration we must take into account in this discussion is the framework of listener expectations in which the distortion is situated. As Kendall Walton argues, it is vital when considering the aesthetic impression of a piece of art that we are able to situate our aesthetic expectations within a relativist schema which can accommodate the context sensitivity of a work. By taking into consideration the art object's aesthetic properties relative to the category context (perhaps in this case usefully thought of as musical genre) to which that object belongs, aesthetic judgements are able to better understand the significance of its fit within the world, and as such offer a more 'weighty' critical judgement in the object's appreciation and evaluation [8].

With distortion occurring across a wide range of genres from commercial pop music to radical experimental electro, each with a whole host of *sui generis* expectations and frameworks, knowledge of function and stylistic context clearly informs our aesthetic judgement around when and how distortion occurs. For example, our threshold for distortion would be much higher at a DIY-style punk concert, where electronic noise and other forms of distortion are not only expected but intrinsic to the sonic language, compared to, say, a DJ set at a major electronic dance music festival, where the faithful reproduction of recordings is a primary aim. By keeping this consideration at the back of our minds throughout this discussion, we are tacitly ascribing to the model of functional aesthetics set out by Glenn Parsons and Allen Carlson, whereby aesthetic success is based on whether an object is well-formed for its function, and as such is perceived as pleasing to 'competent judges', where, within this discussion, a 'competent judge' might be seen as anyone with even rudimentary knowledge of genre expectation [9].

Perhaps the most interesting question that arises from this definition with regards to distortion specifically is how we might define 'well-formed' in the context of a technique intended to modify, disrupt or even rupture a musical experience, and this question forms the basis of the following chapter. Whereas Parsons and Carlson argue for 'well-formed' referring to the satisfactory arrangement of an object's elements within the overall intention and design of that object [9]—for example, that a piece for solo piano is physically playable by the performer and is written idiomatically for the instrument—Panos Paris suggests that we might consider the proper function (or functions) of an object to be distinct from whatever purpose an object happens to serve—for example, that the musical gestures of a piece of piano music designed to be relaxing broadly adhere to the semiotic gestures of being rhythmically gentle, timbrally subdued and dynamically quiet. As such the proper function belongs 'to the object, as opposed to being incidental to it' [10]; a contrast he notes is captured linguistically by the difference between an object 'having function F' and 'functioning as F'.

This distinction is helpful in shifting our discussion away from whether the application or perception of distortion in a musical object is simply 'well-formed' for its function (i.e. successfully fulfils the roles expected of it within a certain schema) toward a fundamental question of distortion's role within music. In this chapter I argue that both micro- and macro-distortion can perform three basic 'proper' functions: (1) to question, through the highlighting of imperfection and difference; (2) to challenge or confront, in the form of violent noise or other modes of alienation; or (3) to oppose as a mode of resistance to normative structures and hegemonies. I will conclude by arguing that, whilst offering distinct modes of aestheticizing, in effect all three aesthetic functions have the same *modus operandi*: they question and challenge in order to reveal, offering the listener a point of enhancement, different or confrontation to the expected norm in a manner which prompts an altered and refreshed form of attending to the sonic object.

17.2   Distortion as Imperfection

The basic tenet that perfection is a fundamental goal of art (and conversely imperfection a failing) is a central and recurring theme throughout much of the history of aesthetics. From the ancient ruminations of Plato—who argued that we instinctively seek to surround ourselves with perfect objects that reflect our own beauty—to the absolutism of Plotinus and other thinkers from the early Christian tradition who believed in art's imperative to imitate the infinite and intrinsic purity of God's beauty through 'proper proportion and pleasant order' [11], the notions of purity and perfection as distinct from ugliness or displeasure had become firmly engrained in Western thought. By the time of the Enlightenment, this thread had taken a decidedly moralistic turn, with Baruch Spinoza questioning, 'if all things have followed from the necessity of the most perfect nature of God, how is it that so many imperfections have arisen in nature—corruption . . . confusion, evil, crime, etc.?' [12]. Perhaps somewhat surprisingly, this view of imperfection being akin to deformity of an ideal state of being has pervaded scholarship until surprisingly recently, with writers like the ecomusicologist and composer Raymond Schafer pushing the notion to its extreme [13].

> The Music of the Spheres represents eternal perfection. If we do not hear it, it is because we are imperfect. . . . Distortion results the moment a sound is produced, for the sounding

object first has to overcome its own inertia to be set in motion, and in doing this little imperfections creep into the transmitted sound. . . . All the sounds we hear are imperfect. For a sound to be totally free of onset distortion, it would have to be initiated before our lifetime. If it were also continued after our death so that we knew no interruption in it, *then* we could comprehend it as being perfect.

(pp. 261–262)

Through his distinction between the perfection of a transcendent, harmonic order and some form of negative, 'worldly' deviance, Schafer locates the very essence of this thread: that perfection is effectively unattainable to mortal beings, who must learn to live with the limitations of inherent imperfections as art, and specifically sound, is brought into being within the world. Sonically then, imperfections and distortions in sound are 'an inevitable consequence of sound's material existence . . . [where] a degree of distortion is simply something that has to be tolerated as sound travels within the earthly, material field of clashes, frictions and mutations' [14]. As soon as we discard any tacit value judgements that the human qualities of the imperfect and distorted are in any way inferior to the 'normative' purity of perfection, we can liberate this idea from its theologically tainted history and see the flip-side: that imperfections are ways of celebrating the human, transient and grounded quality of art as distinct from the aspirational perfection of nature (and indeed the divine). 'A task [is] never perfectly completed, [and] is not perfectible', notes philosopher Hans-Georg Gadamer, 'but this imperfectability is not due to a deficiency in reflection, rather it is ascribable to the essence of the . . . [human] being' [15].

When we move from the abstract to the tangible, this relationship between imperfection and the human becomes clearer. In a music industry so firmly centred around the mediations of digital recording technology, it is no surprise that for many musicians the aspiration of unburnished, perfect performances or recording is often at the forefront of their minds [16]. Whether a violinist in an orchestra or a singer in the studio, the desire to rectify any mistake or distortion in search of perfection is a convincing temptation. But as a substantial body of recent scholarship demonstrates [e.g. 17], for many musicians the effect of this desire is perversely the opposite: to distort reality in such a way as to beautify or homogenise the sound by correcting every perceived mistake dehumanises the music in such a way as to make it sound grotesquely artificial and contrived.

A little too much fixing one way and the looseness of a satisfyingly tight, 'in the pocket' groove becomes overly robotic, whilst too far the other way and we are left with something messy and muddled. A useful comparison here is the Japanese tradition of wabi-sabi, or 'the beauty of simple imperfections', perhaps best known manifest in the artform of Kintsugi or 'golden joinery' (Figure 17.1), where broken ceramics are repaired with striking gold or silver in order to celebrate and display the history of an object, rather than trying to disguise its wear. By celebrating the irregularity, imperfection, incompleteness and insufficiency of objects, the viewer is forced to confront the realism of the object's difficult beauty; beauty in spite of—or indeed because of—real-life usage [18]. In the words of philosopher Yuriko Saito, wabi-sabi is 'the antithesis of profusion, opulence, and exuberance that can be brought about by ornately decorated, gorgeous objects . . . [and as such] represents our entire world and life where the ruling principles are transience, insufficiency, imperfection, and accidents' [19].

Striking the subtle balance between playing in time or in tune and leaving the micro-timing or tuning deviances which give the music its character might be a difficult tightrope for a performer or producer to traverse, but as Charles Kronengold notes, this produces 'a productive tension between the precise way the record seems to be crafted [and the human producing it], particularly in regard to showcasing its unusual sounds, and the accidental, noisy quality of some of the sounds themselves' [20]. Hearing the untidiness of sonic 'accidents' in a recording shines a light on the production process by locating the listener in a real-world context

Figure 17.1 An example of a Kintsugi bowl.
(*Credit:* Marco Montalti).

of music-making—e.g. signifying the physical presence of bodies in the studio enacting their creative labour with realism, originality and agency—but it also helps reminds us of the communicative authenticity of human performer [21]. Think of the distorted, gravelly quality of Tom Waits' voice, for example, singing disturbed stories of social outsiders and misfits or the tender fragility of FKA Twigs' breathiness as she sings of unrequited love and public disapproval, heavy with emotion and seemingly on the verge of breaking with devastation in a track like 'Cellophane'.

Strength and vulnerability seem to be two sides of the same coin here: imperfections fragile enough to convey emotional exposure whilst seemingly being powerful enough to escape the sanitising power of the mediating technologies at work. But this form of distortion can uncover more than just fragility and authenticity at an individual level. Think, for instance, of artists and musicians breaking or manipulating their instruments or equipment in search of a richer sound palette [22], for example, the Kinks infamously cutting open their speaker cones on the 1964 hit 'You Really Got Me' to achieve the track's signature fuzzy distortion. Or perhaps electronica artists like Burial or Disclosure using phonographic imperfections like tape flutter or vinyl hiss on their tracks—maybe even just low-level harmonic saturation to a signal intended to generate a sense of 'analogue warmth'—in order to invoke the cultural memory of past musical cultures by attempting to reproduce an 'authentic' listening experience [23]. We see this tendency particularly vividly in sampled music, where the role of the artist is something of an 'archaeologist of imperfection' [24], navigating and uncovering layers of imperfection—both intentional and accidental—and deciding which to preserve for the sake of the artefact (musically or materially) and which to further warp, glitch or distort for their own artistry. Yet, as Anne Danielsen elaborates, whilst the vulnerability of a human mistake and the glitch of a technological imperfection might strongly resemble each other, they are not the same [17].

> When a sample keeps enough of its character to point toward its original aesthetic universe ... the effect of the 'corrupted' sound file or the imperfection of the loop becomes conspicuous. ... [It is] parasitic on our notion of a pre-existing musical whole—something that was not deformed has been twisted or bent, a whole has been cut up and reordered, something that did not show any sign of failure or defect has been manipulated to come forward as containing a glitch. The perceived nonhuman character of these digital manipulations presupposes a notion of musical humanness—that is, an imagining of what the typically human gesture that has been disturbed or destroyed once was.
>
> (pp. 261–262)

There is something really interesting about our increased acceptance of imperfection in the context of an oblique nostalgia for a non-digital world. In all of these examples, the lo-fi breakdown or malfunction of either a musical performance or a medium's material degradation is given aesthetic value by the artist, enacting 'a break with the metaphysics of digital immateriality' [25] by facilitating a re-emergence of the physical and cognitive human experience 'through [those] technologies of representation, reflection, and recognition ... that are distinct from the machine' [26]. By deploying conventionally undesirable distortions of sound (pops, hisses, crackles, glitches, etc.) distortion becomes a productive creative tool to expose the messy reality and idiosyncrasy of human imagination in response to the increasing standardisation from the normalising force of increasingly algorithmic models of musical production consumption.

In spite of the proliferation of digitally perfected music saturating everyday culture, the development of more open values of listening has created a shift towards a culture of listening that is not only broadly open to distortions and imperfections but also embracing of them as ways for listeners to understand their shared humanity as part of a broader social discourse of fragility [27]. In what Kim Cascone has described as the 'aesthetics of failure' which characterise post-digital music-making [28], we no longer shy away from sonic distortions but embrace them as characterful and even draw the listeners' attention to them, either in the mix or through paratextual material like liner notes or interviews in order to highlight the 'honest' or 'intimate' character of the recording [29]. This process recalls Jean Rancière's notion of *le partage du sensible* ('the distribution of the sensible'): a system of 'self-evident facts of sense perception that simultaneously discloses the existence of something in common and the delimitations that define the respective parts and positions within it' [30]. Drawing a listener toward distortions within a sonic object not only embodies the reality of the human experience, but also connects them with a communal shared experience of fragility; one which works to 'outline a certain cartography of the common world [that] we [can] use to make sense of our world and how we take part in it' [31]. Rather than questioning what is perfect, the shift in our listening attitudes (prompted by an increased presence, awareness and acceptance of distortion-as-imperfection) helps us as a society to focus on the positives of imperfection, and their ability to make us see through the veneer of perfection and embrace the realism of our relationship to the world around us.

17.3 Distortion as Rupture

Although much of the audio engineering literature on this topic differentiates between distortion and noise, noise—in the sense of harsh, unnatural or unpleasant sounds [32]—is often the primary means by which many people encounter sonic distortion. For both the casual listener and the professional mix engineer, a key characteristic of the aural phenomena described here as noise (or distortion-as-noise) is its undesirable quality, with music producer Gary Gottlieb claiming that 'all audio professionals will all agree that a mix needs to be free of noise *and* distortion' [33]. Whether from the accidental misuse of an amplification system resulting in screeching feedback, the too-harsh clipping of an overdriven guitar, or an intentional destruction of amplification equipment (for example, in the 'material action' technique of Japanoise artists Merzbow and Hijokaidan [34]), the effect on a listener is largely the same: the sonic intensity of an increased density of sound, causing timbral offence and ultimately—if pushed too far—a feeling of distress. To locate this in the realms of the physical, David Novak describes his visceral listening experience during a concert of the noise-music band Incapacitants in the following way [35]:

> The low-end vibrations are inside my chest, forcing my lungs to compress as I exhale slightly, involuntarily, along with the blasts of sound. . . . [H]eavy, pounding deep drum sounds, droning moans with electric clatter over it all, and as the atmosphere intensifies, growing louder, the lights begin to come up—white, glaring spots in my eyes as their set crashes to an end. . . . It's so loud I can't breathe—they vibrate the air inside my mouth, in the back of my windpipe, as the volume grows and grows. I fear for my eardrums.
>
> (p. 29)

Whilst there may be a complex tension between conceptions of noise as a technical term and as a socio-political one, the idea of distortion outlined by Novak as a mode of rupture or alienation permeates our society, whether through a sonic function—i.e. a force of interference and disturbance on a soundwave—or a type of sound in itself. Distortion here is creating not just an unwanted quality in the listening experience but (taking a nod from information theory) acting as a destabilising force that fundamentally disrupts the process of communication between artist and listener. This is distortion as a weapon of destruction and rupture, but such a form does not need to necessarily be a negative. Steve Goodman offers an example of this [36]:

> [O]ne everyday use [of noise] relates to distortion in a textural surface. Usually, however, noise is taken negatively, in an essentially psychoacoustic fashion, which states that noise is relative and is an unspecified or unwanted sound. . . . [But if] we take the listening experience of pirate radio, for example, we may attribute part of the affective charge to the dose of static interference, which intensifies the sensation of the music heard, intensifies the conjunction of music and the nervous system, as if the frayed edges of percussive and verbal rhythmicity roughen the skin like aerodynamic sandpaper, making it vulnerable, opening it to rhythmic contagion.
>
> (p. 203)

In the same way a producer might add in distortion on a hip-hop track to signify attitude and rawness, the subliminal association between distortion-as-noise and the subcultural allure makes it a productive way for musical genres to demarcate themselves from the mainstream. Genres of electronic dance music such as techno and industrial have long used distortions of timbre and other techniques of noise to invoke an othered quality, by 'favouring granular sounds and complex sonic textures . . . [in] an auditory experience that indexes haptic encounters . . . [as a] timbral evocation of flesh' [37]. Distortion and noise have a long history of association with underground musical scenes, but whilst some of the more primitive semiotically coded connections have pervaded humankind's development, our social tolerance to distortion-as-noise has gradually risen over the centuries, becoming increasingly 'acceptable' to modern ears [38]. Clearly then there is no 'fixed' state of tolerance towards distortion, but rather a set of sociocultural factors which effect how listeners perceived this form of distortion and what is perceived as distorted by one person or social group might not be distorted for another. But as social theorist Jacques Attali observes, these value judgements are not without complex negotiations of power [39]:

> It is sounds and their arrangements that truly fashion societies. With noise is born disorder and its opposite: the world. With music is born power and its opposite: subversion . . . when it becomes music, noise is the source of purpose and power. It is at the heart of the progressive rationalization of aesthetics, and it is a refuge for residual irrationality; it is a means of power and a form of entertainment.
>
> (p. 6)

Attali's allusion to the power of terminology is important: those in power are more likely to ascribe value-judgement-laden words like 'distortion' for political effect. Think, for example, of the Nazis use of the term 'degenerate' music (*Entartete Musik*) as a framework of control and censorship of art during the Holocaust. The use of this political situating of music as noise appears in the earlier school of *Musique Concrète*—a primarily French type of electronic music, which used recordings and electronics to explore aesthetically distorted 'real world' sounds in the concert hall. *Musique Concrète* stemmed out of futurist art, with Luigi Russolo suggesting that the industrial revolution gave modern audiences a greater capacity to appreciate more complex sounds than existed already in music. Unlike many listeners at the time, Russolo valued recorded sounds of the distortions of the world—for example, those often attached to labour (e.g. factory noises)—as aesthetically positive, using this tension of taste as a means of questioning the taste of bourgeois listeners as a form of social disruption and rebellion.

A more recent example of this sort of disruption can be found with noise-cancelling headphones, often adopted by travellers for the reduction of noise (particularly on airplane flights). As Mark Hagood observes, 'in the face of the discomfort and forced togetherness of travel, people are encouraged to employ noise-cancelling headphones as soundscaping devices, carving out an acoustically rendered sense of personal space that Bose has marketed as 'a haven of tranquillity' [40]. Technology here functions as a cultural mediator, offering a process to direct the listener's personalised aesthetic experience in order to control their mood (the tranquil haven), a method of preservation that helps the neoliberal self—autonomous, reflexive and self-managing—to negotiate the noisy and complex world around them. There's a sort of paradox of distortion here: on the one hand, the headphone wearer is blocking out the unwanted, distorted sounds of the world around them (for example, the various mechanical noises of a mode of transportation) in search for the peace and calm, but on the other hand, they are distorting reality to fit their own desires, using mediating technology to amend and edit the world around them.

In seeking to remove the distortions of worldly noise then, the traveller ruptures the aural environment and the space it inhabits, buying into 'a technological way of being in the world that separates us from things—and people—before we have a chance to know whether or not we want them' [40]. This distortion of experience is noticeably political, the sound of public travel being a heightened manifestation of the forcibly proximal congregation of bodies from diverse cultures and creeds. As the individual tactically aids their navigations through the noisy space of travel, the neoliberal self's individualism negates all difference of the other, suppressing the unfamiliar, idiosyncratic, and uncomfortable world around them. The politics of soundscaping may escape the ardent traveller, but the desire to remove themselves from unwanted encounters probably does not. By eliminating the unwanted, disruptive noise, perceived by the traveller as a distortion of the ideal state of 'peace and quiet', the traveller in fact distorts their encounters with the world, charged with the potential for meaningful encounters. Both versions of distortion therefore are disruptive forces, whose power lies in its ability to challenge and reframe our experience of the world around us.

17.4 Distortion as Resistance

For Belgian surrealist René Magritte, painting was very much a philosophical act. His paintings present familiar objects and scenes in unfamiliar contexts, forcing the viewer to question the relationship between them. John Berger cites Magritte as having 'commented on this always-present gap between words and seeing' [41]. 'We never look at just one thing', he says, rather 'we are always looking at the relation between things and ourselves' [41]. Magritte's piece *The Collective Invention* (Figure 17.2) exposes our relationship with its subject, a fish-woman washed up on shore. Magritte clearly attempts this by reworking common situations in a reality we come to take for granted into events that go beyond that boundary, then using them to highlight contradicting relationships between 'things and ourselves' in reality. His work puts the viewer in a place of self-reflection and rejects the norm of self-affirmation. The fish-woman in *The Collective Invention* is lying slack on a beach. She is a perverse reworking of the mythical ideal of mermaids and sirens. The jarring replacement of fins instead of arms and shoulders gives the impression that the fish-woman's arms are tied behind her back. She does not belong on this beach—is literally a fish out of water—and the image seems to suggest that she is trapped here: on the beach as well as in the half-body of a fish. She is imprisoned in the minimized, anti-idealized impression of her, captured in our (male) gaze. Armless and helpless,

Figure 17.2 René Magritte, *The Collective Invention*, 1934.

the anti-mermaid has not blithely washed ashore by the indifferent sea, but deposited on the sand—not only washed up, but used up.

Magritte does not show the world exactly as we see it or even attempt to represent it as a reality. He does not attempt to capture the essence of things or expose their 'inmost soul'. Rather, he questions the whole act of seeing. Through paradox and juxtaposition he subverts the reality we take for granted by making what we have come to know as familiar, typical or cliché become thought-provoking and strange. *The Collective Invention*'s shocking image, like most of his work, makes the viewer uncomfortable. The fish-woman verges on grotesque. It urges the viewer to question their own relationship with the fish-woman and what she represents. It is a reality that we will never precisely face, and as such shows the limits of the gaze—the fragile tension between what we are able to perceive and what is hidden. As Magritte observes [42]:

> Everything we see hides another thing, we always want to see what is hidden by what we see. There is an interest in that which is hidden and which the visible does not show us. This interest can take the form of a quite intense feeling, a sort of conflict, one might say, between the visible that is hidden and the visible that is present.
>
> (p. 35)

As this example demonstrates, art that distorts our lives has a particular ability to mediate our relationship with, and understanding of, the world around us. To return to an analogy from the information theory approach, by infiltrating the original signal, distortion results in an increased and more varied layer of information being added in a way which may add to, or transform, the original signal's intent. 'Exposure', notes Marie Thompson, 'is an inevitable and necessary component of transmission; a signal has to travel through some form of material medium and this medium will always modify the signal in some way' [14]. More information means more uncertainty, but it also means a greater richness of experience and knowledge. As philosopher Michel Serres theorises, the distortion of ideas that occurs when they are taken outside of their original context is not only a necessary risk, but also—and more importantly—a potential source of invention [43].

> Our living and inventive path follows the fringed, capricious curve where the simple beach of sand meets the noisy rolling in of the waves. A simple and straight method gives no information; its uselessness and flatness (or platitude) is finally calculable.... There is only something new by the injection of chance in the rule, by the introduction of the law at the heart of disorder.
>
> (pp. 127–128)

By allowing space for the distortions of ordinary norms, we open the door for phenomena that interrupt the normative flows of things and disrupt pre-existing relations, revealing potentially unexpected insights that show us alternative ways of understanding the world. Through this mode of revelation, art has the ability to highlight the plurality of experience which we—within the perceptual limitations of our own subjectivity—are unable to see for ourselves. As Maurice Merleau-Ponty observes, 'distortions stem from an attempt to capture the moment when the different perspectives fight for our attention in living vision' [44]. Whilst the creation of any artistic identity or persona might require deviation and distortion to some extent to help 'territorialise' and distinguish itself from its competition, we can see distortion's power to expose pluralities most clearly with more extreme iterations of alterity. Think, for example, of the raw, DIY aesthetic of punk rock in bands like the Sex Pistols or The Ramones, where sonic violence and fracture, dangerously imbued with the same threat of injury as the masochistic pogo dancers, belies the precarity of a generation of working-class Brits hit by national socio-economic crisis. For anarchistic movements like the punks, the sonic (and indeed metaphorical) distortion of expected norms was a primary mode of taking back control, by rebelling against 'symbols of privilege, pillars of society and phony optimism through their abrasive rhetoric, fashion and music' [45] in a direct call to action that Philip Tagg refers to as a sonic 'method of empower[ment] . . . against the oppressors' [46].

But the violence of macro-distortion is not the only path to political agency. At the heart of this form of distortion is resistance of a norm through affect: a highly effective mode of engagement, because it enables the political to become directly felt through individually embodied knowledge. Resistance through affect, argues Brian Massumi, is about intensities of relation that register individually, while directly making a difference in the world. These pertain to the individual's autonomy of expression of its powers to be, but only to the extent that that expression is participatory, directly and dynamically entangled with the outside' [47]. It is clear that the direct, bodily violence of auditory noise can afford one of these modes of resistance, but the potential for radical transformation can also be found in the indirect 'structural violence' of smaller acts of distortion that question assumed systemic knowledge. A particularly important body of examples of this comes from the field of disability studies, where a rich thread in the work of scholars like Laurie Stras and Jennifer Iverson demonstrates that the damaged and distorted voices of performers with disabilities—whose voices '[lay] bare the body's traumatic history; the voice itself is evidence of the damaged body' [48]—offer a challenge to ableist preconceptions of what a body should do and sound like [49]. Examples include singers like Bill Withers, who 'reclaimed' his stutter as a signature vocal technique, or Ian Curtis of the band Joy Division, who 'decided to flaunt the imperfection [of his disabled body] . . . through fit-like movements and granular, uncanny vocal style . . . as a way of kicking against the pricks' [50]; both invoking the alterity of their disabilities to not only reassert control over their everyday life but to make visible medical conditions in defiance of social stigma. As Stras argues [51],

> cultural forms that interrogate or challenge assumptions regarding what's normal and what's not, that allow people to engage and identify with, and possibly even embrace abnormality without threat, work to alleviate both the anxiety of the non-disabled and the alienation of the disabled . . . reception and the comprehensive record of their work allows a privileged examination of how the unacceptable became acceptable, and how what was once considered not fit for purpose could eventually become a 'standard of entertainment' and the inspiration for so much that followed after them.
>
> (p. 317)

As Stras alludes to, the process of revelation can only work when the receiver or listener is aware of what the normative iteration is, and therefore understands the act of distortion as a process of transformation in and of itself. In other words, it is not the distorted object that is politically charged, but our awareness of the process of distortion that enables us—as products of the system—to question and disrupt the (assumed) systemic values. As Jack Halberstam suggests, releasing ourselves from an attachment to the norms that a distorted iteration is being judged against equates to a break from the attachment to hegemonic social expectations of value and beauty that underpin our interactions with the world [52]. In doing so, the highlighting of difference not only induces a new affective state at an individual level, but also serves to create a new order of relations at a systemic level. While visibility may not offer direct transcendence from the situation, it does 'provide small subjects with . . . "a little resistance" in their confrontations with larger systems' [53], whereby the visible distortions of the norm unconcealed in a disabled body challenges our thinking from normalising binary frames. Distortion in this way acts, in the words of Salome Voegelin, as:

> [a] politics of the incomplete, the unfamiliar, the unrecognizable and the unheard; that which we have no words for and that which is incommensurable in relation to current norms, but which presses through a naturalized reality, and impresses on us the need and courage to listen-out for alternatives within.
>
> [54]

17.5 Conclusion

Philosopher and psychoanalyst Jacques Lacan famously proposed that we spend our whole lives trying to find the Real (or ideal) truth in the world. The Real resists representation, so we can never truly know it. It loses its 'reality' once it is symbolized (made conscious) through language, image or sound. Literary theorist Naomi Schor argues that it is through the distortion, which truly shows up in the relationship between the artwork and Real. According to Schor, the micro-level in which detail resides is overlooked as decorative, and as such unnecessary, whilst in actuality it accentuates and enables the idea of the whole work of art. In her analysis, Schor observes that [42]:

> even the portrait painter . . . must flatter . . . the purely natural side of the imperfect existence, little hairs, pores, little scars, warts, all these he must let go, and grasp and reproduce the subject in his universal character and enduring personality. It is one thing for the artist to simply imitate the face . . . and quite another to be able to portray the true features which express the inmost soul of the subject.
>
> (p. 34)

Distortion works in much the same way. It opens up a critical space between how things seem and how we experience them as sensuous beings, where we might see how they truly *are*. Questioning the contradictions and uncertainties in the perceptual gaps in-between allows us to know better. Art's power and influence comes from its ability to give expression to this deep-rooted sense of uncertainty of its world, perhaps sitting at odds to a modern world with strict demands for functionality and communication: distortion in music mirroring the distortion of society around, rather than being a deficiency of representation. As Andrew Bowie asserts, art's worldly function is not to make the world around us 'completely comprehensible' but to provoke us to 'pursue as far as we can the attempt to comprehend it discursively' [55]. But the radicalised and confronting nature of distortion enables us to yield in this way by questioning the polished, 'easy' veneer of perfection and offering the potential for empathy and connection through the creation of a shared liminal situation. Embracing distortion then is the ultimate breaking of the grand narratives we are complicit in; not something to be shunned or feared, but something to be welcomed. Knowledge and truth no longer rest solely on the authority of tradition, but rather upon a critical self-reflection with 'the ability to reject the claim of traditions . . . [and] break dogmatic forces' [56].

Connecting back to music production aesthetics then, distortion, in its various manifestations as either imperfection, rupture or resistance, functions as a mode of revelation, both at an aesthetic and auditory level. With its modes of meaning-making through manipulation breaking down the established norms—whether of an individual listening experience or the cultural imagination of an entire social consciousness—distortion affords new ways of engagement with the world around us. We are inspired, thinking in these terms, not to attempt to eliminate distortion but to learn to think with and even through it: neither adding to sound, nor destroying it, but rather reframing it and morphing it into new pluralities. Much like the care and attention sound engineers and plugin designers spend creating reverb units which faithfully recreate the technical reflections of a space, we should pay equal care to the potential for music to reflect back the artistic and emotional spaces of our lives. Distortion helps us to confront and test the culturally coded norms around us by challenging our pre-existing semiotic relationships with sound to open up a truth beyond the illusions of ideology and hegemony that we take for granted. But perhaps distortion functions best when, as with Magritte's work, it is a reflection of society—as uncomfortable as that may be. If art is fundamentally a lens to deal with society in experience of modernity and technology, then distortion reflects, reveals and empathises with the complexities and imperfections of the modern world in a necessary and worthwhile way. It is precisely through critically listening to—and ultimately accepting—distortion in the sounds around us that we are enabled to break free from the simplistic narratives that are becoming increasingly prevalent in our lives and begin to uncover the hidden truths in the messy spaces between.

References

[1] Provenzano, C.: Making Voices: The Gendering of Pitch Correction and the Auto-Tune Effect in Contemporary Pop Music. *JPMS*. 31, 63–84 (2019). https://doi.org/10.1525/jpms.2019.312008
[2] Williams, J.A.: Intertextuality, Sampling, and Copyright. In: Williams, J.A. (ed.) *The Cambridge Companion to Hip-Hop*. pp. 206–220. Cambridge University Press, Cambridge (2015)
[3] Vessel, E.A., Starr, G.G., Rubin, N.: Art Reaches Within: Aesthetic Experience, the Self and the Default Mode Network. *Frontiers in Neuroscience*. 7 (2013). https://doi.org/10.3389/fnins.2013.00258
[4] Everest, F.A., Pohlmann, K.: *Master Handbook of Acoustics*. McGraw-Hill/TAB Electronics, New York (2009)
[5] Bourbon, A.: Hit Hard or Go Home: An Exploration of Distortion on the Perceived Impact of Sound on a Mix. In: Gullö, J.-O. (ed.) *Proceedings of the 12th Art of Record Production Conference*. pp. 19–36. Royal College of Music (KMH), Stockholm (2019)
[6] Exarchos, M.: Sample Magic: (Conjuring) Phonographic Ghosts and Meta-illusions in Contemporary Hip-Hop Production. *Popular Music*. 38, 33–53 (2019). https://doi.org/10.1017/S0261143018000685
[7] Goldman, A.H.: *Aesthetic Value*. Westview Press, Boulder, CO (1995)
[8] Walton, K.L.: Categories of Art. *The Philosophical Review*. 79, 334–367 (1970). https://doi.org/10.2307/2183933
[9] Parsons, G., Carlson, A.: *Functional Beauty*. Oxford University Press, Oxford; New York (2008)
[10] Paris, P.: Functional Beauty, Pleasure, and Experience. *Australasian Journal of Philosophy*. 98, 516–530 (2019)
[11] Tomaszewski, V.: *A Transcultural and Interdisciplinary Approach of the (False) Dichotomy Subject-Object in Aesthetics. Dialogue Between Western Aesthetics, Sociology of Art and Non-Western Aesthetics*, 391 (2010).
[12] Ratner, J.: *The Philosophy of Spinoza*. Jovian Press, New York (2018)
[13] Schafer, R.M.: *Soundscape: Our Sonic Environment and the Tuning of the World*. Destiny Books, Rochester, VT (1994)
[14] Thompson, M.: *Beyond Unwanted Sound: Noise, Affect and Aesthetic Moralism*. Bloomsbury Academic, New York (2017)
[15] Gadamer, H.-G.: *Truth and Method*. Continuum, Leipzig (1975)
[16] *Brian Eno Explains the Loss of Humanity in Modern Music | Open Culture*, www.openculture.com/2016/07/brian-eno-explains-the-loss-of-humanity-in-modern-music.html
[17] Danielsen, A.: Glitched and Warped: Transformations of Rhythm in the Age of the Digital Audio Workstation. In: Grimshaw-Aagaard, M., Walther-Hansen, M., and Knakkergaard, M. (eds.) *The*

Oxford Handbook of Sound and Imagination, Volume 2. pp. 593–609. Oxford University Press, Oxford (2019)
18. Saito, Y.: The Japanese Aesthetics of Imperfection and Insufficiency. *The Journal of Aesthetics and Art Criticism*. 55, 377 (1997). https://doi.org/10.2307/430925
19. Saito, Y.: *Everyday Aesthetics*. Oxford University Press, Oxford; New York (2007)
20. Kronengold, C.: Accidents, Hooks and Theory. *Popular Music*. 24, 381–397 (2005). https://doi.org/10.1017/S0261143005000589
21. Dolan, E.I.: Musicology in the Garden. *Representations*. 132, 88–94 (2015). https://doi.org/10.1525/rep.2015.132.1.88
22. Kelly, C.: *Cracked Media: The Sound of Malfunction*. MIT Press, Cambridge, MA (2009)
23. Baade, C., Aitken, P.: Still "In the Mood": The Nostalgia Aesthetic in a Digital World. *Journal of Popular Music Studies*. 20, 353–377 (2008). https://doi.org/10.1111/j.1533-1598.2008.00169.x
24. Cubitt, S.: Glitch. *Cultural Politics*. 13, 19–33 (2017). https://doi.org/10.1215/17432197-3755156
25. Born, G., Haworth, C.: From Microsound to Vaporwave: Internet-Mediated Musics, Online Methods, and Genre. *Music and Letters*. 98, 601–647 (2017). https://doi.org/10.1093/ml/gcx095
26. Pettman, D.: *Human Error: Species—being and Media Machines*. University of Minnesota Press, Minneapolis, MN (2011)
27. Clarke, E.: *Ways of Listening: An Ecological Approach to the Perception of Musical Meaning*. Oxford University Press, Oxford (2011)
28. Cascone, K.: The Aesthetics of Failure: "Post-Digital" Tendencies in Contemporary Computer Music. *Computer Music Journal*. 24, 12–18 (2000)
29. Supper, A.: Listening for the Hiss: Lo-fi Liner Notes as Curatorial Practices. *Popular Music*. 37, 253–270 (2018). https://doi.org/10.1017/S0261143018000041
30. Ranciere, J.: *The Politics of Aesthetics*. A&C Black, London (2006)
31. *On the Borders of Fiction | Stoffel Debuysere*, www.sabzian.be/article/on-the-borders-of-fiction
32. Keizer, G.: *The Unwanted Sound of Everything We Want: A Book About Noise*. Public Affairs, New York (2012)
33. Gottlieb, G.: *Shaping Sound in the Studio and Beyond: Audio Aesthetics and Technology*. Thomson Course Technology, Boston (2007)
34. Potts, A.: The Internal Death of Japanoise. *Journal for Cultural Research*. 19, 379–392 (2015). https://doi.org/10.1080/14797585.2015.1065654
35. Novak, D.: *Japanoise: Music at the Edge of Circulation*. Duke University Press Books, Durham (2013)
36. Goodman, S.: *Sonic Warfare: Sound, Affect, and the Ecology of Fear*. MIT Press, Cambridge, MA (2012)
37. Garcia, L.-M.: Beats, Flesh, and Grain: Sonic Tactility and Affect in Electronic Dance Music. *Sound Studies*. 1, 59–76 (2015). https://doi.org/10.1080/20551940.2015.1079072
38. Demers, J.: *Listening Through the Noise: The Aesthetics of Experimental Electronic Music*. Oxford University Press, Oxford (2010)
39. Attali, J.: *Noise: The Political Economy of Music*. University of Minnesota Press, Minneapolis (1985)
40. Hagood, M.: Quiet Comfort: Noise, Otherness, and the Mobile Production of Personal Space. *American Quarterly*. 63, 573–589 (2011)
41. Berger, J.: *Ways of Seeing*. Penguin, London (2008)
42. Schor, N.: *Reading in Detail: Aesthetics and the Feminine*. Routledge, Oxford (2013)
43. Serres, M.: *The Parasite*. University of Minnesota Press, Minneapolis (2013)
44. Theodore, A., Toadvine, J.: The Art of Doubting: Merleau-Ponty and Cezanne. *Philosophy Today*. 41, 545 (1997)
45. Hebdige, D.: *Subculture: The Meaning of Style*. Methuen, London (1979). http://www.tagg.org/articles/xpdfs/dartington2001.pdf
46. Tagg, P., Collins, K.E.: *The Sonic Aesthetics of the Industrial*, 11 (2001)
47. Massumi, B.: *Politics of Affect*. John Wiley & Sons, New York (2015)
48. Howe, B., Jensen-Moulton, S., Straus, J.N., Iverson, J., Holmes, J.A., Bakan, M.B., Dell'Antonio, A., Grace, E.J.: On the Disability Aesthetics of Music. *Journal of the American Musicological Society*. 69, 525–563 (2016)

[49] Stras, L.: The Organ of the Soul: Voice, Damage and Affect. In: Straus, J. and Lerner, N. (eds.) *Sounding Off: Theorizing Disability in Music*. pp. 173–184. Routledge, New York (2006)

[50] Partridge, C.: *Mortality and Music: Popular Music and the Awareness of Death*. Bloomsbury Academic, London (2015)

[51] Stras, L.: Sing a Song of Difference: Connie Boswell and a Discourse of Disability in Jazz. *Popular Music*. 28, 297–322 (2009). https://doi.org/10.1017/S0261143009990080

[52] Halberstam, J.: *The Queer Art of Failure*. Duke University Press, Durham, NC (2011)

[53] Ngai, S.: *Ugly Feelings*. Harvard University Press, Cambridge, MA (2009)

[54] Voegelin, S.: *The Political Possibility of Sound: Fragments of Listening*. Bloomsbury Academic, London (2019)

[55] Bowie, A.: *Music, Philosophy, and Modernity*. Cambridge University Press, Cambridge (2009)

[56] Habermas, J.: *On the Logic of the Social Sciences*. John Wiley & Sons, New York (2015)

Index

Note: page numbers in *italics* indicate a figure and page numbers in **bold** indicate a table.

3% tube-based harmonic distortion 190
12AX7, usage 32, 53–54
12BH7 valves, usage 53
21st-century jazz guitar, impact 237–238
"200 Pound Bee" (The Ventures) 7
1176 (FET compressor): design analysis 54–55; distortion, compression activity *62*; total harmonic distortion (THD) *59*
6386 valves, usage 54

Abbey Road Studio 8; "Sale of the Century" 155
A/B comparison 175, 179, 180
Abercrombie, John 235
A/B online survey 114
A/B testing, usage 117
A/B tests 114; content 114–115
accessibility options 180
accomplishment, myths 191
Ace of Skunk Anansie 180
acousmatic music 218
acoustic cues, identification 247
acoustic distortion, jazz guitar (relationship) 232–233
acoustic phenomenon 95
ADAT, usage 109
Adenot, Paul 40
aesthetic categorisation 262–263
aesthetic, engineers (hiring) 104
aesthetics, progressive rationalisation 269
Agawu, Kofi 246
aggressive distortion 76
AKG C414 EB condenser microphone, usage 185
aliasing, usage 16–17
"All Buttons In" mode 55, 154, 195
Al Ray plugin 105
AmpFarm 195
amplifier: selection, flexibility 181; strengths/weaknesses 181–182; usage/importance 201
amplitude dynamics compression, usage 224–225
amplitude, dyn-space model 204
amplitude modulation (pumping effects), introduction 58, 60

analog modelling 24–26
analogue distortion circuits, overview 129
analogue equipment 150
analogue maximum, clipping 9
analogue soft-clipping 10
analogue-to-digital (A/D) converter 15, 18
And Justice for All (Metallica) 210–213
Aphex parametric EQ, usage 2091
API 3124+ mic preamp 136
API 3124V four-channel microphone amp, usage 70
API EQ, specialness 102
API high gain frequency response *73*
API high gain Hammerstein model *71*
API low gain frequency response *72*
API low gain Hammerstein model *70*
API maximum gain frequency response *73*
API maximum gain Hammerstein model *71*
API phase measurement *74*
API signal path, frequency response 72
API Vision (preamp) 85
API Vision Channel Strip (preamp), analysis 69–72
Arbiter (Dallas/Arbiter) Fuzz Face pedal 8
arctangent distortion effect, harmonic distortion *22*
arctangent effect, waveform/characteristic curve *21*
artistic identity, self-expression/statement 150
attack: design (RC network) *49*; fast attack times, impact 50
attack control (compressor) 49–50; non-linearity 50
Attack Magazine 189
Attali, Jacques 269
Audient ASP880 Hi Z (preamp) 85
Audient ASP880 Low Z (preamp) 85
Audient ASP880 Mid Z (preamp) 85
audio: dropouts, occurrence 50; experience, survey respondents *165*; low-level audio features (compressor) **63**; mixing, distortion (ecological approach) 109; production role *166*; production role, survey respondents *165*; production workflow, traditional/software-emulated

278 *Index*

distortion (comparison) 160; professionals, perspectives 166–167; software (development), distortion (inclusion) 13; usage 67
audio graph, building 29
audio production workflow, harmonic distortion (impacts) 169–170
audio professionals, roles 164
audio recording formats, technical performance (improvement) 189
audio signal: alteration, absence 189; analysis 56–63
AudioWorklet node 31
auditory noise, bodily violence 272
aural frontier 218
aural thinking 216
authorship, accounts 191
autoregressive (AR) model 23
Auto-Tune, glitch 262
Avalon VT 737 (preamp) 80–82; line input Hammerstein model *81*; mic input Hammerstein model *82*; usage 69

Bac to Black (Winehouse) 96
Baker, Ginger 8
Barnard, Matthew 216
Bascombe, Dave 96, 100, 102, 105
bass amp simulation software 15
Beatles, The (band) 8, 251, 254; EMI recording sessions, usage 191–192
BF245A FET, usage 54–55
bias control, idle voltage (correspondence) 39
Bielmeier, Doug 160
Big Muff descriptors 134–136; survey *138*; text mining list *135*
Big Muff, feature 128
Bilinear Transform method 25
bi-polar junction transistors (BJTs) 15
bit-crushing (distortion effect) 10
bit reduction 16
B&K/DPA microphones 190
Blackbird Audio Rentals 190
Black Box HG2, usage 10
black-box modelling 24–25
"Blackened" (Metallica) sound-box Dist-space graph 210–211, *211*, *213–214*
Blake, Tchad 99, 103, 104, 110, 195
Blayer, Archie 6
"Blue Ghost Blues" (Johnson) 233
Blue Matter (Scofield) 235
Bluesbreakers (John Mayall) 8
Boersma, Paul 248
Boogie guitar amplifiers, usage 8–9
Boss DS-1 128; gain, usage 130; tone control 134; usage 9
Boss DS-1 descriptors 133–134; survey *138*; text mining list *133*, *135*
Boss HM-2 Distortion Pedal 201
Bourbon, Andrew 65

Bowie, Andrew 273
Bradley Film and Recording Studio 6
breathing, occurrence 50
Brenston, Jackie 6
brightness, perception 63
"Bring It On Home" (Led Zeppelin) 245
"Brisbane Girl": clean mix *119*; mix 5 *118*; mixes, comparison *122*
broken-up sound 235
Bromham, Gary 3, 93
Brown, Mansur 221
browser-based WebAudio Ecosystem 28
Bruce, Jack 8
Buffa, Michel 28
Bullock, Hiram 235
Burrell, Kenny 233
Bush, Kate 222

cabinet simulator stages, absence *43*
Carlson, Allen 264
Carter, Ron 236
Cascone, Kim 267
Case, Alex 4
Cash, Johnny 6
"casual terms" 216
causal complex 224
Celestion alnico speakers, usage 8
"Cellophane" (FKA Twigs) 266
chance, injection 271
Chandler EMI product, resurgence 105
character, concept 97
Chesterfield radio programme, creation 222
chorus, clean tone 182
Christian, Charlie 231–232, 234, 240
chromatic scale, singing 193
Clapton, Eric 8
Clarke, Eric 103
cleaner tones, preference 119
cleanliness, environment 103
client experience, harmonic distortion (impacts) 168–169
clipped waveform, sine wave (display) *5*
clipping: asymmetry 33; distortion effect 9
clones, usage 156
CNRS 29
Cobain, Kurt 128, 202
Coghlan, Niall 147
Collective Invention, The (Magritte) *270*, 270–271
Collins, Phil 225
colour (sonics) 150–151
colouration 101, 150; absence 88–89; addition 101–102; experience 110–111; studio presence 10–11
"Communication Breakdown" (Led Zeppelin) 255; analysis 251–254; insane episodes *252–253*; songmap *251*; vocal climax *253*
component-level modelling 25

compression: activity (dbx165A) *61*; activity, THD performance *60*; benefits 168; effect 101; usage, reason 47–53

compressor: "All Buttons In" mode 154, 195; attack/release controls 49–50; attack/release design (RC network) *49*; audio signal analysis 56–64; block design *48*; breathing, occurrence 50; clarity (loss), fast attack times (impact) 50; design 48–50; design analysis 53–56; dropouts, occurrence 50; feedback/feedforward compressor designs *52*; field effect transistor (FET) compressor, design analysis 54–55; gain reduction styles 52–53; level detection, usage 50–52; low-level audio features *63*; non-linearity 50; optical compressor, design analysis 52–53; overshoots, occurrence 50; pumping, usage 50; sidechain, function 50–52; total harmonic distortion (THD), input level **57**; valve (tube) compressor, design analysis 54; vocal recording analysis 63; voltage-controlled amplifier (VCA) compressor, design analysis 55–56; waveshaping, fast attack/release (impact) *51*

console, usage 153

content analysis 129–136; method, summarising 177

contour pitches (c-pitches) 202–203, *204*, 208–209

contour segment (c-seg) 203; interpretation *203*

contour space (c-space) 202–203; interpretation *203*

contour theory 202–204

contrasts, noticing 104

control surface, usage 153

control voltage (derivation), sidechain block (usage) 52

converters, hard-limiting 18

Corea, Chick 240

Costey, Rich 195

"Crafting the Signal" (Styles) 189

Cream (band) 8

"creaming the mic pres" process 65

creative abuse 155

creative recording/mixing, low order distortion (usage) 65

creative relationship, ease 99

creativity: accounts 191; stress 182

critical listening skills 98

cross-modal correspondences 95

cross-over distortion, introduction 39

cross-sensory correspondence 98

cross talk, presence 101

CS/N label, denotation 206

Cubase (Steinberg) 186

cubic distortion effect: harmonic distortion *22*; waveform/characteristic curve *21*

cubic distortion, harmonic production 20

cubists 223–224

cultural forms, impact 272

Curtis, Ian 272

cymbals, feeling 87

Dallas "Rangemaster" treble booster 7, 8

danger, distortion (significance) 103

dark space, silent space (equivalence) 222

Davies, Dave 7

Davis, Miles 240; acolytes, impact 234; post-Miles fusion guitar landscape 234–237

DAW AmpedStudio.com, WebAudio amp sim plugins (display) *30*

Dayes, Yussef 221

"Dazed and Confused" (Led Zeppelin): analysis 248–251; restricted vocal pitch *249*; songmap *248*; vocal climax *250*

"Dazed and Confused" (Led Zeppelin) 254–255

dbx165A (VCA compressor): design analysis 55–56; distortion, compression activity *61*

DDMF, Plugin Doctor 68

DeArmond Trem Trol 800 tremolo pedal 7

Deepmind, WaveNet music generator 29

Deep Purple (band) 8

delay node 40

Depeche Mode (band), audio work 99

descriptors, cross-validation 141–142

diatonic descending melodic thirds, usage 250

differences, stacking 103

Diffusion of Innovation (Rogers) 176

digital audio workstations (DAWs): availability 161; deficiency 109; linear time axis 204; multi-track DAW-based workflows, usage 152; usage 13, 31, 105, 112

digital frequency domain 34

digital hardware processors, analog state 168

digital immateriality, metaphysics 267

digital material, transfer 148

Digital Signatures (Hanssen/Danielsen) 10

digital technology, accuracy 194

digital to analog (D/A) converter 15; performance 193

digital to analogue converters (DACs), usage 10

Di Meola, Al 235

diode-bridge, compressor design variation 52

diodes, usage 15

direct-to-disc recording 192

disabled, alienation 272

Discrete Kirchhoff (DK) method 25

distorted guitar, shaping 236

distortion: action 273; addition, danger element 4; aesthetics 262; algorithm, complexity (increase) 21; amplifier, usage/importance 201; analysis 111, 210–213; authentication 238; awfulness 110–111; browser-based WebAudio Ecosystem, interaction 28; compression activity (1176) *62*; compression activity (dbx165A) *61*; compression activity (Fairchild) *61*; compression activity (LA2A) *62*; context 110–112; contrast 103–104; creation, amplitude dynamics compression (usage) 224–225; creation, studio function 201; creative alternative 100; current/future directions 105–106; danger, significance 103; defining 3–5, 96–97; deliberate usage, discussion 97; description 4; differences, stacking (impact) 100; discussion 123–124; ecological approach 109; ecology 102–104; effects (hearing), volume (reduction) 100; excess 119; excitement level, addition 4; experience, role 98–100; generation 49; guitar, relationship 230–231, 238–241; "happy accidents" 99; harmonic distortion, presence 10; history 3; imperfection 264–267; implementation 17; inclusion 13; interpretation 217; jazz guitarist usage, utilitarian purpose 238; labelling 229; levels, increase (perception) 211; listening skills, usage 98–100; low order distortion, usage 65; metaphor 98; methodology/method 113–118; methodology/method, limitations 117–118; misappropriation 99; non-linear audio processing tool, equivalence 4; permutations 220–222; perspectives 97–102; primitiveness 241; properties 230; reframing 241–242; replacement, pink noise settings *116*; replacement, program settings *116*; resistance 270–273; results data 118–123; rupture 267–270; saturated market 185; semantics, understanding 93; sequential dimension 205–206; shapes/forms 9–10; sound staging 103–104; taste/stylistics 104; theory 112–113; timeline 5–9; traditional distortion, software-emulated distortion (comparison) 160; types 9–10; usage 104, 237

Distortion: pejorative consideration 229

Distortion and Rock Guitar Harmony (Herbst) 230

distortion-as-noise, social tolerance 268–269

distortion-as-process, complication 263

distortion pedals: Big Muff descriptors 134–136; Boss DS-1 descriptors 133–134; content analysis 129–136; descriptive survey method 136–143; descriptors, comparison 141–142, **142**; descriptors, list 130–131, *131*; gain setting, results 139; high-gain survey descriptors *141*; lexicon, approach 128; literature review 128–129; low-gain survey descriptors *140*; medium-gain survey descriptors *140*; methodology 129; nebulous descriptors 142–143; ProCo RAT descriptors 134; results 136–139; text mining descriptors *131*; text mining method 129–136; tube screamer descriptors 131–133

Dist-space 204–206, *205*; analysis *206*; analytical tool 202; extension 207; limit 206; model 231; saturation 208; sound-box (relationship) 207–212

DIY recording paradigm 124

DN-023R recorder 192

Domain Specific Languages, usage 31

"Don't Worry" (Marty Robbins) 6

downsampling (distortion effect) 10

dropouts, occurrence 50

drums: recording 103; saturation/edge 119

"Drums Were Yellow, The" (Holdsworth) 236–237

DS-1 descriptors, text mining list *135*

DS-1 pedal, descriptors 137

DSP code, usage 35

DSP routines 171

duty-cycle modulation 33

Dylan, Bob 241

dynamic changes, sequential order 204

Dynamic Range (DR) 115

dynamic range compressors (DRCs): non-linearity (relationship) 47; usage 47–48

dynamic range, reduction 224

dynamic space (dyn-space) model 204

dynamics, sequential dimension *204*

Earthworks microphones 190

EBow, usage 239

ECC83 (preamp valve distortion) 8

Echoplex, usage 155

ecological theory, proposal 113

ecological tropes, application 112

ecology, definition 112–113

editing skills, development 216

Effects and Processors (White) 195

Electric Lady Studios 192

electroacoustic music 218

Electro Harmonix Big Muff 128

Electro Harmonix Big Muff Pi, design 8

Electroharmonix, freeze pedal 232

Electroharmonix POG, usage 237

electronic components, tolerance range 151

Electro Smash website, usage 128

Elen, Richard 195

Ellington, Duke 233

Ellis, Herb 232–234

Ellis, Mark "Flood" 3, 97, 99–100

Elpico AC55 amplifier 7

emotional state, intensification 249

Empirical Discourse Analysis (EDA) 129

EMT 250 reverb 15

enhancer 14

Enlightenment 264

Entartete Musik 269

"EQ for free" 101–102

equipment: combinations, usage 150; levels, pushing 153; misuse 155

Escher, M.C. 219

eternal perfection, Music of the Spheres representation 264–265

even harmonics, production 14

exciter 14

experimentation, requirement 155

Fab Filter ProQ2, usage 115

fade-ins, usage 212

Fairchild (valve/tube compressor): design analysis 54; distortion, compression activity *61*; total harmonic distortion (THD) *58*
Fairchild 670 (Valve) compressor 57, 147
Fairchild 670 Vari-mu compressor 11
Falsetto note, usage 257
Farlow, Tal 232–234
fast attack/release, impact *51*
fast attack times, impact 50
Fast Fourier Transform (FFT) function 160–161
fast level attenuation, accomplishment 48
Fast, Susan 254
FAUST 31, 40; code, usage *43*; distribution 41; implementation 41–44; power amp plugin, display *43*; re-implementation 41–42
FAUST-based power amp re-creation 42
FAUST IDE, WAP GUI Builder (usage) *43*
feedback: result 268; usage 13, 52
feedback circuit *42*
feedback/feedforward compressor designs *52*
Felder, Nir 237
Fender Princeton, modification 8–9
Fender Twin, clean sound 181
field effect transistor (FET) 15; cessation 55; feedback, usage 52
field effect transistor (FET) compressor: design analysis 54–55; feedback, usage 52
figurative listeners 219
filter block, usage 21, 23
Filters (simulation), WebAudio API (usage) *31*
final product, harmonic distortion (impacts) 168–169
Findle, Paul 100
FireWire, usage 162
FKA Twigs 266
Focusrite Scarlett 44
Foo Fighters (band) 191
fourth-order ladder filter, usage 26
Franklin, Aretha 167
freeze pedal (Electroharmonix) 232
frequency dependent distortion 9
frequency response 114
Friedman organic gain staging, usage 181
Frindle, Paul 11
Frisell, Bill 228, 235–236, 240; jazz/music perspective 240
full-wave rectification 18; harmonic series 19
full-wave rectifier effect: harmonic distortion *20*; waveform/characteristic curve *18*
Functional Audio Stream 40
functionality familiarity 154
Funky Junk Ltd. 189–190
fuzz (distortion effect) 10
fuzz pedal, usage 14, 156

Gadamer, Hans-Georg 265
gain reduction 55; method/amount **53**; styles 52–53

Gambale, Frank 235
gate signal, display *44*
germanium transistor-based design, second/fourth harmonic *187*
Germanium transistor circuit, usage 128
"Get Out of My House" (Bush) 222
Gibson Les Paul (1959/1960), usage 8
Gibson Maestroi FZ-1 Fuzz-Tone, usage 128
Gilmour, Dave 128
GML8200, usage 153
golden units 152
Goldman, Alan 263
Gold Star, ambient chambers (usage) 220
Gonzales, Rod 179
Goodman, Steve 268
Goold, Lachlan 109
Gottlieb, Gary 268
Gould, Billy 181
graphic EQ, transfer function curve 40
gray-box modelling 25–26
Grindcore music genre, distortion 134, 136
Grohl, Dave 128, 191
grunge music, distortion 130–131
GUI controls 39
Guild Thunderbird (1966) 9
guitar: 21st-century jazz guitar, impact 237–238; amplification technology, innovation/conservatism 174; amplifiers, stages *32*; distorted guitar, shaping 236; distortion, relationship 230–231, 238–241; early amplified sounds/distortion 233–234; extension/augmentation 239; horn-like quality 240; instrument emulation 239–240; jazz guitar, distortion (effect/creativity/extension) 228; playing, modes 239; post-Miles fusion guitar landscape 234–237; signal paths *32*; sonic identification 240
guitar fuzz, usage 99
Guitar Rig 6, usage 29
guitarscape 231–238
guitar tube amps: distortions, browser-based WebAudio Ecosystem (interaction) 28; real-time simulations, browser-based WebAudio Ecosystem (interaction) 28
GUI thread 35
GULA Studion 195

Hagood, Mark 269
half-wave rectification 18
half-wave rectifier effect: harmonic distortion *19*; waveform/characteristic curve *18*
Hall, Jim 233
Hammerstein model, measurement 80, 82
Hammerstein-Wiener model 23
Hammond C3 organ, Marshall guitar amplifiers (usage) 8
haptic feedback, absence 153

hard clipper effect: waveform/characteristic curve *17*
hard clipper effect, harmonic distortion *14*
hard clipping 9, 14; process 48; rectification algorithms, usage 18–19
hard rock (compositional structures), distortion creation (studio function) 201
hardware: advantages **166**; digital function 168; driving 168; interest 171; simulations 15–16; software (contrast), audio professionals (perspectives) 167–168; units 171; usage 153
hardware-created harmonic distortion **163**
hardware-created THD 160–161
hardware THD, advantages/disadvantages (comparison) 163–164
harmonica, distortion (purposefulness) 5
harmonic content 118; addition 11, 82–83
harmonic distortion (HD) 85, 160, 186; arctangent distortion effect *22*; creation *166*; cubic distortion effect *22*; discussion 170–171; emulation 164; full-wave rectifier effect *20*; half-wave rectifier effect *19*; hard clipper effect *14*; hybrid use 169; impacts 168–169; presence 10; usage 13–14, 168–169
Harmonic Distortion Survey 164–169
harmonicity 253–254
harmonics: addition 5, 101–102; amplitude, decrease 16; analog stage, occurrence 168; production 33–34
heaviness, perception 230
heavy metal (compositional structures), distortion creation (studio function) 201
Helios product, resurgence 105
Henderson, Chris 182
Henderson, Scott 228, 240
Hendrix, Jimi 155
Herbst, Jan-Peter 174, 230
Hersch, Charles 238
"He's a Jelly Roll Baker" 233
Hesse, Chris 181
heterogeneity, impact 231–232
Hetfield, James 201–202
Heubaum, Marco 178
high-gain guitar sounds 139
high-impedance (increase), preamplifier (usage) 34
high-level simulation *31*
high quality (HD) mode 17
Hindemith, Series 2 formation 205–206
hit hardware 147; factors, identification 149; methods 149; technocultural choice 149–150
Hobbs, Revis T. 6
Holdsworth, Allan 228, 235–236
Holiday, Billie 167, 245
Holmes, Jake 248
homogeneity, rejection 240–241
Horning, Schmidt 216
Howlin' Wolf 6, 245
"How Many More Years" (Howlin' Wolf) 6

Huldt, John 179, 180, 182
human mistake, vulnerability 267
Hurst, Gary 7
hypercompression, deleterious effects 111–112
hypothetical worlds, manufacture 219
hysteresis 67

Ibanez Tube Screamer 128
"I Can't Quit You Baby" (Led Zeppelin) 145
Icky Thump (White) 54
iconicity 155–156; investment 156
iconic sounds, replication 152
idle voltage, bias control (correspondence) 39
impedance settings 88–89
imperfection: appearance 265; layers, uncovering 266–267
Incapacitants (band), listening experience 268
incomplete, politics 273
infinite negative feedback, impossibility 52
innovation, diffusion 176
"Input Effects" demo, usage 28
instrumental climax, occurrence 248–249
instruments: emulation 239–2340
instruments, virtual loss 222
internal space, prominence 221
inter-relationships, discussion 95
"In the Air Tonight" (Collins) 225
in-the-box mixing 112, 148
in-the-box recording 148
Iverson, Jennifer 272
"I Want to Hold Your Hand" (The Beatles) 251
iZotope Trash 2 17

Jackie Brenston and his Delta Cats 6
JavaScript: implementation 42; power amp, bypass *43*
jazz: 21st-century jazz guitar, impact 237–238; early amplified sounds/distortion 233–234; examination 229; guitarist distortion usage, utilitarian purposes 238; guitarscape 231–238; post-Miles fusion guitar landscape 234–237; Western scaffolding, dismantling 231
jazz/electric guitar studies 230
jazz guitar: acoustic distortion, relationship 232; authentication 238; extension/augmentation 239; homogeneity, rejection 240–241; instrument emulation 239–240; primitivity, return 241; sonic identification 240; timbral creativity 239; tonal nostalgia 241
jazz guitar, distortion: definitions/scope 229; effect/creativity/extension 228; properties 230; research field overview 229–231; timbre, importance 229–230
JCM800. *see* Marshall JCM800
JCM 2000 series 201
Jenkins, Chris 9
Jensen transformer, usage 74
Jesus and Mary Chain (band) 96

Jethro Tull (band) 207–208
Johns, Andy 100
Johns, Glyn 7
Johnson, Lonnie 233
Johnson, Willie 6
Jones, Gareth 104
Joy Division (band) 272

Kæreby, Ask 185
Kemper, Christoph 175
Kemper GmbH, guitar amplifier launch 175
Kemper profiler: inferiority 180; usage 29
Kemper Profiling Amplifier (KPA): familiarity 24; usage 178
Kemper profiling universe 176–183; adoption 179; attitudes/beliefs 177–179; attributes 181–182; method 176, 178; production benefits 182–183; sound quality 180; website, overview 177
Kessel, Barney 232–234
kick drum, feel 88–89
Kinks, The 7, 96, 266
Kintsugi bowl 265, *266*
Klangheim DC8C 3, usage 115
Knub, Jørgen 195
Korn (band) 201
Kraftwerk (band) 155
Krantz, Wayne 228, 237
Kriesberg, Jonathan 232
Kronengold, Charles 265–266
KT66 (power valve distortion) 8
KT88 (valves) 9

LA2A (optical compressor) 49; design analysis 53–54; distortion, compression activity *62*; go-to option 164; total harmonic distortion (THD) *59*
Lacan, Jacques 273
Lagrene, Bireli 241
Lammert, Lasse 175
Lang, Eddie 233, 241
Laplace Transform method 25
latency, sample-wise accuracy (measurement) *44*
learning curve 181
Lebrun, Jerome 28
Leckie, John 97; distortion description 4; harmonica distortion, purposefulness 5
Led Zeppelin, music (vocal climax) 245, *250*, *253*; restricted vocal pitch *249*; songs 248–259
"Lemon Song, The" (Led Zeppelin) 245
Lennon, John 54
Les Demoiselles d'Avignon (Picasso) 223
level detection, usage 50–52
Lewis, Jerry Lee 6
Lexicon 224 hardware reverb 15
Lifetime (band) 236
Lillywhite, Steve 105
limitation factor 105–106
linear distortion effects 28–29

Liquid product series (Focusrite) 194
listener, subject-position 263
listening experience 266–267
listening out, courage 273
listening skills, usage 98–100
Little Walter, harmonica (changes) 5
Liu-Rosenbaum, Aaron 245
Live (Ableton) 186
localisation 219
Lo-Fi 14
lo-fi aesthetic 97
Lo-Fi algorithm 16
lo-fi deluxe 195
Lo-Fi distortion 15
LoFi-plugin (Avid), usage 192–193
lo-fi sounds, creation 10
Lord-Alge, Chris 110
Lord, Jon 8
loudness: normalisation 85; perception 111, 231
Loudness Units Full Scale (LUFS): measurements 169; usage 115, 117, 120–121, 136
"Love Is the Message" (Dayes) 221
low-cost software emulations 160
low-drive channels 133
lower-level objects, interpretation 219–220
low-gain settings 139
low-gain survey descriptors *140*
low-latency opportunity, usage 28
low-level audio features **63**
low-level harmonic distortion, addition 100
low order distortion 86; focus 66; preamps, relationship 65–68; usage 65
low order harmonic distortion: affordances 65–66; sonic cartoons, relationship 85–87; tools 68
low-order harmonics 110–111
Low Road, The (Scofield) 235

Macbook Pro 16 44
machines, rise 147–148
macro-distortion, violence 272
Maestro FZ-1 Fuzz-Tone 6–7
Maestro RM-1A ring modulator effect, usage 8
magnitude value 18
Magritte, René 270–271, 274
Malcolm Toft 105
Manieri, Marco 195
Mara, Chris/Yoli 191
Marshall distortion, recognition 201
Marshall JCM800 175; annotated schematic *33*; power amp stage, master volume/presence knob control *36*; power head, rear-view (amplification stages display) *32*; power stage 36–39; preamp tubes, transfer functions *34*; tone stack circuit *35*
Marshall, Jim 7
Marshall JTM 45 guitar amplifier 7, 8
Marshall, Terry 7
Martin, George 191–192

Index

Martin, Grady 6
Martin, Remy 207
Massey, Sylvia 65
Massingill, Rem 182
Massive Attack, bit-crushing (usage) 10
master fader, control 100
Matthews, Mike 8
Mayall, John 8
McCartney, Paul 191–192
McDSP Analogue Processing Box (APB) technology 105
McIntyre, Philip 185
McLaughlin, John 228, 235
McLuhan, Marshall 188
Mead, Clive 3
Medica Capture, usage 28
medium-gain survey descriptors *140*
Medium Is the Message, The (McLuhan) 188
Mellor, David 188
Memphis Minnie 245
Memphis Recording Service 6
Merleau-Ponty, Maurice 271
MesaBoogie 2:90, Negative Feedback Loop (NFB)/presence *37*
MESA/Boogies, avoidance 201
Metallica (band) 201–202, 210–213
Metal Music Manual (Mynett) 175
Metheny, Pat 235, 240
Meyer, Leonard 247
micro-cultures, existence 232
microdisruptions 263
microphone preamps: impact 83–85; list **69**
microphones: placement, interpretation 222; positioning 222–223; usage 222–223
Microsoft Forms, usage 117
MIDI sequencing 186
mimetic stereo framing, usage 223
mind's ear, aural thinking 216
"Minor Swing" (Reinhardt) 233
Mist, Alfa 221
mix bus magic 100–101
mixed-methods approach 113–114
Mix Hub plugin, usage 110
mixing: audio mixing, distortion (ecological approach) 109; creative recording/mixing, low order distortion (usage) 65; journey 98; process 100; template 112
Mixing Secrets for the Small Studio (Senior) 194
mixing skills, development 216
modellers, presence 178
modelling 175; analog modelling 24–26; black-box modelling 24–25; component-level modelling 25; gray-box modelling 25–26
mode rejection 66
Moffat, David 3
Momentary Max (M MAX) 118–120
Monder, Ben 232
Montgomery, Wes 233
"Mooche, The" (Ellington) 233
Moog Ladder filter, usage 26
Moore, Allan F. 207
Moore, Austin 47, 128
motivation, myths 191
Moulder, Alan 93, 96, 98, 103
Mountain (band) 9
multi-microphones, usage (multiperspectivity) 223–224
multiple effects, chaining 240
multi-track DAW-based workflows, usage 152
multi-tracking recorders/mixing desks, usage 207
Musa, Tarek 103
music: cognition, information-processing approach 113; construction 153; distortion, timeline 5–9; sensation, intensification 268
musical humanness, presupposition 267
musical reception, distortion impact (evaluation) 83–89
musicians, seating (importance) 222
music-making, real-world context 265–266
Music of My Mind (Wonder) 221
Music of the Spheres, eternal perfection representation 264–265
music production: chance, distortion (presence) 97; distortion, history 3; sound quality, maximisation 192–195; space, distortion 216
music production, semantics 95–96; background/historical context 93–95; crosstalk 95–96
"Music Sounds Better with You" (Stardust) 225
Musique Concrète 269
MXR Distortion+, usage 9
Myer, Bob 8

negative feedback loop 31
Negative Feedback Loop (NFB)/presence (MesaBoogie 2:90): circuit, simulation 38; control 36–37; RC network, display *37*
Negative Feedback Loop (NFB) settings, change/dynamic adjustment *40*
neo-trad artists, timbral development 228
Neural DSP, gears (usage) 29
Neve 1073 (preamp) 78–80, 85; line input Hammerstein model *78*; mic input high input signal Hammerstein model *79*; mic input phase response *81*; mic input reduced input signal Hammerstein model *80*; slowness 87
Neve mixing desk 147
Neve, pushing 101
Newfangled Audio Saturate 17
Newton-Raphson method 25
Nippon Columbia 192
Nirvana (band) 191, 202

noise: disorder 269; usage 268–269
noise floor 47–48; visibility *187*
non-diatonic pitches, preponderance 249
non-disabled, anxiety 272
non-linear audio processing tool, distortion (equivalence) 4
nonlinear autoregressive (NAR) model 23
non-linear distortion effects 28–29
non-linearity: attack/release controls, impact 50; attractiveness 102; dynamic range compressors, relationship 47; perception 60; reduction 55
non-linearity, attack/release role 50
non-magnetic devices, harmonic distortion 85
non-verbal visual thinking 216
non-verbal vocalisations 257
"Nothing Else Matters" (Metallica), dist-space analysis *206*
Novak, David 268
Noy, Oz 237
Nu Metal (band) 201
Nyquist frequency 16

object-based language, usage 21–219
odd harmonics: addition/production 4, 14, 102; distortion 86
odd-order harmonics, production 142
odd sounds, ear candy 102
Okazaki, Miles 232
Omni Channel 110
onset distortion, absence 265
Op-amp-based distortion, usage 9
open multiperspectivity 224
opinion leaders 176
optical compressor, design analysis 53–54
option dilemma 152
Orbison, Roy 6
organology, usage 230
Osborn, Brad 246
outboard equipment, usage 153
out-of-the-box clipping software re-creations 167
overdrive (distortion effect) 10, 14
oversampling, usage 16–17
overshoots, occurrence 50
Overstayer 105
"Oxford Park" 119; clean mix *120*; loudness meter export *120*; mixes, comparison *122*

Padgham, Hugh 225
PAF Humbuckers, usage 8
Page, Jimmy 251, 255
Palladino, Rocco 221
Palmer, Tim 181
paradoxical, disentangling 219
parallel processing 23
Parlophone 191
Parsons, Glenn 264
Pass, Joe 234

Patty, Austin 247
Paul, Les 8–9, 174, 234
Payne, John 179
peaking filters, set (usage) *42*
Peak Level (PL) 115
Peak to Loudness Ratios (PLR) 118–119
perceived volume 100
perception, defining 113
perceptual descriptors 129
percussiveness 268
performance: acoustic fingerprint 232–233
performance, switchability 181
Phillips, Sam 6
Pianist, The (Popova) 223
piano, tuning 186
pink noise settings *116*
pirate ratio, experience 268
pitch-class relationships 230
Pixies, The (band) 202
Plant, Robert (voice) 245–246, 253; phrases, non-verbality 256; scream, extension 244–255
playback medium 112; non-professional identification **121**; professional identification **121**
playback system 117
Plugin Doctor (DDMF) 68
plugins, usage 98
polyphonic texture 255
popular music, Western scaffolding (dismantling) 231
Portishead, bit-crushing (usage) 10
post-Miles fusion guitar landscape 234–237
poststructuralism 219
potentiometers, inclusion 25
power amp output, display *44*
power amp stage: FAUST re-implementation 41–42; implementation, WebAudio high-level nodes (usage) 39–40; master volume/presence knob control (JCM800) *36*
PowerAmp stage, usage 31
power stage 36–39
power valve distortion 8
practice-based manifestations 110
preamps: analysis, measurement 68–83; API Vision Channel Strip (preamp), analysis 69–72; characteristics 32–34; comparison, sonic overview 87–89; low order distortion, relationship 65–68; microphone preamps, impact 83–85; microphone preamps, list **69**; Neve 1073 78–80; SSL E Channel Strip 72–78; switching *32*; transformers, usage 66–67; tube preamps, usage 68; UAD 610 B 82–83; valve distortion 8
pre-existing musical whole, notion 267
presence filter (obtaining), peaking filter set (usage) *42*
Presley, Elvis 6
primitivity, return 241

Prism ADA8 converters, usage 57
ProCo RAT 128; distortion pedal, usage 238; usage 9
ProCo RAT descriptors 134; survey *138*
product endorsement 179
professionalism paradox 149–150
profiler, inferiority 180
PROFILER™ 177
profiling amps, issue 178
profiling, definition 175
program-dependent design, usage 50
program settings *116*
Pro Tools (Avid) 186
Pro Tools peak meter 115
pseudo loudness 100
psychoacoustic processing 111–112
Pultec EQP-1 147, 154
Pultec EQ, TubeTech remake 190
pumping: effects, usage 58, 60; usage 50
PUNISHR 500 series module, usage 105
push/pull power amp, block diagram *37*
"Put It Up to Eleven" (Herbst) 231
Putnam 610 valve console, usage 10

Ramones, The (band) 271
Rancière, Jean 267
Raney, Jimmy 233
Rangemaster 7
RCA BA6A Compressor, usage 105
RC network, display *37*, *49*
real-time simulations, browser-based WebAudio Ecosystem (interaction) 28
real-world experiences 222
record consciousness, concept 233–234, 241
recording: distortion, impact 83–89; experience 216; mic preamp, impact 83–85; process, versatility 181
Recording the Beatles (McCartney) 191–192
recordists: technology, usage 148; vintage processing technologies, relationship 147
record, making 153
rectification 14; algorithms, usage 18–19
REDD console, usage 8
reductive ecological approach, usage 123
Reinhardt, Django 32–233, 241
Reiss, Joshua 3
relative advantage 181
release control (compressor) 49–50; non-linearity 50
release design (RC network) *49*
Remedy, The (Rosenwinkel) 237
residual irrationality, refuge 269
retromania 188–195
retro-trend: explanations 191–192; presence 189
reverb 34–36; effect, creation 121
reverberation 225
"Revolution" (The Beatles) 8
Ribot, Marc 228, 237

Rice, Tom 128
Richards, Keith 6–7
Rig Exchange 177
RME Fireface 802 193
Robbins, Marty 6
Robjohns, Hugh 193
"Rocket 88" (Jackie Brenston and his Delta Cats) 6
Røde microphones 190
Rogers, Julian 3
Roland Space Echo 147
Rolling Stones, The 6–7
room: aggravation 103; effect, adjustability 36
root mean square (RMS) 52; amplitude, ratio 160–161; clean mix average RMS volume, impact 122–123; detector 56; levels, increase 111
Rosenwinkel, Kurt 232, 237, 240
rounding off (waveform change) 9
RStudio, usage 130
"Rumble" (Link Wray) 6
Russolo, Luigi 269

Saito, Yuriko 265
sample-wise accuracy, measurement *44*
sampling approach 24
"Satisfaction" (The Rolling Stones) 6–7
saturated inductor filter, usage 190
saturation 14; level, pushing 99; studio presence 10–11; usage 101
Schafer, Raymond 264–265
Scheps, Andrew 97, 101, 110; mix template guidance 115, 117
Scheps Particles plugin 105
Schoeps microphones 190
Scofield, John 228, 235–236, 240
Scott, Ken 54
Scotto, Ciro 201
ScriptProcessor node, usage 35
second harmonic distortion 86
second-order state-variable filter, usage 26
Senior, Mike 194
Serres, Michel 271
Sex Pistols (band) 272
Shoemaker, Trina 97
sidechain block 48; usage 52
side-chain compression, usage 225
sidechain function 50–52
side-effect sounds, ignoring 162
signal processing 26
signal to noise ratio 190
silent space, dark space (equivalence) 222
silicon transistor-based designs, third harmonic (presence) *188*
similarity-based multidimensional models, comparisons 94
Simulation Program with Integrated Circuit Emphasis (SPICE) 162

Sinatra, Frank 167
sine wave, display 5
single-use descriptors, usage 130
Sixteen Men of Tain (Holdsworth) 236
"Smells Like Teen Spirit" (Nirvana) 202
Smith, Bessie 245
Smith, Julius 34
Smith, Randall 8–9
Sneap, Andy 178, 182
Snoddy, Glen 6
Social and Applied Psychology of Music (Hargreaves/North) 104
social stigma, defiance 272
soft clipping 9, 14; algorithms 19–20; analogue soft-clipping 10; forms 263
software: advantages **166**; types, distortion (inclusion) 13–16
software-created harmonic distortion **164**
software-emulated distortion, traditional distortion (comparison) 160
software THD, advantages/disadvantages (comparison) 163–164
Sola Sound Tone Bender 7
"Songs from the Wood" (Jethro Tull) 207–208
songs, mixes (subtleness) 123
sonic aesthetic 121
sonic affordances 263
sonic attributes 98
sonic cartoons 148, 150–151; low order harmonic distortion, relationship 85–87
sonic events 209
sonic identification 240
sonic object, buzzes/hisses infiltration 262
sonics 150–152; colour 150–151; unpredictability 151–152
sonic signature 141
Sonic Youth (band) 96
Sony Oxford Inflator, design 11
Sony Oxford plugins 100
sound: access, positive effect 181–182; differences, stacking 103; envisioning 216; file, corruption 267; getting 96; iconic sounds, replication 152; imperfection, appearance 265; inspiration 181; quality 180; quality, maximisation 192–195; removal 103; replicability 154; search 150; source, reprojection 219; staging 103–104; value judgements 269
soundboard, gain structure (understanding) 100
sound-box: filling 207–208; saturation 213
Sound-box Dist-space 207–210, *210*
Sound City Studios, opening 191
Soundcloud, usage 117
Soundelux microphones 190
Sound on Sound (magazine) 188
Sound Recording Practice (Elen) 195
source-bonding 219
source signal, odd multiples 15

space: distortion 216–225; fabrication 218; shift 103
Spaeth, Sigmund 242
Sparse Tableau Analysis 25
spatial contract 216, 219
spatial environment, significance 222
spatialization artefacts (swirl) 38
spatial samples, microphones (usage) 222–223
"Spatial Turn" (music production) 217–218
"Speaking of Sound" 95
special effect distortion 148
Spector, Phil 220
spectral analysis 205
spectrogramme, usage 255–257
spectrograms, readings 231
Spencer-Allen, Keith 192
Sphere Electronics 194
SPICE analog circuit simulator, usage 28
Spicer, Mark 246
Spinoza, Baruch 264
SPL Twin Tube, usage 68
squish (temporal lag) 38
SSL 400E, transformer (exclusion) 85
SSL 4000E console, emulation 72, 74
SSL 4000E harmonic performance, measurement 88
SSL 4000E, transformer (inclusion) 85
SSL channel strip, recording 87–88
SSL E Channel 74; frequency response, consistency 78; usage 69, 76
SSL E Channel Strip (preamp) 72–78
SSL Fusion, usage 10
SSL high gain with pad and transformer Hammerstein model *77*
SSL low gain line input Hammerstein model *76*
SSL low gain transformer Hammerstein model *75*
SSL low gain transformerless Hammerstein model *75*
SSL mixing desk 147
SSL, operation 101
SSL overload harmonic analysis *77*
Stardust (band) 225
static interference 268
statistical climaxes 247
Steely Dan, instruments (usage) 167
Stent, Mark "Spike" 97, 101, 102, 106
stereo buss: frequencies, collapse 101; usage 167
Stern, Mike 235, 240
Stewart, Eric 193
Still Life with a Basket of Apples (Cézanne) 223
s-time dimension 204
story-telling, recording tradition 191
Stras, Laurie 272
Streams API, usage 28
studio: distortion, usage 97; equipment, myths 191; harmonic distortion, presence 10–11; modellers, presence 178; saturation/colouration, presence 10–11, 97

"studio speak" 96
Styles, Ashley 189
Suit No. 2 (Gigue) 203
Sun Records 6
Sun Studio 6
sustain, obtaining 237
Swedish Death Metal bands 201
Swettenham, Dick 10
swirl (spatialization artefacts) 38
syntactic climaxes 247
SynthAxe, usage 236, 237

tactile feedback, absence 153
Tagg, Philip 272
tape-based delay unit, output 154
tape machine, usage 153
tape saturation (distortion effect) 10
Tarr, Eric 13
technical competence 176
technical expertise, usage 189
technocultural choice 149
technological development, rejection 174
technological imperfection, glitch 267
technology: rejection 174–175; usage 148
technostalgia 178
Teletronix LA2A (Opto) compressor 57
temporal dynamics real-time adjustment *39*
temporal lag (squish) 38
terminally climactic form (TCF) 246
Terry, Sunny 5
text mining method 129–136
Thermionic Culture 105, 195
Thermionic Culture Vulture, usage 10, 68, 195
third harmonic distortion 79, 87
Thompson, Marie 271
Throughout (Frisell) 236
timbral creativity 239
timbral development 237
timbral qualities, structural mapping 231
timbre: change 4; importance 229–230; perception 94
Timbre (Saitis/Weinzerl) 94
time constants 49, 54; attack/release 48
time-domain simulation 162
tonal camera angles, crafting 222–223
tonal nostalgia 241
tone stack 34–36; circuit (JCM800) *35*; switching *32*
topologies, exploration 66
Topology Preserving Transform (TPT), usage 26
total harmonic distortion (THD): calculation 57; creation 170; emulators, authenticity 161–162; hardware-created THD 160–161; hardware THD, advantages/disadvantages (comparison) 163–164; measurements 114; measurements, inadequacies 111; performance **60**; software, generational affinity 170–171; software THD, advantages/disadvantages (comparison) 163–164; testing 109; usage, audio professional (perspectives) 166–167
total harmonic distortion (THD), input level **57**
totalised image 223
totall recall 182
Townshend, Pete 7
Toy, Raymond 40
traditional distortion, software-emulated distortion (comparison) 160
transformer-based harmonic distortion 190
transformers: compression 67; phase shift/non-linearities 67; usage 66–67
transient smoothing 5
transients, taming 101
Tribal Tech (band) 240
triode valve designs, second harmonic/noise floor (visibility) *187*
Tristan und Isolde (Wagner) 247
tube distortion 10
tube guitar amplifier simulations, understanding 32–39
tube preamps, usage 68
tube rectifiers-based power supply, usage 38
Tube Screamer 131, 137–138
Tube Screamer descriptors 131–133, 139; coding, categories **132**; survey *137*; text mining list *132*
tube simulation (temporal dynamics real-time adjustment), waveshapers (basis) *39*
Tubes (simulation), WebAudio API (usage) *31*
tube transfer function curves, slope 38–39
tunable filter, usage 37–38
typomorphology 124

UA610 A (preamp) 85
UA610B (preamp) 85
UA610 B (preamp) 88
UAD 610 B (preamp) 82–83; gain line input Hammerstein model *84*; line input frequency response *84*; unity gain line input Hammerstein model *83*; usage 69
Universal Audio preamps, usage 68
Universal Audio product, resurgence 105
Universal Audio, Teletronix purchase 185
Urei 1176 147; compressor, usage 161, 186
Urei 1176 Revision D (FET) compressor, usage 56
USB, usage 162
use-case scenarios 151

valve (tube) compressor, design analysis 54
valve distortion 10
Vaughan, Stevie Ray 128
Vectorial Volterra Kernels, basis 24
Ventures, The 7
verbal rhythmicity, impact 268
Vintage Audio Rentals 190

Index

vintage equipment 185; background 185–186; industry reaction 189–191; technical aspects 186
vintage hardware: absence 154; focus 148
Vintage King, establishment 189–190
vintage processing technologies, recordist (relationship) 147
vintage-sounding mixes, re-creation 189
vinyl production, lo-fi aesthetic 97
virtual acoustic space 218
Virtual Analog (VA) filter design 26
virtual environment, real-world experiences 222
virtual pedalboard, FAUST power amp plugin (display) 43
visual reprojection 219
vocal bridge 255
vocal climax (Led Zeppelin music) 245, *250*, *253*, *256*; restricted vocal pitch *249*; songs 248–259; theoretical background/methods 246–248
vocal climax, meaning 245
vocal quality, obtaining 237
vocal recording: analysis 63; clarity/distortion 103; forwardness/excitement 119; tape, usage 101
Voegelin, Salome 273
voltage-controlled amplifier (VCA) compressor, design analysis 55–56
Voltera Kernel 23
Volterra Series 23
Vox AC30, usage 38
VU meter, control 55

Wagener, Michael 178–181
Waits, Tom 266
"Wall of Sound" technique (Spector) 220
WAP GUI Builder, usage *43*
Warm Audio products 190–191
Warm Audio WA76 Discrete FET Compressor, usage 163
Wave Digital Filters (WDF) 25
waveform: rounding off 9; squaring 98; total saturations 208
WaveNet music generator (Deepmind) 29
waveshaper algorithms 17–20
waveshaper curve 17
waveshaping: algorithms 21–22; fast attack/release, impact *51*
waveshaping/filtering: combination 20–23; series combination 21–23
Waves L1 limiter, usage 115
Waves plugin, usage 110
Waves Scheps channel strip, usage 68
Ways of Listening (Clarke) 103
WebAssembly standard 31
WebAudio: amp 34–35; sim plugins, display *30*; FAUST re-implementation 41–42; final implementation *41*; high-level nodes, usage 39–40; implementation 42; nodes, usage 39; plugin (creation), FAUST code (usage) *43*; waveshaper node 33
WebAudio API: high-level nodes, providing 29; implementation *42*; reliance 29, 31; tube transfer function curves, slope 38–39
WebAudio API ecosystem 29–31; designer tool *30*; evaluations 42–44; high-level simulation *31*; implementations 39–42; tubes/filters, simulation *31*
WebAudio API implementation *42*
Webern's Symphonie 203
Weenink, David 248
Weiner model 21
West, Leslie 9
white-box modelling 25
White, Jack 54
White, Paul 188–189, 195
"Whole Lotta Love" (Led Zeppelin): analysis 254–259; arousal, non-verbal signifiers *257*; closing vocal gesture *258*; distorted vocal drone, harmonic "Oh" (usage) *258*; songmap *255*; vocal climax, scream (usage) *256*
"Whole Lotta Love" (Led Zeppelin) 245
Wiener-Hammerstein model 21, 23
Williams, Tom 228
Williams, Tony 236
Wilson, Chris 28
Wilson, Wally 194
windings, ratio 110
Winehouse, Amy 96
Wonder, Stevie 221
Wood, tim 190
workflow 152–155; efficiencies 147; familiarity 154; harmonic distortion, impacts 169; interface/tangibility 153; misuse 154–155; total harmonic distortion (THD) usage, audio professionals (perspectives) 166–1667
Wray, Link 6, 233

x-factor 147
X-parameters 162
X Saturator plugin, usage 68

Yeh, David 34
You Dirty Rat 134
Youlean Loudness Meter software, usage 114
Young, Toby 262
"You Really Got Me" (The Kinks) 7, 96, 266
"You Shook Me" (Led Zeppelin) 245

Zorn, John 232

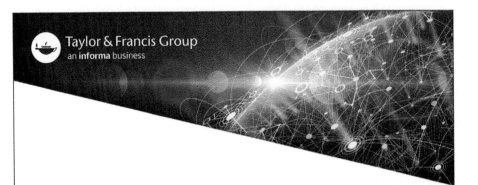